DISENTANGLED

DISENTANGLED

Ethnozoology and Environmental Explanation of the Gloucester Sea Serpent

Robert L. France

Dalhousie University
Ocean Research Group
Department of Plant, Food, and Environmental Science
P.O. Box 550
Truro, Nova Scotia
Canada B2N 4P8

Wageningen Academic
Publishers

EAN: 9789086863358
e-EAN: 9789086868865
ISBN: 978-90-8686-335-8
e-ISBN: 978-90-8686-886-5
DOI: 10.3920/978-90-8686-886-5

Cover: After Captain John Beach Jr. View of Gloucester Harbor with the Sea Serpent of 1817. Engraving. Sheet: 34 x 48 cm. Museum of Fine Arts, Boston. Source unidentified. 2007.976. Photograph © 2019 Museum of Fine Arts, Boston.

Wageningen Academic Publishers
P.O. Box 220
6700 AE Wageningen
the Netherlands
www.WageningenAcademic.com
copyright@WageningenAcademic.com

‘Ecological Entanglements in the Anthropocene’

Title of book by Nicholas Holm and Sy Taffel (2016)

‘No one wanted to tangle with the strange, silent animal in the harbor.’

Gloucester’s Sea Serpent, Wayne Soini (2010)

‘These several conflated [sea monster] creatures are a hassle to disentangle.’

Abominable Science: Origins of the Yeti, Nessie, and Other Famous Cryptids,
Daniel Loxton and Donald R. Prothero (2015)

‘Disentangled and Classified at Last.’

In the Wake of the Sea-Serpents, final chapter title, Bernard Heuvelmans (1968)

‘A more detailed and discriminatory study needs to be made with a view to disentangling the various sightings that Heuvelmans places in his ‘many-humped’ category.’

In the Wake of Bernard Heuvelmans: An Introduction to the History and Future of Sea Serpent Classification, Michael A. Woodley (2008)

‘When they propose to disentangle facts from folklore, cryptozoologists assume that such facts are anomalous ones. I suggest, to the contrary that it is the interpretation which is anomalous.’

Lake Monster Traditions: A Cross-Cultural Analysis, Michel Meurger (1988)

‘Disentangled from the treacherous world.’

Divine Comedy, Paradise, Dante Alighieri (1320)

For Mr Stover

About whom we know nothing else besides that, in 1817, while his fellow citizens of Gloucester were heading out in boats, armed with harpoons and rifles, to kill the sea serpent gambolling about in the harbour, he, on a stroll accompanied by his daughter Susan, and upon seeing the creature within a few feet of the shore, 'took off his hat that he should not frighten him.'

And for Joseph Michael Howlett

Former Gulf of Maine commercial herring fisherman who became a leading practitioner of rescuing whales ensnarled in fishing gear, and who perished while disentangling an Atlantic right whale during the disastrous summer for such in 2017, the two-hundredth anniversary of the Gloucester sightings, and also when this book was being written.

Contents

Literary epigraph

Excerpted from the Prologue to the acclaimed and award-winning work of literary fiction, *The Essex Serpent*, by Sarah Perry (Serpent's Tail, Profile Books Ltd., London 2016):

> A young man walks down by the banks of the Blackwater under the full cold moon...
>
> His breath is slow, the air is full of pins; the marsh at his feet all at once wet, as if something out there has displaced the water. *Nothing, it's nothing*, he thinks, patting about for his courage, but there it is again: a curious still moment as if he were looking at a photograph, followed by a frantic uneven motion that cannot be merely the tug of the moon on the tides. He thinks he sees – is *certain* he sees – the slow movement of something vast, hunched, grimly covered over with rough and lapping scales; then it is gone.
>
> In the darkness he grows afraid. There's something there, he feels it, biding its time – implacable, monstrous, born in water, always with an eye cocked in his direction. Down in the deeps it slumbered and up it's come at last; he imagines it breasting the wave, avidly scenting the air. He is seized by dread – his heart halts with it...Out he looks to the black Blackwater and there it is again – something cleaving the surface, then subsiding...

PREFACE

THE SEA SERPENT AS ENDURING MYSTERY

Sarah Perry's book, *The Essex Serpent,* is a novel about Victorian ideas. The work of literary fiction uses the rumoured presence of a serpent, supposedly haunting the East Anglian shores, as a MacGuffin and then leitmotif to explore concepts of love and feminism, discovery and belief, amateurism and scientism, medicine and quackery, and superstition and religion, prevalent in the late-nineteenth century. The mysterious creature that hovers about, as both metaphor and muse, in the background to the novel is not an invention of fiction by the author. Rather, the creature was described in an 1699 pamphlet entitled *The Flying Serpent, or, strange news out of Essex, being a true Relation of a Monstrous Serpent which hath divers times been seen at a Parish called* Henham on the Mount *within four Miles of* Saffron-Walden. This early and earnest pamphlet – one of the first investigations of what will later be called the field of *cryptozoology* – sets out, as advertised on the front cover of its 1885 reprinting (Figure P.1), to show 'the length, proportion, the bigness of the Serpent, the place where it commonly lurks, and what means hath been used to kill it.' Using a methodology that is prescient to today's cryptozoology, the pamphlet promises to lay out 'the truth of the Relation of the Serpent,' as attested to by a handful of reputable eyewitnesses, whose anecdotes are used to create a composite image of the mysterious, often hidden beast, or *cryptid* (Figure P.2). Also as promised on the cover, in true cryptozoological fashion, is a further 'discourse of other Serpents, and particularly of a Cockatrice killed' in the village (cockatrices, it is worth noting, are mythological animals that resemble a fusion between a serpentine dragon and a giant rooster).

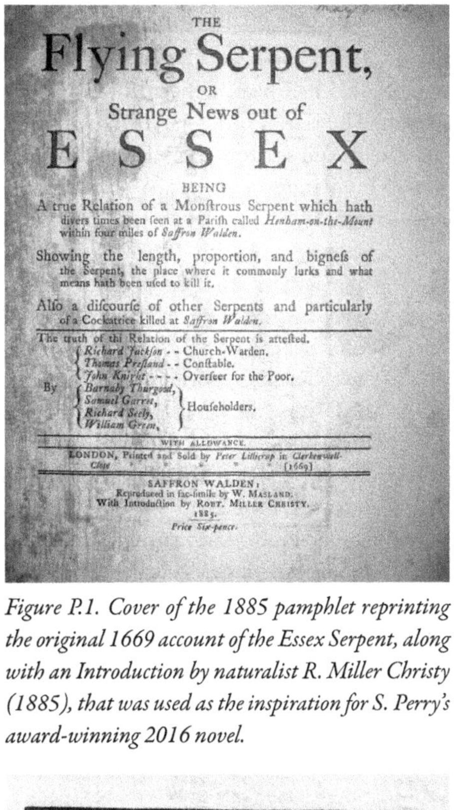
THE

Flying Serpent,

OR

Strange News out of

ESSEX

BEING

A true Relation of a Monſtrous Serpent which hath divers times been ſeen at a Pariſh called *Henham-on-the-Mount* within four miles of *Saffron Walden*.

Showing the length, proportion, and bigneſs of the Serpent, the place where it commonly lurks and what means hath been uſed to kill it.

Alſo a diſcourſe of other Serpents and particularly of a Cockatrice killed at *Saffron Walden*.

The truth of thi Relation of the Serpent is atteſted.

By
Richard Jackſon - - Church-Warden.
Thomas Preſland - - Conſtable.
John Knight - - - Overſeer for the Poor.
Barnaby Thurgood, *Samuel Garret*, *Richard Seely*, *William Green*, Houſeholders.

WITH ALLOWANCE.

LONDON, Printed and Sold by *Peter Lillicrap* in *Clerkenwell-Close* * * * * * [1669]

SAFFRON WALDEN:
Reproduced in fac-ſimile by W. MASLAND;
With Introduction by ROBT. MILLER CHRISTY.
1885.
Price Six-pence.

Figure P.1. Cover of the 1885 pamphlet reprinting the original 1669 account of the Essex Serpent, along with an Introduction by naturalist R. Miller Christy (1885), that was used as the inspiration for S. Perry's award-winning 2016 novel.

Figure P.2. Seventeenth century illustration of the Essex Serpent reprinted in the 1885 pamphlet.

Truth be told, the original seventeenth-century pamphlet by an anonymous local historian is not as interesting as is the Introduction to the Victorian reprinting by the noted Essex naturalist Robert Miller Christy. It is to him whom gratitude is due for his searching out the original document held deep within the bowels of the British Museum Library – apparently the only copy known to exist, and which would later be consulted by the novelist Sarah Perry – and keeping the legend alive. Miller Christy's reinterpretation of the events of two hundred years prior to his time offers an interesting parallel to the present book. Writing from a period of imagined scientific certainty, he comments that 'many of the statements contained in it [the original document] appear absurdly ridiculous in the present day.' He goes on to state that:

> The narrative contained in the pamphlet may well cause us to wonder that our ancestors, only two centuries ago, were capable of such ignorant credulity. Clearly the narrative is set forth in sober earnest by one who did not see in it anything at all impossible; and the fact that its truth is attested by the Church-warden, the Constable, the Overseer of the Poor, and four Householders, indicates that in that day the public mind was quite able to accept seriously such statements as those found in the tract.

The original had been written during the Age of Enlightenment, well past the time when people consulted bestiaries and 'imagination abandoned by reason produce[d] impossible monsters,' as Goya phrased it. And so, from the vantage point of two centuries of accumulated knowledge, Miller Christy offered his own more mundane explanation for what those earlier individuals had purported to have been a serpent:

> We may take it for granted that in this, as in all (or nearly all) similar cases, there was some actual foundation in fact for the tale, which the ignorance and superstition of those days has greatly exaggerated. Nearly all ancient legends, ghost stories, and miraculous occurrences have in them some slight germ of truth; and it seems to me most likely that the accidental appearance at Henham-on-the-Mount of some large bird belonging to the Diver-tribe (*Colymbus*) may have originated the whole of this astounding narrative... These birds, when so situated, do not make use of their wings as a means of escape, and, therefore, would be likely to remain some days near the same spot [as indeed had been the case for the Essex Serpent]. The long, snake-like body and short wings of these birds would, doubtless, greatly assist, in those ignorant days, in starting a rumour that a 'flying serpent' had appeared where-ever one of these birds happened to be carried by the wind.

The suggestion by Miller Christy that the Great Northern Diver (*Colymbus*) or loon, as now featured on the Canadian dollar 'loonie' coin, which can obtain a length of over a metre, might be confused with a serpent, was well ahead of its time. It is also not an unreasonable hypothesis to posit, given that there are many examples of observations made, not just in antiquated, but also in recent, times, of objects in the water that were originally thought to be sea serpents or lake monsters, but which, upon closer examination, turned out to be various species of waterfowl. Miller Christy continued:

> A perusal of the entertaining natural histories of the different kinds of serpents given in the pamphlet will show what absurdities our ancestors were capable of believing in; and there can be no doubt that the numerous mermaids, dragons, and other dreadful creatures, which are recorded to have been caught at sea or thrown up on our shores in bygone centuries, were identical with, and not more dangerous than, the harmless whales and dolphins which are frequently captured or stranded round out coasts now-a-days, and which, in these more enlightened times, do not cause us any terror.

The parallelism to the present book is that I, from the vantage of the twenty-first century and an additional century of accumulated knowledge, will undertake a similar re-examination and reinterpretation of sightings made two centuries earlier, of another mysterious serpent that, much like the Essex Serpent, was viewed by many people, and for which a more parsimonious explanation can also be retroactively invoked. And, in a further instance of parallelism, most of the sightings of this particular sea serpent examined herein occurred in and around the fishing community of Gloucester, a town located, in an intriguing coincidence, also in the County of Essex. However, this is Essex County, not East Anglia, England, but rather that within Massachusetts, New England.

There have been hundreds of sea serpent sightings made from every conceivable corner of the globe over the last four centuries (Heuvelmans 1968; Oudemans 1892). A handful of books exist that focus on sightings made over extended periods of time from general regions, such as New England (O'Neill 1999), the British Isles (Harrison 2001), British Columbia – the Pacific Northwest (LeBlond and Bousfield 1995; LeBlond *et al.* 2014), Scandinavia (Thomas 2011), Newfoundland (Hynes 2012), and Nova Scotia (Hebda 2015). The present book is different in that it focuses on a single corpus of sightings made in a restricted area (mostly Massachusetts) and over a limited timespan (eight years). A 2010 book by Wayne Soini covers a much more limited subset of the same observations.

As an aquatic scientist who has, over the period of more than three decades, published hundreds of papers arising from ecological research conducted in locations from the High Arctic to the tropics, and on organisms from bacteria to whales, and whose work has been cited thousands of time in the peer-reviewed literature, I believe my training and experience is particularly well-suited for undertaking the present investigation. This is an important statement to make, not for purposes of self-aggrandizement, but rather to address the serious criticism raised by many skeptics, as for example Loxton and Prothero (2015), that those individuals involved in cryptozoology have been, for the most-part, woefully unqualified to engage in such avenues of study. My purpose is to use my scientific acumen to reinterpret a limited number of accounts of sea serpents in the light of contemporary knowledge about a particular anthropogenic threat to the conservation of marine animals. In so doing, I suggest that the first widespread documentation of this now recognized, serious threat can be found, occultly, within the nineteenth-century accounts of those whom believed they were observing a 'sea serpent'.

Furthermore, the present book is extremely unusual in terms of its publication. Through reading widely on the subject, I have assembled a personal library containing about a hundred modern contemporary books about sea serpents and lake monsters. The vast majority of these titles have come out in popular presses, with less than half a dozen being of sufficient scholarship to warrant publication by academic houses. And of these latter, most have dealt more with the sociological side of the enigma rather than the scientific. The present book is different in that it may be, if not the only, then certainly one of the very, very few concerning what are termed aquatic cryptids that is published by a specialized and serious scientific publishing house of international renown. That it is a Dutch publisher I have selected for this is no accident. For it can be said that the pseudoscience of modern cryptozoology had its start with the October 1892 publication of Antoon Cornelius Oudemans' *The Great Sea-Serpent* in Leiden. I felt the time was right, now, one hundred and twenty seven years on, for another Dutch publisher to step in and right the wrong of that 'original sin'.

There are other reasons for focusing on this particular group of observations. Most sightings of sea serpents, even the most famous ones such as the 1848 *Daedalus* encounter, were extremely ephemeral affairs, lasting but for a few minutes duration and observed by only a few eyewitnesses. Someone shouts excitedly 'Look, there's something in the water!,' and a handful of people scramble to afford a better vantage on the shore or on one side of the ship to observe the strange phenomenon. And 'Lo!,' there it is! All stare in disbelief at what they are seeing. Some individuals hurriedly pull out their spyglasses or binoculars to obtain a better view. And then, just as quickly as the apparition appeared, it is gone. Forever. Never to be seen again. Leaving the eyewitnesses to confirm, discuss, and possibly reach a consensus about what they had collectively glimpsed. Such is the genesis of many encounters with sea serpents (Heuvelmans 1968; Oudemans 1892). 'Many', but not most, however. For most sightings of sea serpents have actually been made by solitary witnesses with no one present to either corroborate or to discuss the encounter. In contrast, the case of the 'Gloucester Sea Serpent', as the mysterious cryptid has become known to posterity, is unique in the corpus of cryptozoology. Being observed by hundreds, if not thousands, of people, from all walks of life, often for durations of many minutes, and in some cases, even hours, it remains the most sighted sea serpent in history. Indeed, a case can be made that more people saw what they believed to be a sea serpent in Gloucester Harbour and nearby locations in the early nineteenth century, than have seen *all* other aquatic cryptids combined, including those thought to reside in both Loch Ness and Lake Champlain. It is therefore impossible to simply dismiss these sightings as being mass hysteria of the sort that gripped a very limited number of schoolgirls or nuns believing they had respectively seen either witches in Salem or devils in Loudun. Clearly the people in Gloucester at the time did see *something* in the water, which appeared to them to be different from the norm, their confirmation bias kicking in to suggest it to have been a sea serpent.

Another unique feature about the Gloucester Sea Serpent is that it is the most thoroughly investigated such creature of all time. The detailed interviews of eyewitnesses, conducted by legal authorities under the auspices of the region's learned natural history society soon after the bulk of the sightings, means that there is an authenticity to these accounts that is absent in many other contemporaneous anecdotes. So much so in this regard that Paxton (2009), in his analysis of potential biases in eyewitness observations of sea serpents, actually excluded data from the Gloucester case as they are *too* accurate in comparison to the other anecdotal sightings he examined. And it is for this very reason that the Gloucester Sea Serpent forms such an important lynchpin in the theories touted by cryptozoologists.

Fringe scientist H. Bauer (2013), for example, a believer in the Loch Ness monster (who also incidentally denies an HIV linkage to AIDS and regards homosexuality as an aberrant illness), dismissed Loxton and Prothero's (2015) stinging indictment of the pseudoscience of cryptozoology. He accused them of wilfully and deliberately failing to discuss the very strongest case, namely that of the 'Gloucester sightings by innumerable witnesses... [which are] by far the best-documented evidence that sea-serpents exist.' In consequence, there is an established tradition in cryptozoology of using the Gloucester sightings as the touchstone to give an impression of authority to claims made about other aquatic cryptids for which less 'data' are available (e.g. Gould 1934). In other words, 'Look here at the Gloucester example; now take my own sketchy claims more seriously.' As a result, the Gloucester Sea Serpent has, as shown in dozens of web-sites, achieved an exalted status among cryptozoologists: *Ecce Scoliophis atlanticus*, to use, as will be seen, its dubiously ascribed 'official' scientific name. So, if one can offer a different, less sensational, explanation for the Gloucester sightings, this will, I'm sure, not go over well in cryptozoological circles, many of whom are often adverse to accepting mundanity over mysteriousness.

So as to circumvent any possible confusion, particularly among those with a deep-seated longing for the mysterious, it is important to lay the cards upon the table right from the start:

- Sea serpents do not now, and never have, existed in the form commonly perceived in modern cryptozoological lore.
- Almost all sightings of sea serpents can be explained by natural phenomena.
- And where they cannot, the possibility exists that the sighting represents a hitherto undocumented species new to science, most likely a cetacean (whale).

Dutch zoologist Oudemans began his book with an epigraph from Melville in which the novelist expressed his belief that 'one day the question will be satisfactorily solved.' It is my hope that the present book represents a step toward achieving that goal.

Acknowledgements

Research for this project was aided by the welcoming and accommodating personnel at the Cape Ann Museum Archives, the Mystic Seaport Museum Archives and Collections, the Wedgeport Tuna Museum and Interpretive Centre, and Historic Ipswich, in addition to the helpful staff at the Dalhousie Library System. Thanks to Wayne Soini for permission to use his hand-drawn map of harbour sightings, and to The Center for Coastal Studies, The Mystic Seaport Museum, J. O'Neill, Penguin and Fortean Zoology Press for permission to use illustrations/photographs. It is a blessing to live in a region whose maritime heritage is preserved and so celebrated through its many fine museums, such as those visited herein for this research project: the Maritime Museum of the Atlantic (Halifax, Nova Scotia), the Fisheries Museum of the Atlantic (Luneburg, Nova Scotia), the Wedgeport Tuna Museum and Interpretive Centre (Wedgeport, Nova Scotia), the Trinity Museum (Trinity, Newfoundland), the Cupids Legacy Centre (Cupids, Newfoundland), The Rooms (St. John's, Newfoundland), the Mystic Seaport Museum (Mystic, Connecticut), and of course the Cape Ann Museum (Gloucester, Massachusetts). Finally, I would like to thank Mike Jacobs and the entire editing and production team at Wageningen Academic Publishers.

Chapter 1.

BACKGROUND

1.1 SEA SERPENTS AND NATURAL HISTORY

During the eighteenth and nineteenth centuries, the study of the environment, what was then termed natural history, was a strange beast indeed. It was a curious cocktail that blended proto-scientific accuracy and rigid factualism with a sense of wonder and celebration of mystery (Berger 1983; Holmes 2008). Strongly influenced by John Ray's 1691 classic *The Wisdom of God Manifested in the Works of the Creator*, for many, the purpose for studying nature was to discover and marvel at God's benevolence. Nature was regarded as the handiwork of God, and its patterns disclosed His wisdom, power, and goodness. The study of nature was therefore invested with both an aesthetic appreciation and especially a theological significance. It is no coincidence that the 1789 classic of enduring popularity, which became the standard to be emulated by many, *The Natural History of Selborne,* was produced by a country parson, Gilbert White, a man who rarely travelled more than a few kilometres from his childhood home and village church. Natural history had its roots in 'natural theology', as popularized in William Paley's highly influential 1802 book, *Natural Theology; or, Evidence of the Existence and Attributes of the Deity Collected from the Appearances of Nature.* This work was based on the idea that there existed an overall design in nature, a rank and order in the great chain of life, and regularity in the operation of laws, all of which being evidence of a transcendent guiding intelligence of a caring Creator (Berger 1983). For this reason, the practice of natural history was regarded as a morally uplifting enterprise that elevated the mind and expanded the heart (Barber 1980).

Gilbert White's book, with its emphasis on field studies over museum visits, became the inspiration for every amateur naturalist. Natural history societies sprang up all over for the collecting and descriptive activities of such amateurs, and operated as social clubs for self-improvement of the educated classes (Berger 1983). Boston's American Academy of Arts and Sciences, one of America's oldest learned societies, was established in 1780 and is still in existence. To satisfy the insatiable demand for knowledge by avocational naturalists – those engaged in the pursuit for personal pleasure – universities and lyceums sponsored lecture series (Arment 2004). The result was a mania for forming collections which would be prominently displayed in curiosity cabinets (Mason 2017). By the middle of the nineteenth century, as Barber (1980) states, 'there was hardly a middle-class drawing-room that did not contain an aquarium (Figure 1.1), a fern-case, a butterfly cabinet, a seaweed album, [or] a shell collection...And at the same period it was impossible to visit the seaside without tripping over parties of earnest ladies and gentlemen…standing knee-deep in rock pools (Figure 1.2).' Natural history became an obsession. Every newspaper ran a column on the subject and natural history books became bestsellers, only marginally less popular in England than works by famous novelists such as Dickens. Privately owned curiosity museums became the rage (Benson 1988). The noted portrait artist Charles Wilson Peale, for example, satisfied his fascination with natural history by operating one such popular museum in Philadelphia from 1784 to 1845 (Figure 1.3).

Figure 1.1. 'The Aquarium', Weekly Welcome *Magazine, 1879 (from Barber 1980).*

Often the emphasis was on the strange and exotic (Barber 1980). With titles that included phrases like the 'wonders of' or 'marvels of', popular natural history books consistently focused on the more bizarre types of natural phenomena at the expense of the mundane. For the more serious collectors, the most gratifying achievement of all was to find a previously undescribed and therefore 'new' species. Indeed, the *modus operandi* for this type of dedicated natural history was to go into the field, find a new species, kill it, and bring it to a museum specialist to verity its novelty. Perhaps, if one was especially fortunate, the species would be officially christened with one's own name. For all of eternity. What could be better than that? And it wasn't like there was a shortage of undescribed species to go around. All of China, most of South America, and much of western North America, in addition to Australia, Russia, Japan, India, and Africa were still zoologically uncharted, offering the promise of new discoveries to those intrepid enough to travel there (Barber 1980). Newspapers could be assured of record sales under a banner headline of 'Important New Scientific Discovery of….' And every year brought a new sensation: 'an endless list of ever bigger and better marvels to fire the public imagination' (Barber 1980).

Figure 1.2. Sea-side studies by the gentry, 1868 (from Barber 1980).

The great French comparative anatomist and palaeontologist Georges Cuvier set the tone for the first half of the nineteenth century. Under his guidance, natural history as a fledgling professional discipline had little to do with the study of the inner workings of nature, or what we would today call ecology, but rather everything to do with the discovery, description, and formal classification of species (Evans 2010). The race to find new species became an obsession. And if you couldn't find a new species, you could always follow the lead of the French-German-American naturalist, Constantine Rafinesque (who will figure prominently in the Gloucester Sea Serpent story), and make one up. Author of a definitive

Figure 1.3. Charles Wilson Peale unveiling natural curiosities in his museum in 1822 (Pennsylvania Academy of Fine Arts, Philadelphia; from Barber 1980).

1820 text about fishes from Ohio, Rafinesque was cruelly spoofed by noted wildlife painter John James Audubon into inventing a suite of new species from faux information that had been provided to him. The result was ridicule, with Rafinesque going on to become a medical quack and entrepreneur before dying so destitute that his body had to be sold to a medical school to pay his outstanding debts (Barber 1980). An ignoble end to an individual who was one of the biggest believers in sea serpents and whose other, more legitimate, natural history comments concerning the process of speciation actually presaged Darwin by several decades. The end result of this mean-spirited jest by Audubon was far-reaching in that it caused European naturalists to regard all newly described species from America, including, as we will see, the Gloucester Sea Serpent, with the deepest suspicion.

In antebellum America, natural history combined an intellectual fascination about strange forms of life with an intense interest in exploiting new resources (Berger 1983). As in England, collectors were mostly amateurs assembled from the professions of physicians, apothecaries, ministers, merchants, and teachers, all sharing an interest in curiosities and anomalies (Evans 1993; Stearns 1970). For the most part, zoological observations and the acquisition of specimens were brought about by random encounters rather than through serious surveys made of native fauna (Arment 2004). During the nineteenth century, those participating in what would become the formalized profession of natural history, underwent a progression from being observers, then collectors, then explainers, then systemacists, to finally experimenters, more diligent examiners, and proto-scientific researchers (Arment 2004; Elman 1977).

The assertion made by the eminent French naturalist Georges Louis LeClerc de Buffon that due to impoverished environmental conditions existing in the New World, species there were degenerate and of a correspondingly smaller size than those in Europe, was an insult to what would become the dominant creed of American exceptionalism. No less a personage than President Jefferson, himself an accomplished amateur naturalist, produced a series of comparative tables to send to Buffon, along with a stuffed moose, to challenge the Frenchman's biased assertions (Barber 1980). What American naturalists most hoped for and eagerly sought was the discovery of a truly giant American species that would clearly put an end to the slander from the Old World (as will be shown, such a confrontational attitude was to play a major role in the uncritical acceptance of the existence of the enormous sea serpent observed in Gloucester). Enter Charles Willson Peale, he of the family-run museum of curiosities, and an active member of the Philadelphia Philosophical Society, who, in 1801, led an expedition to the then remote wilds of upstate New York. Five months later, Peale returned home in triumph with the bones of not one, but *two* mastodons. These were soon mounted, creating what were to that time, only the second paleontological reconstructions in existence. One model remained in Philadelphia to generate revenue for his family museum, and the other was sent to tour Europe – where presumably it would set 'those Frenchies' straight. Crowds flocked to the Peale Museum to see the fossil skeleton that, in Barnumesque fashion, was advertised as being 'the LARGEST of terrestrial beings,' and 'the ninth wonder of the world' (Barber 1980). Finally the emasculation of America was put to an end once and for all. As Emling (2009) explains: 'The mastodon was far more than an exciting exhibit; it was a natural rallying cry for a new nation... The discovery of the mastodon actually helped to redefine America's image as a more powerful nation.'

The public marvelled at both the size of the creature as well as the knowledge that it was extinct (though Jefferson, for one, thought otherwise, believing that living mastodons might still exist in unexplored regions of the country, and charging Lewis and Clark to keep an eye out for them during their 1803 expedition). And if giant mastodons had once, and possibly still, existed, why not then giant sea serpents as well? Finding one of *those* would *really* establish that the new nation was a major player on the World stage.

Popular natural history in late modernity existed in a peculiar and outmoded vacuum far removed from reality (Barber 1980): its biology was obtained from Lamarck (species are fixed but individuals could inherit acquired characteristics passed on from the actions of their parents), its philosophy from Paley (diversity of life was demonstration of God's benevolence to be understood through natural theology), and its cosmology was from Archbishop Ussher (instance of Creation empirically dated, based on a careful reading of the Old Testament in terms of who begat whom, to preciously 9 am, October 23rd, 4004 BC).

Loxton and Prothero (2015) follow Lyons (2009) in noting the 'truly spectacular coincidence' wherein sea serpent sightings increased in the nineteenth century in parallel to new fossil discoveries that suggested that giant marine reptiles had once plied the world's ancient oceans: 'As the world reflected on a recent rash of sea serpent sightings on the East Coast of North American (especially in and around Gloucester, Massachusetts, in 1817), astonishing fossil finds shone a warm light of plausibility on the modern serpent legend – a spotlight under which the sea serpent thrived.'

In 1811, Mary Anning, a 12-year old working-class girl, whose life continues to interest novelists today (e.g. Chevalier 2010; Thomas 2010), found the first fossil of a complete ichthyosaur while walking along the towering sea-cliffs near Lyme Regis in Dorset, southern England (this is the location where discoveries of new ichthyosaur species continue to occur – Nature 2019). Anning, the inspiration behind the tongue-twister 'She sells sea-shells on the sea-shore...,' would go onto become the greatest fossil hunter of her day, selling finds in order to support her family, to many museums, where they remain on prominent display. In so doing, Anning's discoveries, such as of the first complete plesiosaur in 1824, directly contributed to the understanding of geologic 'deep time' (Fuller 2001), as well as the theory of evolution. It is not without justification that Emling (2009) subtitled her biography of Anning: *Dinosaurs, Evolution, and the Woman Whose Discoveries Changed the World.* Her prologue begins: 'Pre-Victorian England exemplified a powerful period in the history of science, a time when one never-before-seen monster after another was being cajoled from its Jurassic tomb, drawn out into the light of day where it could blow holes through the Biblical account of the earth's history.'

The term 'monster' is not out of place since the enormous eye sockets of the ichthyosaurs certainly did make them 'look devious, even evil' (Emling 2009). But whereas ichthyosaurs resembled dolphins, plesiosaurs were unlike anything seen before; as such, they were truly monstrous in appearance. Imagining an ocean teaming with such creatures, such as envisioned by Mary Anning's friend Henry De la Beche (Figure 1.4), was a terrifying prospect indeed. Another early supporter of Anning, the geologist William Buckland, who was the first to describe a dinosaur species from fossils (*Megalosarus*), and whose son Francis would become a noted zoologist and ardent contributor to later Victorian debates about sea serpents, thrilled audiences with his depictions of these oceans of antiquity (*in* Emling 2009):

> During this period of monsters...huge lizards, their jaws like crocodiles, their bodies as big as elephants, their legs like gateposts and mile-stones, and their tails as long and as large as the steeple of Kilington or Long Habro. Take off the steeple of either church, lay it in a horizontal position, and place legs on it, and you will have some notion of the animal's bulk. These stories look like fables, but ask not your indulgence to believe them. There the monsters are, and I challenge your incredulity in the face of the specimens before your eyes. Disbelieve them if you can.

Indeed, presented with such panache, who could disbelieve otherwise.

The race to discover bones from which to identify new species of prehistoric animals which led to battles between ambitious and opinionated fossil hunters in the pages of professional journals, also carried

Figure 1.4. 'Duria antiquior, *a more ancient Dorset,' an 1830 painting by geologist Henry De la Beche depicting the primeval seas off the Lyme Regis coast filled with Mary Anning's remarkable discoveries.*

over into the popular press (Jaffe 2000). 'Dinomania' ensued (Torrens 1991). The public could not get enough from viewing the few plaster reconstructions that existed in select museums (Rea 2001). The void was filled by artistic imaginations (Rudwick 1992), such as the *Duria antiquior* depiction of Mary Anning's discoveries. But no representations were more famous than the sculptures by B. Waterhouse Hawkins, under the direction of Sir Richard Owen, conceiver of the term 'dinosaur' and the best-known naturalist in England, that were shown at the Crystal Palace complex at Sydenham for the 1854 Great Exhibition (Rupke 1994). Many of these sculptures can still be seen today. It is but a small leap to go from the fantastical drawings in *The Book of the Great Sea Dragons* (Hawkins 1840), to the rise in popular 'prehistoric fiction' by Verne, Conan Doyle, Rice Burroughs, and others a half century later (Angenot and Khouri 1981), and then to 1933's *King Kong* and the first sightings of the Loch Ness Monster immediately thereafter (Loxton and Prothero 2015).

Sea serpents, including the one sighted by large crowds around Gloucester, became wrapped up in the professional maturation of the discipline of paleontology. In 1827, the great botanist Sir William Hooker wrote: 'The recent discoveries of Plesiosaurus and Megalosaurus have made demands upon our powers of credence far greater than the serpent, the descriptions of the latter animal have received very little trust, and even much ridicule and contempt,' before he concluded that sea serpents must now be given their due credence and 'be assumed as a sober fact in Natural History.' Six years later, Robert Bakewell, in his influential textbook, *Introduction to Geology*, stated 'I am inclined to believe, that the ichthyosaurus, or some species of a similar genus, is still existing in the present seas' (*in* Loxton and Prothero 2015). To which the Yale University professor of chemistry, Benjamin Silliman (an individual who played an important role in the Gloucester Sea Serpent story), commented that it was more likely the sea serpents were to be plesiosaurs, given the latter's elongated necks.

The idea of the sea serpent being a relic plesiosaur that had escaped extinction was swiftly taken up by other naturalists, a century before the Loch Ness phenomenon (Loxton and Prothero 2015). In 1835, geologist John Ruggles Cotting wrote that 'the sea serpent, which has frequently visited the waters of New England, is supposed to belong to the genus Plesiosaurus. Its existence has been so often attested by thousands of competent witnesses, that its identity is no longer problematic.' Fourteen years later, fellow Bay State citizen and Harvard professor Louis Agassiz, most well-known for his geological work, echoed a similar belief (*in* Loxton and Prothero 2015):

> I have asked myself...whether there is not such an animal as the Sea-Serpent. There are many who will doubt the existence of such a creature until it can be brought under the dissecting-knife; but it has been seen by so many on whom we may rely, that it is wrong to doubt any longer. The truth is, however, that if an naturalist had to sketch the outlines of an Ichthyosaurus or Plesiosaurus from the remains we have a them, he would make a drawing very similar to the Sea-Serpent as it has been described...I still consider it probable that it will be the good fortune of some person on the coast of Norway or North America to find a living representative of this type of reptile, which is thought to have died out.

Newman (1848), like many naturalists, was in agreement, writing that 'the descriptions of the Northern sea serpents [i.e. those observed in Scandinavia and New England] agree with the supposed appearance of some of the antediluvian species.' The eminent naturalist Philip Henry Gosse, who would later be lauded by such distinguished evolutionary biologists as Charles Darwin and Stephen Jay Gould, was even more definitive. In the immensely popular *Romance of Natural History* of 1861, he writes to 'express my own confident persuasion that there exists some oceanic animal of immense proportions, which has not yet been received into the category of scientific zoology; and my strong opinion, that it possesses close affinities with the fossil Enaliosauria' (i.e. a type of plesiosaur).

Evolutionary theory played a complex role in the discussion of sea serpents and sea serpents a complex role in the early discourse on evolution (Lyons 2009; Regal 2011). The big problem for theology was Cuvier's assertion that some monstrous animals had become extinct, their presence known today only from their fossilized remains left behind. Throughout the nineteenth century, the theological controversy over fossils and extinct animals really had more to do with the concept of time than it did the existence of monsters. This instilled the unsettling feeling of what McGowan-Hartmann (2013) terms 'temporal vertigo': 'Imagine waking up from one's cosy six thousand years of history to the sudden realization that vast ages, hitherto undreamed of, had existed in a past that stretched back an unthinkable distance.' Furthermore, if fossils existed of antediluvian beasts, might not it be that in lands deemed remnant of the primordial world, in other words, the vestigial Eden that many imagined the New World to be (Merchant 2003), that living forms of such creatures might still roam undetected?

Throughout the nineteenth century, the discovery of new and extraordinary species continued to fuel the public imagination that all things might be possible. Europeans were thrilled and perplexed by exotic creatures reported from North America. Some species, such as the moose, seemed so outlandish they were considered myths (Regal 2011). And no animal was more persistently sighted and eagerly sought after than the Great Sea Serpent. No wonder that the enigmatic creature is mentioned in the opening paragraph of Barber's 1980 comprehensive overview of natural history, along with the gorilla, another beast whose legendary existence had been suggested since Classical times. Because of this, Bolster (2012) is therefore correct in his statement that 'no marine environmental historian worth his or her salt can afford to ignore early nineteenth-century sea serpents.' Plesiosaurs, after all, were no less fantastic than the hippogriff of Greek mythology (Regal 2011), a being well-represented in medieval bestiaries (Barber 1999; White 1984), and one which Loxton and Prothero (2015) believe to have been the inspiration behind the modern invention of the Great Sea Serpent during the late Renaissance. Indeed, the nascent science of palaeontology was infused with mythical dragons (Mayor 2000; McGowan-Hartmann 2013), no more so than the gothic reconstructions of the dinosaurs for the 1854 Crystal Palace exhibition.

During the great 'age of wonder' (*sensu* Holmes 2008), the efforts of amateur naturalists and learned professionals alike, concerning sea serpents, were given equal and due attention by an enraptured lay public. Indeed, it was unlikely for any natural historian at that time, especially one with marine interests, *not* to have had an opinion about the existence of sea serpents. The study of sea monsters was not only a legitimate, but also a meritorious field of scholarship (Lyons 2009). Harvard professor Jacob Bigelow, for example, in an 1820 article entitled 'Documents and Remarks Concerning the Sea Serpent,' published in the *American Journal of Science and Arts*, considered the sightings in Gloucester three years prior to represent 'the most interesting problem in the science of natural history.' Many of the leading scientists of the day, as for example Richard Owen, Charles Lyell, Louis Agassiz, Joseph Banks, M. (Georges) Culvier, J.D. Hooker, Constantine Rafinesque, Samuel L. Mitchell, Benjamin Silliman, Thomas Say, and William D. Peck, all felt drawn to comment upon the subject, filling the pages of august journals such as *The Zoologist, Nature, American Naturalist, Quarterly Journal of Literature, Science, and the*

Arts, the *American Journal of Science and Arts*, and others, with their speculations and theories. Indeed, 26 papers about sea serpents were published between 1877 and 1899 (Lyons 2009), including 19 in *Nature* (Westrum 1979), in addition to a flurry of contemporaneous books: *Sea Monsters Unmasked* (Lee 1883), *Mythical Monsters* (Gould 1886), *Monsters of the Sea: Legendary and Authentic* (Gibson 1890), *The Great Sea-Serpent: An Historical and Critical Treatise* (Oudemans 1892). Sir Richard Owen, himself, kept a file called 'Sea-Serpentiana' (Rupke 1994), and was actually deputed by the Admiralty to investigate sea serpent sightings made from British naval vessels (Barber 1980). His lively engagement in debating the 1848 *Daedalus* sighting of a reputed sea serpent, led to him being dubbed the 'sea-serpent killer' by the Prince Regent, Albert (Regal 2012).

By the end of the nineteenth century, many naturalists and much of the general public accepted as a given that some mysterious sea monster had indeed haunted the coastal waters of New England decades before. The word 'monstrous' derives from the Latin *monstrum*, meaning portent or prodigy (Regal 2011). In point of fact, imaginary sea monsters did play a portentous role in the culture and maturation of late Victorian natural science. Fearful humanoid monsters in Victorian fiction, such as Dr Frankenstein's patchwork creature, the alter-ego of Mr. Hyde, and Dr Moreau's hybrids, served to raise questions, in the face of Darwinian theory, about what it meant to be human, as well as to provide a warning about the dangers of unshackled science (Schneider 2015). From a zoological perspective, to be regarded as a monster means being either: (1) an atypically large size (e.g. Darwin uses the phrase 'monstrous as a whale' in his 1859 *On the Origin of Species*); (2) an hitherto unknown creature (Westrum 1979); or (3) an individual displaying teratological abnormalities or unusual inter-species hybridizations (Dixon and Ruddick 2013).

Lyons (2009) considers the moniker of the Victorian Age as being the 'age of science' to be better shifted to the 'age of contradictions' or the 'age of transitions' in consequence of the climate of uncertainty that was generated by the promulgation and widespread discussion of competing theories and beliefs. In this regard, the subject of sea monsters played an important role in Victorian discussions about issues of naturalness, novelty, deep time, and professionalism.

There is a long history of human-animal hybrids contributing to defining what humans are and what their relation to the natural world is in terms of either our uniqueness from or union with other animals (Dinnerstein 1976; Ritvo 2010). The fascination with sea monsters as composite creatures was prevalent in Victorian times, most notably in the form of mermaids, which fuelled scientific discourse about Darwinian theory and the chain of evolutionary life; i.e. could these half human-half fish beings be the sea-to-land missing link (Brink-Roby 2008)? Descriptions of sea monsters pushed the boundaries of what was possible. The statement by Gosse (1861) that 'nature has an awkward way of mocking at our impossibilities,' was certainly born out in the form of the platypus and other Australian animals which were then becoming more widely known (Ritvo 1997). If such odd creatures existed, why not strange monsters from mythology? For some, such as noted naturalist Charles Gould (1886), Darwin's theory gave credibility to the possible existence of composite creatures such as griffins, satyrs, hippogriffs, and mermaids. In 1819, for example, when New England was preoccupied with the mysterious serpentine creature seen along its coast, explorers discovered a species of tapir in Asia, a surprising find given that the genus had hitherto been thought to be restricted to South and Central America. However, the presence of tapirs in south-eastern Asia shouldn't have been a complete surprise since their existence was well-established in Chinese and Japanese legends. And what a weird looking chimera it turned out to be: a bear-like body divided equally into a black front and white rear section, sporting the trunk of an elephant, the eyes of a rhinoceros, the tail of a cow, and the brown-coloured feet of a tiger (Heuvelmans 1990). If such oddities were possible, why then not the Dr Moreau-esque 'Moha-moha' sea monster of legend from Australia, which was certainly no more bizarre looking than its neighbour, the chimerical platypus (France 2017)?

To Victorians, sea monsters 'embodied alternative modes of understanding about the natural world' (Ritvo 1997). Because of this, by their very nature, or possibly better, their 'un-nature', sea monsters, especially those that so obviously seemed to be hybridizations of recognized animals, challenged Linnaean classification. By destabilizing taxonomic categories, sea monsters raised questions about the ability of experts and therefore scientific authority (Lyons 2009; Ritvo 1997). Furthermore, ascribing Latinized names to conveniently classified animals forms the very foundation of natural science, and in terms of metaphysics, is a way of maintaining the perception of humankind's superiority (in the form of stewardship or dominance) over the rest of the natural world. To taxonomically name and classify an animal is to domesticate and in a sense to own it. Therefore, if composite creatures such as sea monsters defy categorization, then our place of wisdom at the apex of Creation must be re-examined, something that fed into the unease raised by Darwin's revolutionary theory. In short, to Victorian naturalists and scientists, sea monsters always remained the most frustrating of fauna.

As previously mentioned, the discovery and recognition of the fossilized bones of ancient 'monsters' throughout the 1800s had a profound effect upon Victorian society. In regard to composite creatures, the noted palaeontologist William Conybeare (1824) described the plesiosaur as being 'a snake threaded through the body of a turtle.' The discovery of fossil remains of hybrid monsters from the past, such as 'fish-lizards' (i.e. ichthyosaurs) and 'almost-lizards' (i.e. plesiosaurs), provided both a scientific basis for mythological sea monsters as well as making plausible the possibility that they might have survived to historic times (Mayor 2000). The result was, according to Lyons (2009), that such Victorian paleontological discoveries gave a level of scientific legitimacy to sea monsters that had hitherto not existed; nor has it since.

Given that the seas were relatively unexplored, it was completely plausible that strange and as yet unknown creatures, possibly living fossils, might inhabit the depths (Newman 1848). For did not the old adage state *In mari multa latent* ('in the ocean many things are hidden')? Sea monsters therefore became the subject of impassioned debate as high profile naturalists attempted to understand what such creatures were rather than simply dismissing them out of hand. In a chapter entitled 'Swimming at the Edges of Scientific Respectability,' Lyons (2009) argues that the sea serpent became the locus for several boundary disputes in the culture of Victorian science. The existence or nonexistence of sea serpents was used to promote alternative views of Earth history and thus the legitimacy of the fledgling sciences of geology and palaeontology. Also the fact that most sightings were made by amateurs, provoked a big debate about the credibility and character of eyewitnesses and indeed the role played by professional scientists in their goal to become the sole voice of authority on all matters of natural history; in short, the definition of what counted as being valid evidence. The anti-establishment opinion was to trust the reports of salty sea dogs observing live animals rather than museum landlubbers studying dead fossils (Newman 1848). In contrast, Sir Richard Owen insisted that no number of eyewitness testimonies counted in the absence of physical evidence confirmed by a museum-based authority (Regal 2012). In other words, legalistic endorsement of eyewitness testimonies, which as we will see became the backbone for the natural history investigation of the Gloucester Sea Serpent, and which can be influenced by many subjective factors (Champagne 2001; Paxton 2009; Paxton and Shine 2016), is no way to go about doing objective science. All the magistrates in the world were not as important as a single specimen; and the opinions of amateur natural historians, many of whom were aristocratic gentlemen, did not count in the eyes of those who were part of the maturing discipline of true science. However, it is worth noting that what we now recognize as constituting real science as practiced by today's professional zoologists and botanists, is very different from the root of those disciplines in proto-scientific natural history as conducted by amateurs in the nineteenth century (Clifford *et al.* 2006; Peters 1980). At that time, a wise amateur naturalist, as for example, Henry David Thoreau (Dean 1993), might very well be more correct about the workings of nature than a university professor such as Louis Agassiz.

Concerning the 1877 *Osborne* encounter with a sea monster (France 2017), Richard Owen scribbled the following in the margins on his copy of letters sent to him directly by the ship's commander: 'The object or phenomena may have been unknown to the observers [but they] were not necessarily caused by a 'monster'' (Regal 2012). And in his marginalia on his copy of a newspaper article from that time entitled 'The Sea-serpent Again,' in which the anonymous author states that 'there is not the slightest doubt that there are monsters in the sea of various kinds yet unknown to man,' Owen simply states 'what proof of their monstrosity?' (Regal 2012). And so by the end of the nineteenth century, the relationship between science and sea monsters coalesced to one of opposition and hostility (Gage 1896). It would be many decades later before amateurs regained the podium with the advent of cryptozoology, largely disseminated by the un-refereed internet, and often undertaken through a desire to prove the experts wrong (Dendle 2006), most of whom avoid sea-serpentology anyway due to its unsavoury contemporary reputation of being an embarrassing pseudoscience (Schembri 2011).

1.2 SEA SERPENTS AND CRYPTOZOOLOGY

In 1829, the English geologist Gideon Mantell published the seminal paper 'Age of Reptiles' in which he discussed not only the extinction of species but also their slow progression from primitive reptiles to advanced mammals over a period millions of years. All this, fully decades before Darwin. This might be considered as one of the benchmarks in the birth of the science of palaeontology. But the idea that God would go to all the trouble to create species that would disappear before humans were around, was an affront to orthodox belief. It was absurd and must be challenged. Enter Reverend William Kirby, he of 1836's influential *On the Power, Wisdom, and Goodness of God as Manifested in the Creation of Animals and in their History, Habitats, and Instincts*, whom would have none of it: 'Who can think that a being of unbounded power, wisdom, and goodness should create a world merely for the inhabitation of a race of monsters, without a single rational being in it to glorify and serve him!' (Emling 2009). Ergo: all species that ever existed *must* still be in existence...*somewhere*. In other words, all species of dinosaurs identified by their fossilized remains have to be around, perhaps hidden from common view. The truth was out there...somewhere. Kirby's book might be considered as one of the benchmarks in the birth of the pseudoscience of cryptozoology.

Cryptozoology is the 'investigation of purported animal species which are undescribed by traditional zoology because their alleged existence is supported by insufficient anecdotal or physical evidence' (Arment 2004). A suite of professional-looking journals on the subject has been produced, most notable perhaps for their short-lived existence due to lack of support from established scientific funding bodies:

Cryptozoology (1982-1996), *The Cryptozoology Review* (1996-2004), *Kraken: Archives de Cryptozoologie* (2008-2011), and the *Journal of Cryptozoology* (2012-present). As Rossi (2016) found in his survey of the subject matter of 119 papers and field reports published during the entire run of *Cryptozoology*, cryptozoologists have a particular penchant for sea serpents and lake monsters, second only to their fixation with 'relict hominids' such as Bigfoot and the Yeti, etc.

Bernard Heuvelmans (1988) provides a glimpse into the world of cryptozoology that he was largely responsible for creating and promulgating. The assembling of testimonial evidence from newspapers creates a corpus of data, the abundance and concordance of which, it is believed, gives weight to the assertions made. In this, he states, 'the whole process of cryptozoological work can best be compared with legal procedures, with apparently unknown animals put on trial.' By so doing, cryptozoology satisfies its major objective of proving the existence of mythological beasts. Cryptids are therefore ethno-known animals for which concrete evidence 'does not yet exist' (Arment 2004). For many critics, it is the insertion of that single word 'yet' that sidetracks the entire process from orthodox science into the woolly realm of science fiction.

Cryptozoology operates on circumstantial evidence as the best marker of legitimacy in support of its claims. For example, to Heuvelmans (1988), the fact that so many large cryptids of similar description inhabit lakes in the same climatic zone around the world, means that they share the same uniform ecological niche, and that this in turn, 'is, of course, the best evidence one could wish for its [i.e. the creature's] existence.' Yes, the very 'best' evidence. No need to follow centuries of established zoological protocol and the requirement that there should exist at least a single representative specimen of a proposed species that is registered, catalogued, and studied by specialists in evolutionary biology, anatomy, and taxonomy at an academic institution. Such a requirement – i.e. 'show me the body,' as made from legions of naysayers (e.g. Shermer 2003) – is, according to Heuvelmans (1988), 'totally arbitrary' and unneeded for cryptozoology to function. That is of course true, but it certainly is not science.

Over the years, many have justifiably taken issue with Heuvelmans' methodological approach, as practiced in his 1968 classic *In the Wake of Sea-Serpents* (translated from the original French of three years before). Meurger (1988), for example, became infuriated by Huevelmans' attempts at scientific legitimacy, accusing him of displaying sloppy scholarship in assembling so-called evidence from newspapers, incredible selective bias and manipulation of facts, and a complete failure to identify overt hoaxes and obvious misinterpretations of natural phenomena in his desire to construct an irrational classification scheme of sea serpents that is delusional at best, or outright dishonest at worst; something that, to use the parlance of the moment, is a house of cards assembled from a Trumpian world of 'alternative facts.'

One measure of what can be regarded as 'good' science is an adherence to the Law of Parsimony, or what is commonly referred to as 'Occam's Razor', named after the fourteenth-century English Franciscan, William of Occam. The precept of this philosophy of praxis – *Pluralitas non et ponenda sine necessitate* (plurality must be posited without necessity), or in other words, no more assumptions should be made than are necessary to explain a certain occurrence (Das 2009) – is that when positing scientific explanations for any phenomena, especially for those in which there is some uncertainty, it is necessary to first consider the simplest explanation possible, and if that does not prove sufficient, only then to progress to alternative hypotheses of increasing complexity. Occam's Razor then is a systematic and heuristic approach to eschew preconceived biases and self-fulfilling opinions, a symbolic razor to 'shave away at the ambiguous metaphysical concepts that either cause confusion or misunderstanding in people' (Konar 2009); in other words, a way to prune theories of embellishments (Das 2009). Avoiding the practice of parsimony compromises the scientific process, and strays dangerously into the disreputable wilderness of what is commonly referred to as 'pseudoscience'.

Pseudoscientists often work alone and without financial support, and can be characterized by a lack of formal education in the subject of investigation, a general ignorance of common scientific praxis, a blind assurance in the accuracy and infallibility of their work, and a stubborn resistance to others' contrary opinions, thereby fostering feelings of being discriminated against and leading to a tendency toward paranoia (Das 2009; Shermer 2010). In contrast, real scientists collect empirical data to test and validate falsifiable hypotheses based on refutability toward an end of reaching consensus (Schembri 2011). Pseudoscience tries to appropriate the prestige of science, and although it may mimic the outward trappings and protocols of the former, it falls short of meeting the accepted standards of the practice of science (Bunge 1984). In consequence, pseudoscience lies outside the mainstream of legitimate science and its proponents are frequently the object of ridicule. Although pseudoscientists claim to do science, in reality they are faith-driven believers who are unwilling to modify their ideas on the basis of new and contradictory evidence (McKinney 2013a).

Despite noble attempts by Arment (2004) to salvage the bruised reputation of cryptozoology by writing about how it should follow established scientific protocol, when examined through such a lens of scientific legitimacy, cryptozoology fares rather badly due to its reliance on anecdotal over physical evidence as the basis from which to draw conclusions. In other words, what is lacking in terms of tangible evidence is supplanted by irresolute assumptions. The result is that, rather than considering the simplest explanation first, cryptozoologists immediately jump to conclusions about the existence of new or mythical creatures that are unwarranted by the nature of the data obtained. As Konar (2009) states, the motivation of cryptozoologists 'to discover these purported creatures is so strong that they will create complex, hypothetical situations stating that this alleged creature is alive and researchable. However, they fail to see the crucial details that might negate their findings or contradict their theories.' By failing to follow a logical examination of hypotheses, from the simplest to the most complex, cryptozoology, Konar believes, 'is not actually a science at all, and should be considered merely a pseudoscience.'

For Hill (2011), cryptozoology is 'sham science', whose major *modus operandi* is confirmation bias; i.e. the interpretation of new evidence to support one's *a priori* beliefs. Cryptozoologists err from the very get-go by assuming a mystery creature is out there waiting for them to find. They begin by asking the wrong question: not 'What is that?', but 'Is that a cryptid?', and by so doing, narrow the range of possible and parsimonious interpretations (Hill 2011). In this, cryptozoologists are remarkably similar to many Ripperologists who force evidence to shore up their own pet theory about which, usually historically eminent Victorian, was 'Saucy Jack.' As one real criminologist once stated, expressing his frustration over all the unfounded theories, arm-waved about in hundreds of popular books, concerning the identity of the Whitechapel murderer, should the unlikely day ever came when the truth would be finally and convincingly revealed, all the armchair amateur theorists would be expected to let out a collective and exasperated cry of '*Who*?!'. For Hill, a scientist, cryptozoology is 'a jumble of poor-quality evidence, a ton of hype, rampant speculation, and unfounded assumptions, even conspiracy theories and too often paranormal explanations.' Cryptozoologists can often be disturbingly naïve, egregiously superficial, wholly amateurish, prone to enormous nonsensical assumptions, ad hoc theories, stubbornly arrogant, frustratingly ignorant, unyieldingly biased toward their own reasoning, unwilling to countenance alternative hypotheses, unethical data manipulators, and as a group, demonstrate a lack of critical analysis, conservatism, and restraint that is normal in sound scientific research (Naish 2006, 2011, 2012a; Rossi 2016; Woodley *et al.* 2012). Cryptozoology is haunted by cranks (Naish 2011) who promulgate alarming and bizarre claims (Woodley *et al.* 2012) in articles published in magazines written for laypeople rather than scholars (Naish 2006), and is simply 'terrible' science, whose conclusions depend more on the imagination and preconceptions of its proponents than anything to do with actual biology (Naish 2012a). As such, cryptozoology is unworthy of serious consideration, and its very name should be abandoned and replaced by 'mystery animal research' (Naish 2011, 2012b).

McKinney (2013b) believes that, for the most part, the label 'cryptozoology' is unneeded as the practice overlaps with established fields of scientific discovery. It is unnecessary to coin a new term and imagine a separate scientific discipline since much of what cryptozoology is, already exists. Once a newly discovered species becomes formally recognized and an accepted reality, it enters the purview of established scholarship. This is simply the normal process. By losing its object of study in this way, cryptozoology becomes redundant. What cryptozoology is, is simply a return to nineteenth-century natural history based on the exploration and discovery of animals known only from hearsay; it is really nothing other than a physically adventurous form of ordinary science.

But McKinney (2013b) goes on to state that there is one critical way that cryptozoology is in fact unique, and that is in its dogged researching of something that simply does not, or could not, exist. The stubborn belief in the existence of a cryptid, the evidence for which is slim to non-existent, based on nothing more than anecdotal stories, is 'sheer pseudoscience.' As he states: 'It may be great fun, and adventurous, and every once in a while it may lead to a real discovery, but none of those things makes it science.' In this regard, the sole sense that cryptozoology should be retained as a useful term, is in a pejorative sense; i.e. its referral to a set of pseudoscientific beliefs and pursuits.

Because of 'expectant attention' (*sensu* Gould 1934) and resulting 'psychological contagion', Loxton and Prothero (2015) consider cryptozoology to be the ecology of imagination, and sea serpents and lake monsters to be creatures of the mind, not of nature. Although not the first to make the connection, they build a case that the Loch Ness Monster owes its origin to the media hype and success of the film *King Kong* which served to create a visual illusion of prehistoric creatures in a modern context. So just as sea serpents in the nineteenth century arose out of art, were altered in relation to changing fashions – notably the discovery of dinosaur fossils, Nessie – a catchall term for anything puzzling observed in the loch – was born of magic: a fusion of the kelpie of Scottish folklore and the movie magic of Hollywood (Loxton and Prothero 2015). Aquatic cryptids are therefore not real species; they are social phenomena. As such, cryptozoology is the study of manufactured legends, and has little or nothing to do with zoology per se.

Cryptozoology, due to its: (1) ignoring of the need for burden of proof to assuage reasonable doubt, especially in the light of the dictum that extraordinary claims necessitate extraordinary evidence; (2) failure to adhere to the scientific creed of falsifiability, and in consequence, its inability to be self-corrective in the generation of a body of cumulative evidence; and (3) bypassing of the established peer-review process, produces results deviant to orthodoxy and has, in consequence, more in common with the belief systems in religion and perhaps politics than it does mainstream science (Schembri 2011). Loxton and Prothero (2015), in their seminal *Abominable Science*, a book as well-regarded by scientists (e.g. Hoch 2014) as it is loathed by pseudoscientists (e.g. Bauer 2013), make the claim that most monster hunters, including the illuminati of famous cryptozoologists, are amateur enthusiasts ignorant of the rules of science, and who customarily have no advanced degrees and relevant training in the basics of ecology, palaeontology, or field biology. As a result, their anecdotal tales of gripping narratives masquerade as being real science.

And certainly the reputation of cryptozoology is not improved any due to its failure to emphatically divorce itself from 'paracryptozoology' or 'supernatural zoology', which is concerned with wild claims of the existence of paranormal or mythical animals such as unicorns, merbeings, dragons, centaurs, and the like, and often affiliated with such epiphenomena as alien abductions, crop circles, demons, ESP, the Bermuda Triangle, etc. (Mclendon 1984; Rossi 2016). Because of this, the best that can be said of cryptozoology is that it is a transitional field situated between mythology and zoology (Schembri 2011). Against all evidence to the contrary, the pseudoscience of cryptozoology is kept viable, not by real data and actual facts, but by the public's enduring 'interest and fascination in the possibility of the existence of fairytale creatures' (Das 2009). He continues:

> Cryptozoologists make assumptions to initiate and advocate their claim that these elusive and sometimes fictional animals are currently living. These rare organisms are supposedly scarce enough to be able to evade detections, but are common enough for humans to have been purported eye witnesses of these animals. Several images, stories from supposed first hand eye witnesses, and archaic tales do not make these animals seem true, but the validity of this [sic] data is immersed in ambiguity. The promulgators assume that these animals have been able to breed and survive without scientists ever able to find a corpse, skeleton, or remains of these creatures.

Meurger (1988) considers that aquatic cryptids are 'real' only in the sense that they are mental constructs that have their origin in folklore and exist within a mythological landscape. The whole idea of cryptozoology in transforming such fables into zoological speculations is wrong, based as it is on naturalistic naivety. To focus on zoological or natural interpretations of historical descriptions of ancient monsters – a common practice among cryptozoologists to create the impression of precedence as a strategy to give weight to modern contemporary sightings, and also of skeptics desirous to disprove such claims as being merely commonplace and misidentified recent phenomena – is to miss the point. Folklorist belief, insists Meurger, should not be so naturalized; for doing so entraps one in a web of illusory facticity. Cryptozoology, in this regard, belongs to the purview of folklore, *not* of zoology.

Despite its unsavoury reputation, cryptozoology, according to Dendle (2006), is of significance, if only from a psychological rather than a zoological perspective. Modern cryptozoology provides several psychological benefits. One such role played by cryptozoology in contemporary society is to recapture a sense of wonder, mysticism, danger, and surprise in a world that is now fully charted and over-explored (Dendle 2006), and in consequence, deemed to be somewhat boring. Many cryptozoologists operate in a haze of romanticized nostalgia for the bygone era of scientific discovery, when much of the world was an unknown wilderness. They often anachronistically envision themselves to be the last stalwarts of those noble naturalist explorers of bygone times who sought out and brought back evidence of hitherto unknown species; in other words, zoological treasure seekers cut in the mould of Indiana Jones. Not for cryptozoologists the emasculated world of ivory tower laboratory science; instead, they – and almost all of them are male – dream of discovering their own Skull Island through dogged determinism and a sense of their own rugged, *Boys' Own* adventurism. Furthermore, by identifying with antecedent naturalists from the nineteenth century, cryptozoologists relive the Us-against-Them, amateur versus academic, mindset that characterized that earlier time in the development of proto-science (Regal 2011).

Dendle (2006) believes that a second psychological role played by cryptozoology in contemporary society is to provide a vehicle for articulating resentment of, and defiance against, a scientific community perceived as monopolizing the pool of culturally acceptable beliefs. To be a cryptozoologist is to proudly position oneself as a deliberate challenge to mainstream science; it is to be iconoclastic, a way to put those highfalutin Oxbridge-Harvard professors, with their know-it-all arrogance, in their place because *you* know, Agent Mulder-like, where they don't, not only that the truth is out there, but that you yourself can explain those truisms. In this mindset, a good theory or belief system can persist not only despite the absence of confirmatory evidence, but actually *in* spite of the presence of contradictory evidence. In this regard, cryptozoology operates as a branch of parascience (*sensu* McLendon 1984), joining such other fringe beliefs as denying that AIDS is related to HIV, or alternatively, insisting that autism is related to vaccines. In this regard, Bernard Heuvelmans' (the father of cryptozoology) confrontational attitude to the mainstream of those whom he referred to as the 'dictators of science', reflected his personal hatred of authoritarianism, something fostered by his by living under Nazi occupation (Regal 2011).

In an era immersed in Baconian rationalism, wherein even ecology, frequently that most problematic of scientific disciplines, endeavours to ascribe to a Popperian empiricism (Peters 1991), belief in the existence of sea serpents and their freshwater cousins, lake monsters, seems romantically anachronistic at best, or outright delusional at worst.

Certainly the case for the existence of such aquatic animals has taken a number of jarring hits over the last several decades, notwithstanding the self-admitted in one case, and near certitude in the other, of hoaxes perpetrated in the form of the iconic photographs of those plesiosaur pinup models 'Nessie' and 'Champ' (Bartholomew 2012; Radford and Nickell 2006; Williams 2015). Contemporary scientists have explained many sightings in terms of natural phenomena that went unrecognized at the time. Merbeings have long been variously identified as sirenians or pinnipeds (Hutchins 1968), as have krakens, the giant squid (Ellis 1994). Famous sightings of individual sea serpents, such as that by the crew of the *Daedalus* in the southern Atlantic Ocean in 1848 or that by Egede off the coast of Greenland in 1734, may be explained by surface skim feeding or the waving about of an excited penis of particular cetaceans (Galbreath 2015; Paxton *et al.* 2005). Putative sea serpent carcasses washed up on beaches or pulled up in fishing nets, which are referred to by the wonderfully onomatopoeic word 'globsters', are inevitably found to be decomposing basking sharks, giant octopi, or other known animals. And my own work, based on empirical relationships existing between animal body size and trophic position in food webs, as measured by stable nitrogen isotope analysis (France *et al.* 1998a), indicates the ecological impossibility of apex predators of the sizes commonly ascribed to sea serpents (France in prep.). Many other such refutations exist beyond these few examples noted here.

When contemporary biologists publish papers related to the mystery of sea serpents, it is: (1) as empirical investigations of the likelihood discovering new species of cetaceans or other large marine animals (Paxton 1998; Raynal 2001; Woodley *et al.* 2008); (2) as screenings of the most likely (i.e.

non-prehistoric or anthropomorphic) candidates (Paxton and Holland 2005; Woodley *et al.* 2011); (3) as empirical studies of eyewitness accounts (Paxton 2009; Paxton and Shine 2016); or (4) reinterpreted sightings of unidentified marine objects as entangled cetaceans or *chelonians* (marine turtles) (France 2016a,b, 2017, 2018; see Deedy 2017). On the other hand, some contemporary zoologists, despite the fragmentary, at best, or complete absence, at worst, of accommodating evidence, and sometimes in spite of the overwhelming evidence to the contrary, persist in believing in the existence of sea serpents (LeBlond and Bousfield 1995; Woodley 2008). No surprise then, that most of the serious and recent academic scholarship on the nature of these imaginary sea creatures has been by social scientists, including anthropologists, folklorists, mythologists, psychologists, and historians (Bartholomew 2012; Brink-Roby 2008; Brown 1990; Burns 2014; Cheezum 2007; Dendle 2006; Lyons 2009; Meurger 1988; Regal 2012).

Within the field of cryptozoology, it is important to distinguish between the publications of scholars (e.g. Heinselman 2001; Mangiacopra and Smith 2007; Radford and Nickell 2006), the efforts of some of whom, albeit not holding academic degrees in aquatic zoology, limnology, or oceanography, should still be treated seriously, but of course critically (Naish 2011), and the expansive corpus of fringe-published or web-based missives of variable absurdity along with the press reportage of ridiculous hyperbole. Overlying all is the traditional view of mainstream and narrow-minded scientists (eggheads) pitted against untrained but passionate amateur cryptozoologists (crackpots) that parallels the situation in the nineteenth century between amateur and academic natural historians (Regal 2011).

But it is not as simple and clear-cut as all that. Take Charles Lyell for example. He was the preeminent geologist of his day, whose writings in his seminal 1829 book *Principles of Geology* about deep time and the progressive development of the Earth and its inhabitants would inspire many, including Charles Darwin. Lyell's thesis, 'uniformitarianism', is based on the belief that 'the Earth has been shaped into its present contours by processes which are still at work today, and which can be observed' (Barber 1980). (This is in contrast to the theory of 'catastrophism' or the belief the Earth was shaped by singular events such as the Biblical flood.) Lyell argued that if the principle of uniformitarianism applies to geology, it must also do so to biology. Or as Emling (2009) phrases it, quoting Lyell: 'Impressionable minds were intrigued by his [Lyell's] conclusion that, because the processes that alter Earth are uniform through time, 'huge Iguanodons might reappear in the woods, and the ichthyosaurs in the sea, while pterodactyl might flit again through umbrageous groves of tree ferns." Sounds very *Jurassic Park*-ish. But the difference is that the reappearance of such creatures was not to be brought about through genetic engineering and the creation of new beings, but rather through the discovery of existing beings that have remained hidden somewhere out there in the then incompletely explored World. So what we may now distinguish as being either 'science' or 'speculation', to use Arment's (2004) convenient dichotomization, was very much conflated in the nineteenth-century world of natural history. And may also linger and be so today as well.

I believe that arguments about whether cryptozoology is or is not a true scientific discipline (e.g. Rossi 2016) may be somewhat off the mark. True, because cryptozoology does not meet the rigors of generating and testing falsifiable hypotheses, it therefore should not be considered a part of science of ecology, as outlined and defined by Peters (1991). But then, using those same criteria, neither should evolutionary biology, natural history, and some restoration ecology really be considered to be true sciences as well (France, in prep.; Peters 1980). And so, cryptozoology, if undertaken by knowledgeable individuals, needn't be automatically relegated to the rubbish heap of fringe or crackpot pseudoscience. Instead, it perhaps may be possible to regard cryptozoology as an anachronistic form of natural history which operates under different evaluating criteria than those used to judge the modern science of ecology.

For example, I have published close to two hundred peer-reviewed papers, most of which deal with some aspect of ecological, often aquatic-based, science. But about a half a dozen of my papers fall under the rubric of what I would consider to be natural history. The ecological science papers use both field experiments of manipulated aquatic systems, and measurements and observations made of populations of aquatic organisms in natural settings. My handful of natural history papers derive from when I was a participant on the Trans-Ellesmere Island Ski Expedition in the Canadian High Arctic. That research was entirely based on rapidly acquired observations made under less-than-ideal circumstances, and involved both quantitative analyses of animal sightings in some cases, and simple anecdotal documentation of behavioural observations in others. (Paper titles include 'The Lake Hazen Trough: A Late Winter Oasis in a Polar Desert'; 'Polynyas as Centers of Organization for Structuring the Integrity of Arctic Marine Communities'; 'Comparison of Avian Species Richness and Abundance of Localized Oases on Ellesmere Island'; 'Non-sterility of High Arctic Icefields on Ellesmere Island'; 'Newly Reported Colonies of Ivory Gulls on Southeastern Ellesmere Island'; 'First Record of the Rough-legged Hawk form Ellesmere Island, Northwest Territories'; and 'Aerial mobbing of a Gryfalcon by Glaucous Gulls', all reprinted in France 2010). One of these publications was actually selected by the journal *Conservation Biology* as being one of best papers in that discipline in the 1990s. But neither this paper, nor the others cited above from the expedition, are science *per se*; rather, they are natural history, which is an entirely different undertaking (France, in prep.; Peters 1980, 1991). (It is worth noting that I also did publish six true scientific papers from the expedition that dealt with trace contaminants in sampled plants, in addition to several environmental papers concerned with litter and equipment abandonment by other polar researchers and the army.) The point is, that all these publications, be they science, natural history, or environmental, have their own particular merit. As long as one is aware of the limitations of conducting natural history based on the collection of observational and anecdotal data in the field, as was I (e.g. citing many studies about the strengths and weaknesses of line-transect sampling methodology), or as is Paxton (2009) in terms of the likelihood of detecting animals based on their respective sizes (for example, his citing of papers concerning line-transect sampling methods), then the investigation of unidentified marine objects or mystery animals can indeed be a valid pursuit. What one must just always keep in mind is that it is natural history, *not* science *per se*. That said, as I have written about previously, there is of course plenty enough bad aquatic zoology science in its own to go around (e.g. DeMelo *et al.* 1992).

1.3 SEA SERPENTS AND FORENSIC HISTORICAL ECOLOGY AND ETHNOZOOLOGY

In the Anthropocene, humans, nature, and their social-ecological interactions have become closely 'entangled' (*sensu* Holm and Taffel 2016), sometimes irreparably so. There is a cardinal need for historical interpretations to become *de rigueur* in conservation biology, lest the practice and its interpretations and proscriptions be considered incomplete (Crumley 1994; Meine 1999). In particular, past societal attitudes and actions are very much ecological driving forces shaping current and future ecosystems (Cronon and McDonnell 1993; Szabo and Hedl 2011). Furthermore, and relevant to the present investigation, as Alexander *et al.* (2017) correctly state, 'historical ecology is a forensic pursuit.'

So under what discipline does the historical forensic study of sea serpents, sometimes referred to as 'sea-serpent*ology*' (as distinguished from 'sea-serpent*ism*,' which pertains to held belief systems), reside? Remember that McKinney (2013b) argued that cryptozoology as a label for the investigation of mysterious animals was unneeded as the practice simply overlaps with established fields of scientific discovery. And I made the case that that field of discovery was not really scientific in the way we now recognize such, but rather an anachronistic return to the traditions of nineteenth-century natural history, which is a different kind of pursuit altogether from that of the modern science of ecology (see Peters 1980, 1991). So, if not science *per se*, what then *is* the study of cryptids (given that both Microsoft Word and Google recognize that term)? Also remember that Meurger (1988) made the case that the study of imaginary animals al à cryptozoology belongs to the purview of folklore, *not* of zoology. However this is not quite as dichotomous an either/or choice as he makes out. For in the three decades since the publication of Meurger's seminal book there has been recognition that the worlds of folkloric beliefs and zoological realities can indeed fall under the rubric of a single and emerging discipline, that of *ethnozoology*.

Ethnozoology is part of the mature discipline of *ethnobiology*, the study of the dynamic relationships between people, biota, and the environment in both past and present times (Anderson *et al.* 2011). Unlike the ephemeral publication history of fly-by-night cryptozoology journals, more than half a dozen ethnobiology journals have become established since the 1980s, with more being added every few years as the discipline continues to expand in recognition and popularity.

The sociological investigation of the relationships of animals to human culture has long been a respected field of study (e.g. Franklin 1999; Mayor 2000; Ritvo 2010). One aspect of scholarship has dealt with the manner in which different societies have addressed the question of establishing classification systems for cataloguing zoological biodiversity (e.g. Burnett 2007; Ellen 1993a; Forth 2011; Ritvo 1997). What is now regarded as ethnozoology, the first general description of the field only being published recently (Alves and Albuquerque 2017), although a few earlier publications exist having used that word (Ellen 1993b; Henderson 1914; Pattazy 2011), formally connects scientific methods with cultural beliefs. Ethnozoology then is a hybrid discipline, structured from both the natural and social sciences (Wikipedia). It is interdisciplinary, subsuming elements from conservation biology, anthropology, archaeology, environmental history, and psychology, in addition to its foundational roots in zoology and ethnology. It focuses on understanding the roles – not just utilitarian but also cultural, religious, artistic, and philosophical (e.g. Posey 2000) – played by animals in human society. An earlier moniker for this type of study was 'folk biology' or 'folk zoology' (e.g. Franklin 1990; Hunn 1978; Medin and Atran 1999), which dealt with people's everyday understanding of the biological world in terms of how they perceive, categorize, and reason about fauna.

Most importantly, ethnozoology entails the ecological, cognitive, and symbolic study of fauna, regardless of whether those animals are either existing or imaginary (e.g. Da Silva Vieira *et al.* 2017; Tidermann and Gosler 2010). In this regard, the present investigation joins previous serious academic studies about aquatic cryptids, including, for example, books by Lyons (2009) and Bartholomew (2012), unpublished theses by Burns (2014) and Cheezum (2007), and journal articles by Stothers (2004) and Starkey (2017), as all being part of the recently formally described discipline of 'ethnozoology'. Contributions to the field by zoologists, as for example by Paxton (2009), Woodley *et al.* (2008), and my own recent efforts (France 2018), all published in peer-reviewed professional (not fringe) journals, are important in providing a scientific perspective, adding to the immediately aforementioned works by social scientists. In so doing, the ethnozoological study of imaginary, mystery, and unidentified animals will advance based on equal parts in consideration of scientific acumen and cultural cognition.

Chapter 2.

NINETEENTH-CENTURY MARITIME CONTEXT

2.1 THE ECO-CULTURAL SEASCAPE

From my present home at the head of the Bay of Fundy it is but a short drive to view the expanses of this dynamic body of water into which millions of litres from the Gulf of Maine – a component of which conceivably at onetime being part of Gloucester Harbour in Massachusetts – are funnelled twice daily to create the world's largest tides. Not far from where I live, the largest wooden ship in the British Dominion of Canada was built in one of the many shipyards that dotted the coastline in the nineteenth century. But today, it is a rare sight to see any ships out in the Bay. The fishing industry is but a vestige of what it was during the 'Age of Sail'. Whales, once so numerous they dominated the churned up waters to feed on the abundant plankton, can today still be seen on special whale-watching tours but in numbers that are but a fraction of what they once were in previous centuries. When such creatures occasionally wash up dead on beaches they are so unrecognized as to be mistaken by the general public and a sensationalist press for being carcasses of sea serpents (Hebda 2015). The overall impression left from gazing at this seascape is one of emptiness of both human and large marine life.

To the south, to where I used to live, coastal New England, or what my Nova Scotian neighbours quaintly refer to as the 'Boston States', is similarly but a muted remnant of the maritime hustle 'n bustle and unbelievable munificence of aquatic life that once characterized the Gulf of Maine. One has only to stare seaward while standing on the wharf beside the famous New England Aquarium at the head of the region's largest port, Boston Harbour, and compare the view with late nineteenth-century photographs of the same (Bunting 1971), to realize how, despite being coastal dwellers, our modern lives are focused on the terrestrial landscape rather than the seascape for sustenance, livelihood, commerce, and transportation. Today, our use of the sea seems to be largely devoted to it being regarded as little more than a recreational resource. Turning to look at the adjacent Aquarium, one is also faced with the grim reality that there may actually be more regional species living inside that building than now persist as residents of the urbanized and depauperate harbour (France 2013) on which the building sits.

Early in Mark Kurlansky's 2008 book *Cod*, a Newfoundland fisherman remarks, while looking out toward the contemporary oceanic horizon, that he sees 'not a whale, nothing.' This depressing state of affairs – that the sea is not immortal – is captured in the title of W. Jeffrey Bolster's (2012) *The Mortal*

Sea, and is a consequence of what Farley Mowat (1997) referred to in the title of his book, as the *Sea of Slaughter*. Oh, but to have had the opportunity to be aboard one of the multitude of ships that plied these waters in the days when the explorers, then itinerant fishermen, and finally colonists, visited and settled the region, when the wealth of the sea's bounty there may have been unmatched anywhere else on the planet.

As he sailed down the coast of Acadia (later Maine) in 1605, Samuel Champlain saw porpoises 'in such abundance that I can guarantee there was never a day or night during which we did not see or hear more than a thousand pass alongside our pinnace.' Captain John Smith noted during his 1616 sail through the region he christened 'New England' that 'scarce any place, but Cod, Cuske, Holybot, Mackerell, Skate, or such like, a man may take with a hook or line when he will.' On one version of his map a vast school of fish is depicted under a ship off the Massachusetts coast. In 1629, a Salem minister stated the abundance of sea fish to be almost beyond belief. At this time, smelt runs would be so numerous that the cascading bodies of the anadromous fish would actually damage fishing weirs. So great was the multitude of whales seen from the shore in 1635 that Massachusetts Bay colonists likened the air being filled with the exhalations of the cetaceans to a dense town cluster of spewing chimneys. Indeed, tens of thousands of baleen whales are thought to have existed in the Gulf of Maine at the beginning of the seventeenth century. Even as late as the eighteenth century, a churchman could comment that 'there can be but few places on Earth where is to be seen such a manifestation of the fecundity of His Creation.' When the Renaissance Europeans first visited the Western Atlantic boreal ecosystem they were simply flabbergasted by the diversity and abundance of marine life compared to the diminished and degraded state of their home waters. In a way, these explorers were heading backward in ecological time (Bolster 2012). As Mowat (1997) details, there were astronomical aggregations of seabirds, such that a single musket shot could kill a bushel-basket full of Eskimo curlews. John Cabot famously, and perhaps apocryphally, mentioned fishing by means of simply lowering a weighted bucket into the sea. Another explorer remarked that 'Cods are so thick by the shore that we hardly have been able to row a boat through them. I killed of them with a Pike.' Whales too were so numerous that they posed a threat to early transportation. More than three thousand striped bass could be caught in the setting of a single net. Schools of mackerel could be nine kilometres long by two kilometres wide, and ten meters thick, containing an estimated seven hundred and fifty million fish. During the seventeenth century, one to three-metre long lobsters, codfish, and sturgeons were routinely caught. So many alewife were captured that they were used, along with lobsters, to fertilize agricultural fields. Seal colonies of tens of thousands were common, and spring whelping patches could cover tens of square kilometres of sea-ice and contain millions of individuals. Remarkably, some species which today we think of being exclusively restricted to the arctic, were onetime residents of temperate Maritime Canada and New England. 'Polar' bears are thus misnamed in the current parlance, as their Latin name is *Ursus maritimus* not *U. polaris*, as a consequence of them once occurring as far south as Maine. Up to three-quarters of a million walruses once inhabited the region, with offshore islands supporting colonies totalling hundreds of thousands. Before it became a valuable resource, tons of ivory tusks that had been abandoned on the killing grounds of these islands were used as ship ballast. The last walrus seen in Massachusetts was killed in 1754, and was probably a straggler who wandered southward, though still within the species' former range. Simply put, when one reflects on the sheer total mass of life that once occupied this particular, tiny region of the planet, compared to the residual remaining today, it completely boggles the mind, and of course, greatly saddens the heart.

The Gulf of Maine stretches from Cape Cod, Massachusetts to the southern tip of Nova Scotia. The coastline is predominantly rocky, glaciation having stripped away much of the sedimentary rock, such that beaches and saltmarshes are less abundant here than in the Eastern Seaboard of the United States, further south. The underwater feature of the outer banks, where the seafloor can rise up to within twenty to five metres of the surface, serves to make the Gulf a semi-enclosed area (Bigelow 1927). The watershed is about one hundred and fifteen thousand square kilometres and includes all of Maine, 70% of New Hampshire, 56% of New Brunswick, 41% of Massachusetts, 36% of Nova Scotia, and 1% of Quebec. The extraordinary bounty of marine life in the undisturbed, pre-European northwestern Atlantic was attributable to it being ground zero for the collision of two major sea currents.

Standing on any seashore, one might be forgiven for believing that the movement of water was mostly to and fro, in and out, due to the crashing and retreating of waves in the short-term or to the more gradual rising and receding tides twice a day. However, as we know from the scattering of debris originating from the Fukushima tsunami and the disappearance of Malaysia Airlines Flight 370, the world's oceans are also characterized by major lateral currents. A large gyre, for example, exists in the Atlantic, moving water from continent to continent in a clockwise direction due to the Coriolis force. Olaus Magnus' famous 1539 map of Nordic waters depicts, interspersed among a menagerie of sea monsters, a series of large swirling eddies, which we now recognize to be the Iceland-Faroes Front (Rosby and Miller 2003). On an aside, and germane to the present book, this cartographic fact led Ulanksi (2008) to postulate that 'if the artist rendered the swirls accurately, then can we accept the existence of sea monsters roaming the North Atlantic?' But back to the topic in hand.

Several smaller currents conjoin off Florida to flow northward, hugging the coast, until reaching Cape Hatteras, North Carolina, where the underwater river is deflected eastward to the outer continental shelf (Ulanksi 2008). At four knots per hour, this Gulf Stream is one of the swiftest ocean currents, and moves fifty seven million cubic metres of water every second. As it continues flowing northward, it begins to snake about like a meandering terrestrial stream, carving off enormous rings or vortices of water much in the way a river will create oxbow lakes. The surface water temperatures of the Gulf Stream may be fifteen degrees warmer than outside the path of the current. And it is this difference that fosters the exceptional biological productivity of the waters of New England and Atlantic Canada.

Ecosystems often have fuzzy boundaries wherein one system's biota overlaps with that of an adjacent system's. Such interfaces may be determined by topography or vegetation, such as between land and water (the riparian zone), or between forest and field, mountain and valley, etc., or they may be caused by thermal differences such as in microclimatic arctic oases or resulting patterns of ocean circulation such as in *polynyas* (open-water areas surrounded by pack-ice) (France and Sharp 1993, 1995). Frequently these areas are biological hotspots. Nowhere is this truer than in the northwest Atlantic where the warm waters of the northward flowing Gulf Stream collide into the southward flowing cold, nutrient-rich waters of the Labrador Current, all being supplemented by the discharge of organic matter and nutrients, originating from the planet's largest bodies of freshwater, the Great Lakes, through the mighty St. Lawrence River. The frontal boundary among these mixing currents, whether near the coast or out over the banks, becomes an incredible incubator of life. Thermal upwelling of the mixed water into the surface *euphotic* (solar irradiated) zone drives algal growth which in turn supports considerable zooplankton and larval fish as well as the giant filter-feeders, baleen whales, in addition to higher trophic groups of pelagic fish like herring and mackerel, which in turn sustain *pinnipeds* (seals) and apex predators such as sharks, billfish, and tuna (Fry 1988). Organic matter rains down to the seafloor to support the coastal *benthic* (bottom) community making use of extensive seagrass beds as a nursery habitat whose food web is fuelled by the attached algae growing there in consequence of the abundant nutrients (France 2013). No wonder that it was the sea, even more so than the land, that became the *raison d'etre* for the European colonization of the region.

Massachusetts has more than one thousand kilometres of coastline, with the 'rocky fist' of Cape Ann to the north and the sandy spit of Cape Cod to the south being 'two giant limbs thrown seaward, like the wings of a fish-weir, to guide sea-borne commerce into Boston's fruitful embrace' (Morison 1921). One of my biggest surprises when I moved to Massachusetts was to find out how divided the State...sorry, the Commonwealth (some are particular about that distinction), is in terms of those who were either North or South Shore people. Indeed, during my explorations I would occasionally take along a Boston or Cambridge resident whom had lived their entire life in Massachusetts and whose family owned vacation property on Cape Cod but had never once visited the North Shore communities or areas to where I was mostly drawn. Yes, it *is* difficult to be a tourist in one's own backyard, but to never experience the spooky-New Age ambiance of Salem, the natural beauty of the Plum Island beach or Essex estuary, or the maritime spirit and historic architecture of Marblehead, Ipswich, Newburyport, or Gloucester, is to have missed the some of the most interesting parts of the region.

One of my favourite pastimes during the decade I lived in Massachusetts was to deeply immerse myself (*sensu* France 2003a,b) into the seascape by paddling an ocean kayak along the coast. Join me for one such trip. Starting very early in the morning and leaving the great harbour of Boston behind, we paddle northward, past Lynn Harbour, location of sightings of Sea Serpent in the early nineteenth century (Figure 2.1A). We progress seaward to round the narrow spit of Nahant (Figure 2.1B) which, in 1819, was the location of many sightings of the Sea Serpent believed to have previously visited Gloucester, and which, incidentally, was the location of my first foray into conducting marine ecology research in the region (France 2011a, 2013; France *et al.* 1998a,b). Crossing Nahant Bay, we make for the rocky point of Marblehead where, should the wind be blowing, we can find refuge in its small harbour, the birthplace of the American Navy, and today filled chock-a-block with luxury sailing yachts. Continuing around the point, we next encounter the much larger harbour of Salem, whose legacy of being the leading port for the China Trade is reflected by the enormous wharves that jut out into the now, mostly silent, bay. Crossing the mouth of the Crane River, we paddle northward past several islands and pass by the tiny harbour of Manchester-by-the-Sea and its wave-battered beach of 'singing' sands. Making good time, another several hours of paddling finds us at the mouth of Gloucester Harbour. For the moment, lets pass it by and continue our kayak journey northward around the tree-clad shores of Cape Ann (Figure 2.2).

Conscious that this can be a dangerous coast whenever a nor'easter is blowing, we quicken our pace and shoot by Milk and Thatcher islands, the latter the scene of an early maritime disaster. Nearby, on the mainland, is Loblolly Cove, a small and otherwise nondescript embayment which, however, played a critical role in the 1817 sea serpent saga in that it was the location where the putative offspring of the creature was caught (Figure 2.1C). Paddling on, we pass the tiny harbour of Rockport, being careful to avoid interfering with the busy to-and-fro boat traffic transporting tourists on whale watching cruises.

Figure 2.1. Massachusetts locations related to the nineteenth-century Gloucester Sea Serpent sightings: (A) Lynn Harbour; (B) Nahant Peninsula; (C) Loblolly Cove, Cape Ann; (D) Sandy Bay, Cape Ann; and (E) Ipswich Bay.

Once across the sea serpent sighting location of Sandy Bay (Figure 2.1D) and rounding the rocky outcrop of Halibut Point we leave Cape Ann behind and enter Ipswich Bay (Figure 2.1E), where the mysterious creature was supposedly captured. Here are the protected areas of the extensive Essex and Great saltmarshes, and the sand beaches of Crane and Plum Island, in my mind the most beautiful part of the entire Massachusetts coast, an opinion held by others, as the area has the distinction of being the first wilderness in the entire country to be praised for its scenery (Nash 1982). Sadly, as the hour is getting late, we cannot linger but must progress northward through the Plum Island Sound and navigate our way through the complex saltmarsh of the Parker River/Essex Bay Area of Critical Environmental Concern to enter the mouth of the Merrimack River to arrive at our first day's destination of the onetime bustling but now quiet harbour of Newburyport.

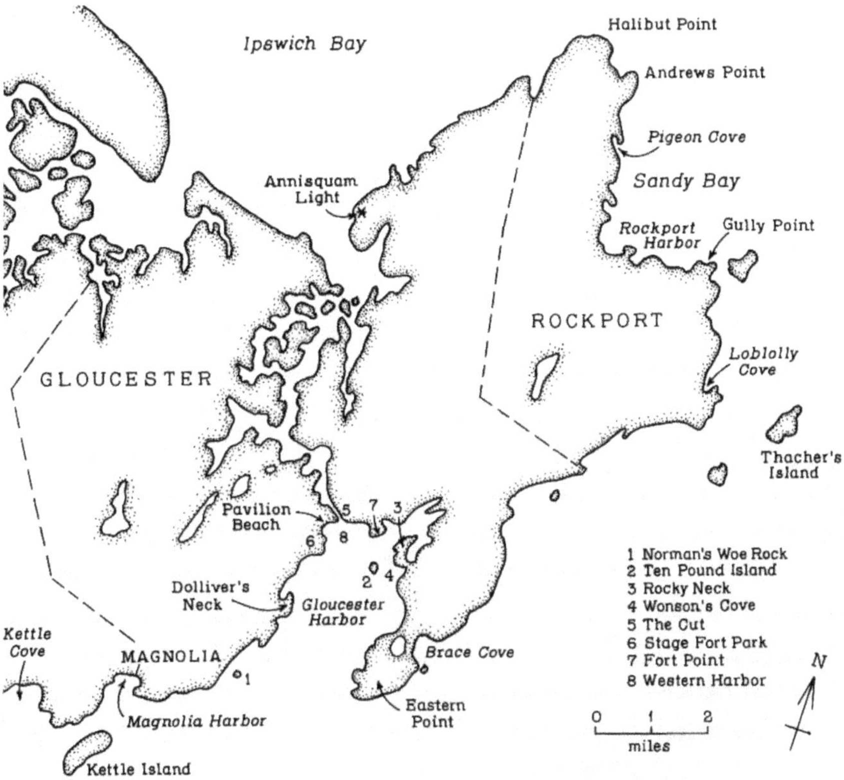

Figure 2.2. Map of Cape Ann from J.P. O'Neill's 1999 book The Great New England Sea Serpent: An Account of Unknown Creatures Sighted by May Respectable Persons Between 1638 and the Present Day.

For another day's imagined outing, we can return to explore the topographic features of Gloucester Harbour (Figures 2.2 and 2.3). Let us begin our paddle at the western side of the harbour mouth, at the rocky islet called Normans Woe (Figure 2.4A), which figures prominently in Longfellow's *Wreck of the Hesperus*. From here it is several kilometres across the mouth of the harbour to the tip of Eastern Point on the other side (Figure 2.4B), and five kilometres to the streets of the city of Gloucester (Figure 2.4C) at the base of the inner harbour. Paddling along the western shore we pass Hammond Castle (Figure 2.4D), nearby the location where on August 6th, 1817, two anonymous 'ladies' would be the first to sight the Sea Serpent, and from where the Sea Serpent was seen out near Normans Woe eight days later by several children whom fled terrified from it to the safety of their homes. Continuing on our clockwise tour of the harbour, we visit other locations specifically mentioned by those interviewed by Justice of Peace Lonson Nash for the official Linnaean Society report and by C.L. Sargent for his own investigation (see Chapter 3). Rounding a small headland, we paddle into the protected embayment of Freshwater Cove (called Webber's Cove back in 1817), and then pass the small islets of Field Rocks (Piper's Rocks back in 1817), and arrive at the site of the old military fort, today Stage Fort Park (Figure 2.4E), from where one eyewitness became much alarmed when she spotted the creature swimming near Piper's Rocks on August 14th. Here too is Half Moon Beach (Figure 2.4F), where many sightseers gathered in 1817, hopeful to be afforded a glimpse of the famed Serpent 'gamboling' about out in the harbour. Leaving the Stage Head promontory behind, we paddle across the opening of the canal that technically makes Cape Ann an island, the Cut, both sites mentioned by eyewitnesses in depositions and interviews as locations from where they spotted the creature, the latter being near where a ropewalk was laid out. Just before entering Inner Harbour, we can turn and look out across the panorama of the main harbour proper and see Ten Pound Island (today Babson Ledge; Figure 2.4G), a reference point mentioned in many of the eyewitness accounts, and referred to as 'the sea serpent's playground' by Soini (2010). The other side of Inner Harbour (Figure 2.4H) is formed by the small promontory of Rocky Neck (Figure 2.5) from where, on August 14th, several locals saw the Serpent making rapid progress toward Ten Pound Island. We continue our paddle along the length of Eastern Point, which forms the eastern side of the protected harbour, and rounding it progress several kilometres along its outer edge (Figure 2.4I) toward Salt Island near where, on August 28th, crew members of the *Laura* observed the Sea Serpent heading out to the Atlantic, in what would be the last Gloucester sighting of 1817.

Had we been in a larger sailing ship and not kayaks, we would have been wise to read through the United States Harbor Commission's warnings about navigating Gloucester Harbour:

> Gloucester Harbor and approaches have very broken ground and many rocks and ledges, some of them unmarked; careful navigation is necessary, especially in thick weather. Dog Bar, on which the breakwater is built, extends 100 yards westward of the end of the breakwater where it is marked by a buoy. Round Rock Shoal, a rocky ledge about 400 yards in extent northeast to southwest and covered 13 feet, extends from 0.15 to 0.3 mile westward of the breakwater light. It is marked on its northeastern edge by a buoy and on its southwestern edge by a lighted buoy. Green Rock, 175 yards eastward of Tenpound Island, is marked by a daybeacon. The passage between Tenpound Island and Rocky Neck is shoal and foul, and should not be attempted, especially by strangers. Ten Pound Island Ledge and Mayflower Ledge, on the eastern side of the approaches to the

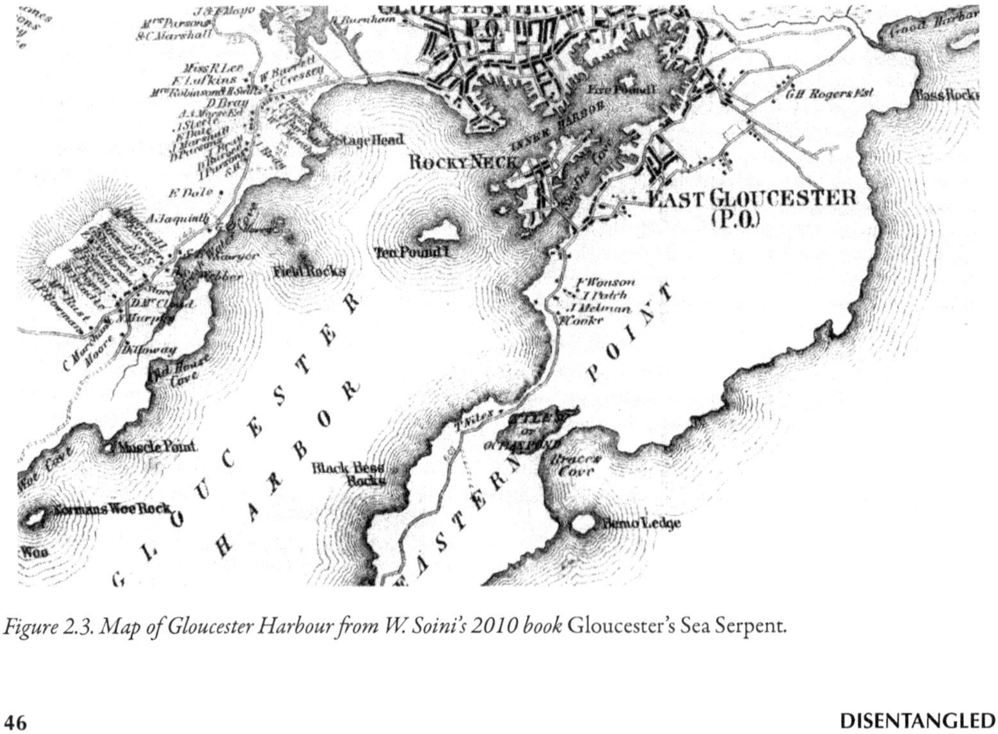

Figure 2.3. Map of Gloucester Harbour from W. Soini's 2010 book Gloucester's Sea Serpent.

Figure 2.4. Gloucester Harbour locations related to the nineteenth-century Gloucester Sea Serpent sightings: (A) Norman's Woe; (B) Eastern Point, (C) Gloucester city; (D) Hammond's Castle; (E) Stage Fort Park; (F) Half Moon Beach; (G) Ten Pound Island; (H) Inner Harbor; and (I) outer edge of Eastern Point.

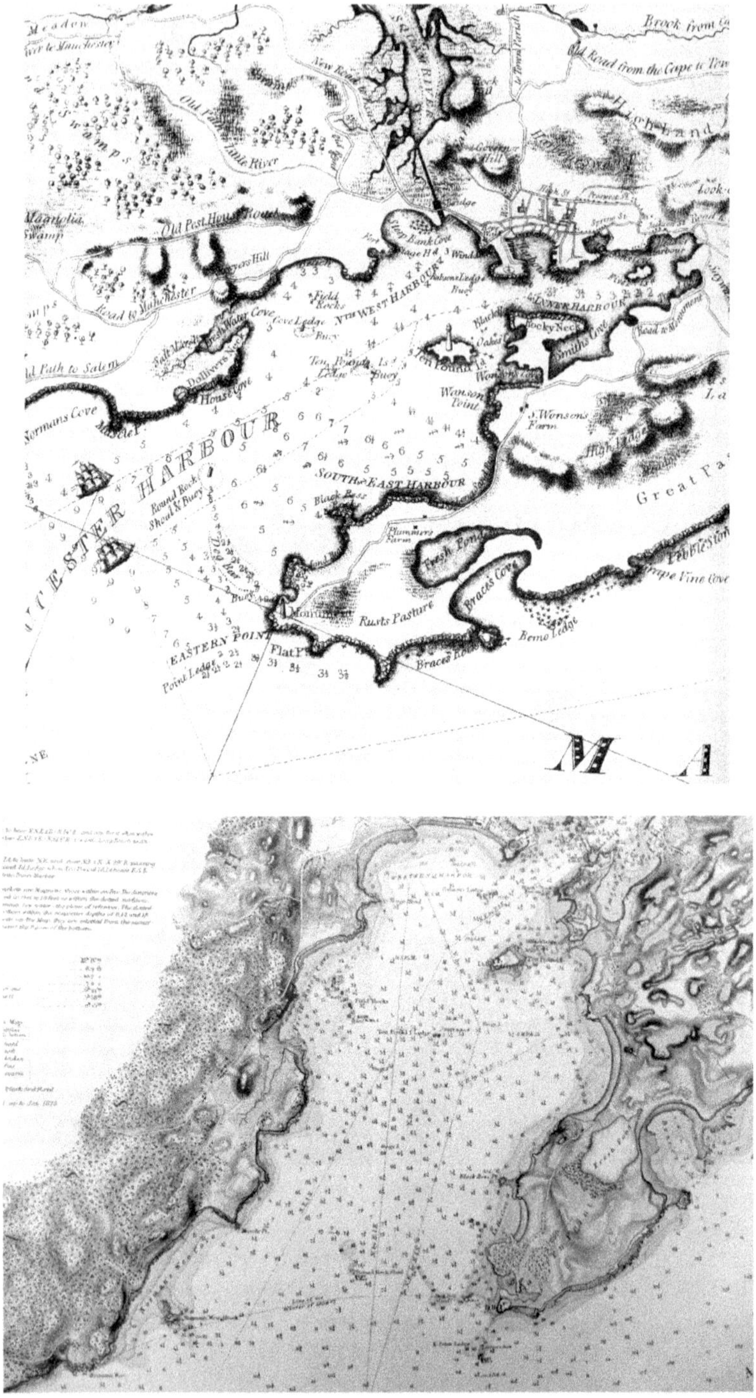

Figure 2.5. Bathymetry maps of Gloucester Harbour (Upper: from 1831, John Mason, Cape Ann Historical Association; Lower: from 2016, MassGIS, Commonwealth of Massachusetts, U.S. Geological Survey, USDA Farm Service Agency)

> Inner Harbor, are covered 19 and 16 feet, respectively; a buoy is off the northwest side of Mayflower Ledge. Two shoal spots, covered 16 and 19 feet and marked by a buoy, are about 220 yards southwestward of Tenpound Ledge. There are also unmarked 18-foot and 19-foot spots about 230 yards north-northwestward of Tenpound Island Light. Prairie Ledge, on the western side of the approach, is covered 4 feet and marked on its eastern end by a lighted buoy. Babson Ledge, 500 yards north of Tenpound Island, is covered 13 feet and marked on its south side by a buoy. Rocky Neck, a high and partly wooded island on the east side at the entrance to Inner Harbor, is connected with the easterly shore by a causeway. Black Rock, about 100 yards off the western end of Rocky Neck, is marked by a daybeacon.

Obviously, none of this was an issue to dissuade the Sea Serpent from swimming into and, according to more than one eyewitness, 'gamboling' about the harbour, during 1817. Still, however, the creature, like any marine animal, would have had to pay attention to its surroundings in terms of water depth so as not to become stranded in relation to the tidal range for the harbour, which is between one and two metres.

Bathymetric measurements shown on the nautical map of the harbour reveal it to be generally quite shallow (Figure 2.5). This is a very important point which will be used to help in the screening of candidate animals for the purported sea serpent (see Chapter 5). The area between Rocky Neck and Ten Pound Island, for example, ranges in depth between three and seven metres. The area around Piper's (Fields) Rocks, another location where the creature was observed, is only three metres feet deep. Shoreward from these two locations, a transect from Stage Head on the west to Fort Point on the east at the entrance to Inner Harbour (which itself has an average depth of three metres for non-dredged sections) shows depths of three, seven, five, five and four metres. Moving seaward, a transect across the two kilometre-wide mouth of the outer harbour from Normans Woe Rock to the Eastern Point Lighthouse reveals depths of six, 10, 14, 16, 17, 11, 12, nine, six, 11, 12 and seven metres.

Finished with paddling for the day, let's pull up our kayaks on one of the wharves and make our way into the town of Gloucester, itself.

2.2 THE TOWN BUILT BY FISH

Despite Kurlansky's (2008) belief in Viking visitations to Cape Ann, professional historians put no weight in such assertions. What we do know for sure is that in 1606, during his second visit to what would become New England, Samuel Champlain entered a harbour on southern Cape Ann, referring to it as *La Beauport*, or the beautiful harbour (Figure 2.6). Native inhabitation of the region and fishing of the near shore waters goes back over ten thousand years (Cronon 2011). At the time of Champlain's visit as many as four thousand Amerindians may have resided on the peninsula. But in one of history's greatest mortalities (White 2012), by the time that the first European settlement was established a few decades later, only a handful of the original inhabitants had survived eradication by European diseases such as smallpox.

From the start, the early European seasonal visitors and later permanent settlers, many from Gloucestershire, were impressed with the nearby supply of large codfish (Figure 2.7). One individual complained that they were so numerous that they 'pestered' his ship (Babson 1860; Pringle 1892). Early

Figure 2.6. Champlain's 1606 map of La Beauport *– the good harbor (online from the Sawyer Free Library, Gloucester: The Book of the Three Hundreth Anniversary Observance of the Foundation of the Massachusetts Bay Colony at Cape Ann in 1693 and the Fiftieth Year of the Incorporation of Gloucester as a City 1924).*

explorer and colonist Captain John Smith claimed that 'the seas is better than the richest mine known' (Kurlansky 2008). As described by Morison (1921): 'God performed no miracle on the New England soil. He gave the sea. Stark necessity made seamen of would be planters...Massachusetts went to the sea, then, not of choice, but of necessity.'

In 1623, a fishing plantation was established in the harbour by the Dorchester Company of merchant adventurers. This would last for three years until it was moved down the coast to Salem, which offered better prospects for agriculture than the predominantly granitic Cape Ann. Roger Conant, an influential member of the Plymouth Plantation at this time, and responsible for relocating the Cape Ann Plantation, was by all accounts a warm-hearted individual, not the dour fellow portrayed on a Salem statue (Figure 2.7). The first permanent settlement of Cape Ann occurred with the reoccupation of the harbour in 1633 or 1639 (depending on sources), which by 1642 was incorporated into a town christened as Gloucester. Much of the early decades were spent in harvesting the rich bounty of the sea and characterized by squabbles with the Plymouth Plantation over fishing rights to various coastal locations in the Massachusetts Colony. Soon much of Cape Ann was denuded to provide timber for ship building, and in 1688 land grants began to be awarded. The witchcraft hysteria of 1692 did reach Gloucester but there were no executions. Like Venetians, those from Gloucester were practically focused on making a living, having little time for theological arguments. One early resident in response to being berated by a Salem clergyman for his lack of religious enthusiasm, stated 'Reverend, we come not to pray but to catch fish,' an attitude that still exists in the form of local bumper stickers that proclaim 'In COD we trust' (Brigham 2007). Several skirmishes with His Majesty's ships occurred in and around the harbour during the Revolutionary War. At this time, townsfolk experienced deprivation, leading to penury and the threat of starvation. In consequence, many turned temporarily to privateering. The following years were little better since, with United States independence, the cod-fishery was deprived of its greatest market, the British West Indies. The sea dominated all ways of life with the majority of men being engaged in fishing or sailing. By the beginning of the nineteenth century, over forty fishing vessels were registered in the town. Since 1784, a wooden carving of the 'sacred codfish' (*Gadus morhua*) has hung in the Massachusetts State House in Boston in homage to the importance of the fish to the welfare of the Commonwealth (Morison 1921). A channel was cut in the eighteenth century to join Ipswich and Massachusetts Bays, so as to avoid the dangerous conditions of having to round Cape Ann, but this became clogged with sediment until its restoration several decades after the dramatic events of 1817.

Kurlansky (2008) describes Gloucester's history as being one of periodic crises. Certainly, the decade preceding the Sea Serpent sightings was a time of strife for Gloucester residents. The town, like many New England coastal communities, vehemently opposed the Embargo Act of 1807 which curtailed all trade with Britain and her North American colonies. The prevention of any American vessel to clear an American harbour bound for a foreign port, had serious implications for Massachusetts given that the Commonwealth was responsible for 37% of the total United States foreign trade, with its fishing fleet being 55% of the nation's total, and its revenue equalling that of the entire Federal Reserve (Morison 1921). Their petition to President Jefferson to be exempted had the opposite effect in that instead of relief they received the Enforcing Act. With a larger patrolling fleet than would be mobilized for the war half a decade later, the American Government shut down the lifeblood of maritime New England, interrupting fisheries and nearly destroying commerce. In response, Gloucester fishermen turned to smuggling in order to survive. Due to suffering severely during the Embargo, residents opposed the War of 1812 which they deemed to be inopportune, unnecessary, and unjust. The most exciting event to have ever occurred in the harbour up to that time was a battle with a British ship. By 1817, the post-war population of Gloucester was about six-and-a-half thousand. In the following year, the mackerel and cod fisheries were hampered by a British treaty restricting American access to their one-time fishing grounds in the Gulf of St. Lawrence.

Figure 2.7. It was the great abundance of marine life, such as codfish (model from The Rooms) that attracted European settlers, such as Roger Conant (statue in Salem) to the region.

Morison (1921) describes a prosperous and thriving town at the start of the nineteenth century, one 'boasting a bank with a vault carved out of solid rock, a schoolhouse with a cupola, and a two-story 'artillery house' or armoury, with four field pieces and a bell procured from Denmark,' whose residents, to quote Dr William Bentley, the Samuel Pepys of the time and region, 'excel in their parties, their clubs, and also in their military parades.' Babson's (1860) description of Gloucester four decades after the events of 1817 is the closest written description we have to what the town looked like at the time of the Sea Serpent sightings:

> The principal harbor is on the south side of the town, and is formed by Eastern Point, – a strip of unequal width, extending in a south westerly direction about three miles, in a rock-bound coast that defies the violence of the surging waves dashing almost constantly against it, and affording within its friendly embrace safe shelter and anchorage for a large fleet. In the course of the year, many hundred vessels, mostly of the coasting and fishing classes, find here a refuge from the dangerous easterly gales and storms by which the coast is so often visited; and in autumn, when a favorable wind and clear weather follow an easterly blow, it is not an uncommon sight to see a fleet of two or three hundred sail

> working out of the harbor at the same time, and presenting a scene of surpassing interest and beauty.
> The surface of the town is uneven; and its peculiar character strikes every beholder, at first sight, with astonishment. Bald, rocky hills, bold and precipitous ledges of rock, with acres of bowlders of various sizes, in many places scattered thickly over the surface, combine to present a rare scene of naked ruggedness and complete sterility.

Today, examples of Gloucester's architectural heritage from the nineteenth century persist (see Appendix One), such that it is still possible to capture a flavour of the maritime ambiance of those bygone days.

Gloucester owes its existence to fishing (Brigham 2007) and bills itself as being 'America's Oldest Seaport' (Kurlansky 2008). One advantage over the northern waters around Newfoundland was that fishing in Massachusetts was a year-round enterprise. That, and the fact that the cod achieved larger sizes, gave an advantage to Gloucester. Cod was prized because with virtually no fat, once salted and dried, it would be a concentrated source of protein that would rarely spoil (Kurlansky 1997). Not only did cod last longer, it also tasted better than other fishes one rehydrated and unsalted. Three grades of prepared cod existed, with the poorest quality being shipped to Caribbean slave plantations. Cod fishing and human slavery therefore became closely linked commercial enterprises, a stain upon Gloucester's memory. Atlantic/Boston mackerel (*Scomber scombrus*) was a favourite target species with a single jigger manned by a crew of eight being able to harvest over a thousand barrels of the prized baitfish during a single cruise. Following a procedure unchanged since medieval timers, small two-man rowboats or dories (Figure 2.8) would set out from shore or their mother ship, sometimes a barque, and spread out to drop their lines for these or other fishes. Originally, each line would sport a single hook (Figure 2.9), but the nineteenth-century development of long-lining meant that to the heavy or ground line were fastened shorter lines every half metre. The strength needed in pulling such lines up from productive, fish-filled waters, hour after hour, day after day, would have been considerable. The mackerel 'jig' (a hook around the shank of which was cast a plummet of lead or pewter (Figure 2.9) which would be baited with herring, menhaden (*Brevoortia tyrannus*), or immature mackerel) was invented in 1812 (Bolster 2012). A school of mackerel would be attracted to the surface by the over-side broadcasting of chopped baitfish. The fish were caught on baited jigs, each member handling several short lines, and 'dexterously snapping his mackerel into a barrel with the same motion that jerked him out of water' (Morison 1921; Figure 2.9). The fish would be dressed and salted on board and made ready for distribution throughout the country. The mackerel fishery was dependent on an ample and cheap supply of baitfish such as alewife herring (*Alosa pseudoharengus*), referred to by McPhee (2002) as 'the founding fish.' Herring were also critically important to colonial Massachusetts, and in particular to granitic-soiled Cape Ann, as a source of nutrients for agriculture. In a practice learned from the Amerindians, and known as 'spot fertilization', a few kernels of corn would be planted inside of a herring (Maddigan 2014). Through decomposition, the fish would provide organic matter and help to neutralize the acidic soil. During spring spawning runs, tens of thousands of excess herring would be scattered about and ploughed into the fields as a fertilizing agent. Although certainly not the most pleasant fish to consume, local herring became an inexpensive and easy-to-obtain source of protein for the relief of the poor, including widows and orphans of those lost at sea (Maddigan 2014), as well as for slaves on plantations in the Southern States.

Figure 2.8. Illustrations of nineteenth-century fishing methods: (A) baitfish barrels, and (B) line fishing from dories (Goode et al. *1884).*

Gloucester, Brigham (2007) states, was 'populated by simple men with modest means and without affectations, but who walked with an aristocrat's belief that they were a special breed.' It was a stratified society with many fishermen becoming indebted to ship-owners who doubled as shopkeepers in supplying fishing gear in addition to all living staples. A fisherman's son was predestined to follow his father to the sea (Morison 1921):

> As soon as he could walk, he swarmed over every [boat]...that came into port, began 'hand-lining' for cunners [i.e. small fish] off wharves and ledges, and begging older boys to teach him to row. At six he was already some aid in curing the catch, and he helped his mother with the household work, in order to qualify as sea-cook. Boys of nine to

> twelve years did the cooking...for Gloucester fishermen [at sea] at this period. After a voyage or two he handed over his cooking utensils – a single iron pot and large spoon – to a younger brother or cousin, became an apprentice, learned the secrets of luring codfish to hook, and the art of heading, splitting, and salting with quick precision. A strong boy of fifteen or sixteen might be as accomplished a fisherman as any...To save enough to acquire a fishing vessel, and live ashore on her earnings, was his highest ambition. Otherwise he grew gray in the service of the sea. When rheumatic arms could no longer haul on sheet or cable, and eyes grew dim from straining through night, fog, and easterlies, he retired from deep waters, and puttered about with lobstering, shore fishing, or clam-digging.

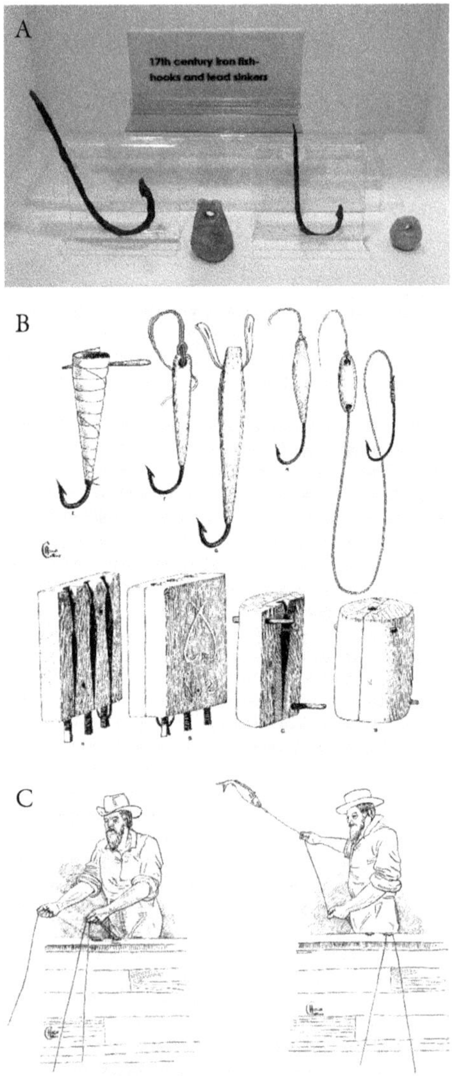

Figure 2.9. Iron hooks (A) were originally used (Cupids Legacy Centre, Newfoundland) until being replaced by the mackerel jig (B), which subsequently improved catch during the process of 'jigging' (C) (Goode et al. *1884).*

Gloucester developed the most self-sufficient fishery in all of New England (Brigham 2007). By the end of the eighteenth century, one-quarter of Massachusetts seaport residents relied on fresh fish. In addition to selling to local markets, the catch was exported to the West Indies and to Europe. From as early as 1623, fish was exported to Bilbao at a time when sixty percent of the fish eaten in Europe was salt-cured cod (Kurlansky 2008). By the time of the Sea Serpent sightings, more than two dozen Gloucester ships were making regular voyages delivering fish and timber to Surinam. The first vessels used by early settlers were ketches, shallops, and pinnaces until 1713 when a revolutionary form of sail rigging led to the invention of the schooner by a Gloucester native (Pringle 1892). This would become the defining symbol of the town (Figure 2.10). By the early nineteenth century, vessels began travelling out to the Georges Bank. But the bulk of the fishery was still inshore, some of it taking place as far away as the Gulf of St. Lawrence. The days of multi-week trips to the outer fishing banks to collect massive stocks of codfish or the rapacious plunder of coastal mackerel due to the invention of the purse-seine were still decades to come. The inshore fishery was simply so good there was no need to brave the more tempestuous offshore waters.

The romantic allure of the fishing life of Gloucestermen became the subject for a flurry of novels as, for example, Joseph Reynolds' 1856 *Peter Gott, the Cape Ann Fisherman*, William Adams' (Oliver

Optic) 1869 *The Story Flag, or the Young Fisherman of Cape Ann,* James Connolly's series of fishermen's tales starting with 1902's *Out of Gloucester*, and Elizabeth Phelps' series from the 1870s, such as *Jack the Fisherman* and *A Singular Life*, the latter filled with moralizing about hard drinking ways of Gloucester's fisher-folk (Brigham 2007). Only two works of fiction, however, have stood the test of time, such that they could foster the maritime dreams of a boy like myself growing up in the central plains of the continent, far from any seashore. Rudyard Kipling said of his 1897 *Captains Courageous* that 'I wanted to see if I could catch and hold something of a rather beautiful localized American atmosphere that was already beginning to fade' (Brigham 2007). And who can forget, when standing for the first time beside the castle of inventor John Hammond and gazing out toward Gloucester Harbour to the rocky outcrop of Norman's Woe emerging from the sea, those words of Longfellow from his 1841 *Wreck of the Hesperus* that, as Brigham states, 'made palpable the hazards that crews of fishing schooners faced in winter storms along the Cape Ann shore': 'And fast through the midnight dark and drear, / Through the whistling sleet and snow. / Like a sheeted ghost, the vessel swept / Tow'rds the reef of Norman's Woe.'

Figure 2.10. The schooners Thomas E. Lannon *(upper) and* Ardelle *(lower) in Gloucester Harbour.*

Schooners were built for speed, but with so much sail they could become unstable in gale-force winds and turbulent seas. The realistic tone of Kipling's and Longfellow's classics underscores the perilous conditions faced by a life on the sea, as testament to the lengthy listing of names on Gloucester's famous Fisherman's Memorial (Figure 2.11), inscribed with the quotation from Psalms 107:23: 'They That Go Down to the Sea in Ships.' Indeed, between 1830 and 1892, a staggering 576 vessels and 3,224 lives were lost. As Kurlansky (2008) states: 'The sea would take thousands upon thousands of Gloucestermen and the resulting pain shaped the town of Gloucester as much as did the dense, grey-spotted granite or the fine, white-flaked cod.' In several instances, more than a hundred men were lost in a single storm. Because multiple generations of men from the same family would fish together, if the ship went down, the devastation would be on par with that of the 'Pals

Figure 2.11. The Fisherman's Memorial, Gloucester Harbour.

Brigades' in the Great War a century later when an entire village's men could be wiped out at a single time. Gloucester, it was said, was a town full of widows and orphans, with total estimates of as many as ten thousand fishermen being lost at sea (Brigham 2007).

Cape Ann, as a result of developing separately from the world around it (Brigham (2007), was incredibly insular, with many residents of the seventeenth to early nineteenth centuries spending their entire lives there without ever leaving the peninsula except, of course, to travel out to sea (Kurlansky 2008). At three times in its history, Gloucester has broken free of its sequestered existence to become the focus of international attention and consequent tourist visitations. Most recently, this occurred with publication in the late 1990s of Sebastian Junger's best-selling book, *The Perfect Storm*, and the subsequent movie starring George Clooney, part of which was filmed in the city. The saga of Captain Billy Tyne and his mates, portrayed as flawed but Homeric heroes, and their ill-fated fishing trip aboard the *Andrea Gail* during the Halloween nor'easter of 1991, transfixed the world. This was coincident to my moving to Massachusetts, such that when I first visited Gloucester, its waterfront was filled with tourists, some arriving in large motor-coaches, browsing through pop-up souvenir shops, taking harbour tours on fishing vessels, or visiting the hard-drinking establishment that figures prominently in the book. For the curmudgeonly salty-dogs of Gloucester, the last time the city had experienced such a hullabaloo was way back in 1935 when Spencer Tracey and his Hollywood cronies were in town to shoot scenes for the movie version of *Captains Courageous*. However, before Tyne and Junger, before Tracey and Kipling, the event that made Gloucester the subject of global attention – from meetings held in the shadowy gas-lit halls of academe by learned, top hat-apparelled, gentlemen, brandy snifters in hand, to animated discussions occurring in dockside watering holes by opinionated labourers slamming tankards on the table for emphasis, or to more genteel coffee houses frequented by other members of society, sipping exotic roasts while aspiring to upward class mobility, all armed with fresh information obtained from the latest local broadsheets or periodicals – there was the Great Sea Serpent craze of 1817.

2.3 THE GREAT NEW ENGLAND SEA SERPENT

The conception of the Gloucester Sea Serpent did not suddenly appear *de novo*. When Gloucester residents and sightseers visiting from elsewhere in New England gazed out at the creature gambolling about in the harbour during those dog-days of August 1817, they were unhesitant in identifying the type of animal being seen. The only real mystery was not what the creature was, but rather why it had taken so long to make itself known in their own particular patch of the world. In other words, the surface of Gloucester Harbour was far from being a *tabula rasa*; instead, it was a canvas, or perhaps more accurately, a palimpsest, upon which eyewitnesses could leave (layer) their own imprint as part of a long established tradition. In short, eyewitnesses *expected* to see a sea serpent, and that is indeed exactly what they unhesitatingly believed they observed.

Of those contemplating the existence of a sea serpent along the New England and New York shores, the more erudite would have drawn upon their education in the Classics and known that the ancient Greeks had populated their 'wine-dark seas' with all manner of fabulous beasts (Jaffe 2013; Papadopoulos and Ruscillo 2002). From classical art, they would have known of the hippocamp, that chimera-like fusion of the front-end of a horse with a serpent's coiled tail, that was commonly used to decorate Hellenic and

Hellenistic vases and tiled floors, and which, after adoption and promulgation by the Romans, would later become a ubiquitous feature in medieval bestiaries. This particular artistic tradition of the hippocamp may in turn have been the inspiration behind the cultural invention of the Great Sea Serpent in the Western World. In a case of (imagined) life imitating art, the vertical undulations – which, as we will see, is a trait mentioned by most observers of the Gloucester Sea Serpent – may derive from the Classical artistic solution to picturing wiggly movement in two dimensions (Loxton and Prothero 2015).

Stothers (2004) believes that the medieval and later modern image of the Great Sea Serpent owes its origin to three main sources of information from antiquity: primitive myths propagated through classical literature, inferences drawn from the chance discovery of giant prehistoric fossils, and ancient historical observations of rare snakes or other animals. From classical literature, some New Englanders would therefore have known of 'The Bagradas River Incident' concerning the giant water serpent that swallowed Roman troops after crushing them in its coils during the Punic Wars; and perhaps also known of the giant snake that emerged from the water to devour cattle before being captured and presented to Alexander (Sweeney 1972). The Roman encounter, cribbed from Charles Rollin's seminal, eighteenth-century *Ancient History*, was printed in a popular 1817 pamphlet (Anonymous 1817; Bowen 1817). Entitled *A Particular Account of a Monstrous Sea-Serpent, the Largest Ever Seen in America, which has Lately Made its Appearance in Gloucester Harbor, Cape Ann, (Mass.) and has been Seen by Hundreds of Respectable Citizens. With an Exact Representation of the Wonderful Monster,* this pamphlet seemed to have been designed to shock and instil a level of worry among coastal dwellers, similar to what the book and movie *Jaws* would do a century-and-a-half later:

> In the meantime Regulus, everywhere victorious, led his army into a region water by the river Bagradar, near which, an unlooked for misfortune awaited them, and at once affected the Roman camp with considerable loss, and with apprehensions still more terrible, for a serpent of prodigious size attacked the soldiers who were sent for water, and while they were overwhelmed with terror and unequal to the conflict, engulphed several of them in its enormous mouth, and killed others by writhing round them with its spires, and bruising them with the strokes of its tail...It caused so much trouble to Regulus, that he found it necessary to contest the possession of the river with it, by employing the whole force of his army; during which a considerable number of soldiers were lost, while the serpent could neither be vanquished nor wounded...Several discharges were made against it without success, till its back was broken by an immense stones, the formidable monster began to lose his powers, and was yet with difficulty destroyed; after having diffused such a horror among the army, that they confessed they would rather attack Carthage itself than such another monster.

For the majority of the 1817 Gloucester community, however, being descendants of devote Puritan colonists, it must have seemed as if the observed creature had been ripped from the pages of their well-thumbed Bibles. Did the arrival of the ominous monster have anything to do with the 'summer that never was' of the preceding year, which we now know to have been caused by the 1815 Tambora eruption and consequent changes in global climate (see Chapter 5)? (Incidentally, the frigid weather of 1816 forced many to stay indoors, including several writers holidaying on the shores of Lake Geneva, thereby giving birth to the two monster classics *Frankenstein; or the Modern Prometheus* and *The Vampyre.*) Certainly, those whose lives were so closely tied to the sea would have remembered King David singing, in Psalm 104, about the 'great and wide sea, wherein are things creeping innumerable, both small and great beasts.' Was not the great beast seen in the harbour, therefore, but some manifestation of the Leviathan, that 'twisting serpent' of *Isaiah*? Indeed, for those whom would later read through the eyewitness testimonials in the Linnaean Society's report (as will be seen in Chapter 3 and 4), the parallels to *The Book of Job* would have been striking:

> Canst thou draw out leviathan with an hook?...I will not conceal his parts, nor his power, nor his comely proportion...His teeth are terrible round about. His scales are his pride, shut up together as with a close seal. One is so near to another, that no air can come between them. They are joined to one another, they stick together, that they cannot be sundered. By his neesings [i.e. sneezing or snorting] a light doth shine, and his eyes are like the eyelids of morning...The flakes of his flesh are joined together; they are firm in themselves; they cannot be moved...When he raiseth up himself, the mighty are afraid...The sword of him that layeth at him cannot hold: the spear, the dart, nor the habergeon...The arrow cannot make him flee: slingstones are turned with him into stubble. Darts are counted as stubble: he laugheth at the shaking of a spear...He maketh the deep boil like a pot: he maketh the

Figure 2.12. Destruction of the Leviathan (A) by Gustav Doré, Illustrations for the Bible *(1866) ('In that day the Lord with his sore and great and strong sword shall punish Leviathan the piercing serpent . . . and he shall slay the dragon that is in the sea' – Isaiah 27:1). Note how the Biblical image was later transformed into that of a sea monster in an 1873 edition of the works of Rabelais (B) and another book by Buel (1887; see also www.historicinpswich.org 28/11/2018 (C).*

> sea like a pot of ointment. He maketh a path to shine after him; one would think the deep to be hoary. Upon earth there is not his like, who is made without fear.

These words would have been well-known to the town's churchgoing mariners without need for Gustav Gore's later interpretations for a less Biblically-literate audience of Victorian landlubbers (Figure 2.12).

Such classical and Biblical exposure laid the foundation for the genesis of what would be regarded as the Gloucester Sea Serpent. And to reinforce this interpretation, there were dozens of widely-circulated illustrations and writings of sixteenth- to eighteenth-century Scandinavians who, almost single-handily, invented the modern, stereotypical concept of the sea serpent.

During the Age of Discovery, as whalers pursued their quarry to more and more remote waters and brought back bones of unknown creatures, 'sea monster stories started to circulate widely and became so numerous that people began to make them a matter of record' (Sweeney 1972). One such encounter, told in Hans Lilienskiold's 1694 *Speculum boreale* (*in* Loxton and Prothero 2015), is remarkably similar to the prose used in many of the later Gloucester sightings, in regards to describing the sea serpent, in this case of a prodigious (and impossible) length of 80 metres, lying curved about in quiet seas and looking like 'a number of ox-heads [i.e. barrels] have been thorn into the water.' Another account is from historian Jonas Ramus's 1689 *Nori Regnum* (*in* Sweeney 1972): 'A large Sea-snake was seen by many people in Dramsfiorden; and at one time by eleven persons together. It was in very calm weather; and so soon as the sun appeared, the wind blew a little, it shot away just like a coiled cable, that is suddenly thrown out by the sailors; and they observed that it was some time in stretching out its many folds.' Remember the specific words from these two particular sightings – i.e. *ox-head barrels* and *a coiled cable*, both as if thrown out upon the water by sailors – when I introduce my explanation for the Gloucester Sea Serpent in Chapter 5.

Figure 2.13. Close-ups of the Sea Worm (A) and Sea Serpent (B) from Olaus Magnus' 1539 Carta Marina, *and a serpent (C) from his 1555* A Description of the Northern Peoples.

No piece of cartography from the *Hic Svnt Dracones* ('here are dragons') period of the sixteenth century is more famous for its depiction of a menagerie of fabulous sea monsters as the cleric Olaus Magnus' representation of northern waters, the *Carta marina*

Figure 2.14. Image of sea serpent inspired by Magnus, from Konrad Gessner's 1551-1558 Historia Animalium.

of 1539. Nigg (2013) describes it as the 'largest, most detailed, and most accurate map of any part of Europe up to that time,' whereas for Van Duzer (2014) it also contains the 'fullest and most influential collection of sea monsters of the sixteenth century.' Magnus obtained much of his knowledge of marine life from seafarers' tales. As shown in a series of side-by-side close-ups from the map compared with modern photographs, the fabulous beasts depicted can easily be identified today as recognized animals (Nigg 2013): the *Rockhas* is a ray; the *Polypus*, a lobster; the Sea unicorn, a narwhal; the *Prister*, a sperm whale; the *Zephius*, an orca; the Island whale, a blue whale; the *Balena*, possibly a northern right whale; the *Rosmarus* (and also perhaps the Sea swine and the Sea cow), a walrus; the Kraken, a giant squid; and the ten to thirteen-metre-long Sea worm (Figure 2.13A), perhaps either an oarfish or an eel. However, when it comes to finding a known candidate for the impossibly large (65 metre) Brobdingnagian Sea serpent (Figure 2.13B), which he states is the most notorious, copied (e.g. Figure 2.14), and influential of all of Magnus' marine monsters, Nigg (2013) is at a loss for explanation, unsatisfactorily filling the known animal box with an illustration of a plesiosaur, as a sort of place-holder pending further information.

As well as being a cartographer, Magnus was also an author. In his 1555 *A Description of the Northern Peoples*, which would become the standard reference on Scandinavia (Loxton and Prothero 2015), he writes:

> They who, either to trade or to fish, sail along the shores of Norway, related with concurring evidence a truly admirable story, namely that a very large serpent of a length of upwards of 200 feet, and 20 feet in diameter lives in rocks and holes near the shores of Bergen; it comes out of its caverns only on summer nights and in fine weather to devour calves, lambs and hogs, or goes into the sea to eat cuttles, lobsters and all kinds of sea-crabs. It has a row of hairs two feet in length hanging from the neck, sharp scales of a dark colour, and brilliant flaming eyes. It attacks boats, and snatches away me, by raising itself high out of the water, and devours them.

The accompanying illustration captures this sobering scene (Figure 2.13C).

Several years later, Swiss naturalist Konrad Gessner, regarded as the father of modern zoology, published another volume to his *Historia Animalium* in which he takes the supernatural riparian snake of Magnus, the *lindorm* (Loxton and Prothero 2015), and transports its encounter with the hapless sailors to the open water (Figure 2.14) with the following description:

> There is another sea serpent, a hundred or two hundred feet long, which sometimes appears near Norway in fine weather, and is dangerous to Sea-men, as it snatches away men from their ships. Mariners tell that it incloses ships, as large as our trading vessels... by laying themselves around them in a circle, and that the ship is then turned upside down. It sometimes makes such large coils above the water, that a ship can go through one of them.

Another illustration depicts what Loxton and Prothero (2015) refer to as the canonical representation of a sea serpent floating 'with its coils cartoonishly out of the water,' and which Heuvelmans (1968) snidely dismisses given that 'the way the animal floats almost entirely out of the water would be mechanically impossible for anything but a balloon.' The answer as to how a purported sea serpent could have its co-called coils be impossibly at the level of, or above, the water lies at the core of the explanation I propose in this book.

But if there is a single source responsible for popularizing sea serpents to the extent so as to predispose Gloucester eyewitnesses to believe they saw the same, it is Erik Pontoppidan, the Bishop of Bergen's, seminal 1753 work, *The Natural History of Norway*. Following its translation into English in 1765, it, according to Loxton and Prothero (2015), 'caught the imagination of the world...[and] took off as a global popular mystery... igniting a scientific debate that would rage' until today. Indeed, if Dickens can be said to have been the man to have invented the modern concept of Christmas (Standiford 2011), a similarly bold statement can be made with regard to Pontoppidan and the modern invention of sea serpents. With very few exceptions, the Great Sea Serpent was to remain an exclusive Nordic creature during the eighteenth century.

Pontoppidan tackles the three most prevalently sighted sea monsters, correctly concluding that mermaids equate with dugongs and seals, and that kraken are giant squids. About sea serpents, however, he believes them to be real and neither mythological nor misidentified creatures. Pioneering the methodology that would later come to characterize cryptozoology (e.g. Heuvelmans 1968; Oudemans 1892), he focuses on the aggregate description obtained from 'creditable and experienced fishermen, and sailors, in Norway, of which there are hundreds, [who] can testify that they have annually seen [the sea serpent].' One of the most famous of these sightings was described in a letter from Captain Lorenz von Ferry that was read before a court magistrate and two sworn witnesses in Bergen. Such a practice was to become the *modus operandi* when assessing later sea serpent sightings, such as those in Gloucester, based on the belief that such a legalistic procedure imparted extra validity by transforming an eyewitness account into eyewitness testimony.

> In the meantime the sea-snake passed by us, and we were obliged to turn the vessel about in order to get nearer to it. As the snake swam faster than we could row, I took my gun, which was loaded with small shot, and fired at it; on this it immediately plunged under water...The head of this sea-serpent, which it held more than two feet above the surface of the water, resembled that of a horse. It was of greyish colour, and the mouth was quite black, and very large. It had large black eyes, and a long white mane, which hung down to the surface of the water. Besides the head and neck, we saw seven or eight folds, or coils, of this snake, which were very thick, and as far as we could guess there were a fathom's distance between each fold.

Pontoppidan describes another sighting: 'A great sea serpent was seen several times by several persons together. The weather was calm, but as soon as the sun set and the wind began to blow, it left the fjord, and like one who runs out a coil of rope can know the length therefore, so one could see how long it was, before it had wound off all its coils, and stretched itself at full length.' An even more interesting account was obtained from Governor Benstrup, as it was accompanied by an illustration (Figure 2.15). One of the other eyewitnesses remarked on the 'difference between the body and the tail of this creature as to thickness,' continuing 'It appears that this creature does not, like the Eel or Landsnake, taper gradually to a point, but the body, which looks to me as big as two hogsheads [i.e. barrels], grows remarkably

Figure 2.15. The sea-worm from Erich Pontoppidan's 1753 The Natural History of Norway.

small at once where the tail begins.' It is hard to reconcile this description with the illustration, despite Heuvelmans's (1968) belief that the two pieces of information are compatible. For himself, Pontoppidan, states, in a tiny gem of prose that presages my thesis about the Gloucester creature and many similar 'multi-humped' sea serpents in general, that 'These [the alleged serpent's bends] at a distance appear like so many casks or hogsheads floating in a line, with a considerable distance between each of them.' And at risk of stealing my own thunder, to this statement of Ponotoppidan, Heuvelmans (1968) sideswipes the truth when he comments 'In future I shall speak of such sea-serpents appearing 'as a string of buoys,' and we shall find all sorts of theories put forward to explain these strange lines of humps, which, whatever they are, cannot be part of a true snake.'

Pontoppidan includes other, similar anecdotes. To Sweeney (1972), what is important is that they recur again and again, thereby giving weight to the belief that something was seen which 'looked like a serpent, and which would appear on the surface only if the weather was calm.' Oudemans (1892) too, stresses the replicability of the Norwegian sightings: 'I beg the reader to fix his attention on those apparently insignificant statements, as it will be seen that they are given several times independent of one another.' In this regard, a case can reasonably be made that the Pontoppidan evidence, because it is derived from eyewitness accounts across a broad expanse of both distance and time something that is not the case for the Gloucester Harbour events is probably less prone to inherent confirmation bias and visual contagion.

From Pontoppidan, we are left with the impression that sea serpents are an accepted facet of the Nordic identity. The good bishop wraps things up as follows:

> Before I leave this subject, it may be proper to answer a question that may be put by some people, namely, what reason can be assigned why this Snake of such extraordinary size, etc., should be found in the North sea only? For, according to all accounts from seafaring people, it has never been seen anywhere else. Those who have sailed in other seas in different parts of the globe, have, in their journals, taken particular notice of other Sea monsters, but not one of them mentions this.

For Loxton and Prothero (2015), it is precisely this geographic idiosyncrasy or cultural specificity that is the key problem for the cryptozoological belief in the existence of sea serpents. But Pontoppidan was to be proven wrong in the last decades of the eighteenth century when the sea serpent he did so much to popularize would spread beyond its Nordic homeland. For as the new century approached, reports of sea serpent sightings became more frequent as they were shared in the wharf-side conversations among seafarers and maritime merchants in English, the global *lingua franca* at the time.

The 1815 to 1824 sightings of the Great Sea Serpent in Gloucester Harbour and environs (Chapter 4) did not occur within a regional vacuum. June Purbach O'Neill, in her comprehensively researched *The Great New England Sea Serpent: An Account of Unknown Creatures by Many Respectable Persons Between 1638 and the Present Day*, documents twenty-one regional sightings that occurred before 1815, when I take up the story in the present book. In fact, from the start, it seemed that Cape Ann was a location that was particularly favoured by the elusive and mysterious creature or creatures. A 1793 letter to the *Salem Gazette* mentioned that citizens of Marblehead had been mustered to guard their harbour from a sea creature that had been seen 'a short distance this side of Gloucester' (O'Neill 1999). Earlier, in 1641, Obadiah Turner recorded seeing a creature off Lynn, Massachusetts that was 'as big around in ye thickest part as a wine pipe; and they do affirm that he was fifteen fathoms or more in length [i.e. thirty metres]. A most wonderful tale....We have likewise heard yt Cape Ann ye people have seene a monster like unto this, which did there come out of ye sea and coile himself upon ye land mch to ye terror of them yt did see him' (O'Neill 1999). As well, in what is one of the first mentions by a European of mysterious aquatic animals in the New World, Josselyn (1674) recounts hearing of an incident in 1639 that happened at Cape Ann: 'they told me of a sea-serpent or snake, that lay coiled upon a rock at Cape Ann; a boat passing by with Englishmen on board, and two Indians, they would have shot the serpent, but the Indians warned them, saying that if he were not killed outright, they would all be in danger of their lives.'

Several interesting historical images exist of the Great New England Sea Serpent. One is a twentieth-century map that anachronistically depicts one of those *hic svnt dracones* drawings popular in cartography of three centuries before. Here, off the north coast of Cape Ann can be seen a monster sporting a gapping mouth, a spiny dorsal fin, and a long, curly-cued body terminating in a whale-like tail (Figure 2.16). That the monster is shown to be of a size comparable to those of several of the fishing vessels, is more than a little worrisome. More alarming in this regard is the illustration (Figure 2.17) showing, admix a scene of early colonists steering their boats past an Edenic landscape filled with wildlife and Natives, no less than three multi-humped sea serpents. One, in the background, seems to be making its way toward an unassuming deer nibbling submerged plants; another, in the middle distance, has its predatory designs on a large fish; and, horror of horrors, the one in the foreground has actually clutched a human in its jaws. The message from these images is clear: the Yankee coast is not something to be trifled with lightly by either effete Europeans or dissolute Southerners.

Josselyn's anecdote is also of significance through indicating that the Amerindian peoples of coastal New England were familiar with giant aquatic monsters before the arrival of Europeans. At first blush it seems that these might provide an antecedence to the sightings of the Great New England Sea Serpent. However, although giant aquatic beings are known from folklore traditions of many aboriginal cultures of North America, and are depicted on rock art throughout the continent (Dewdney and Kidd 1973), including in New England (Lenik 2010), often as serpentine creatures that do resemble what Europeans describe as being sea serpents, it would be wrong to fall into the cryptozoological trap, in implying that aboriginal renderings can be related to later European sightings. Amerindian mythology is distinct from European imagination (Meuger 1988). In other words, serpents seen in shamanic visions are different from those thought to be glimpsed in the water.

Figure 2.16. Coulton Waugh's 1927 map displaying historic parts of Cape Ann and the North Shore.

The shamanic rock art in New England dates from the Middle Woodland period, more than a thousand years ago, and is found on prominent bedrock outcrops overlooking coastal rivers and bays, as well as on portable flat stones and ornaments. Images range from simplistic zig-zag or wavy glyphs pecked

Figure 2.17. Illustration from Johann Ludwig Gottfried's 1655 Newe Welt und americanische *of the New World's sea serpent haunted shores.*

into rocks or incised onto stones, to more elaborate depictions of serpents curled about or flicking their tongues (Figure 2.18). Myths tell of the Native god Glooscap coming down from his home in the Bay of Fundy to do battle with a dragon in the Kennebec River in Maine: 'The waters became stagnant and foul. And as he moved along he saw a serpent at a long distance, as it lay on the surface of the dark water. Upon nearing the monster, it raised its head and began to run out its fir-like tongue rapidly at him; by this action he was well aware that this was another deadly enemy, so he steered his canoe directly for the monster...' (Strong 1998). A petroglyph of just such a serpent (Figure 2.18) exists on the banks of the Kennebec River (Strong 1998), possibly drawn during a vision quest in homage to the legend (and not because the artist necessarily saw something in the river). And then there is the legend of a black widow-like squaw whose husbands kept dying mysteriously (Leland 1884), which was famous enough to inspire an illustration on a piece of birch bark (Lenik 2010). The protagonist of the myth is her sixth husband of the year, who spies on her when she is supposed to be berry picking:

> A great foam, or froth, rose to the surface of the water. Then in the foam appeared the tail of a serpent. The creature was of immense size. The woman, who had laid aside all her garments, embraced the serpent, which twined around her, enveloping all her limbs

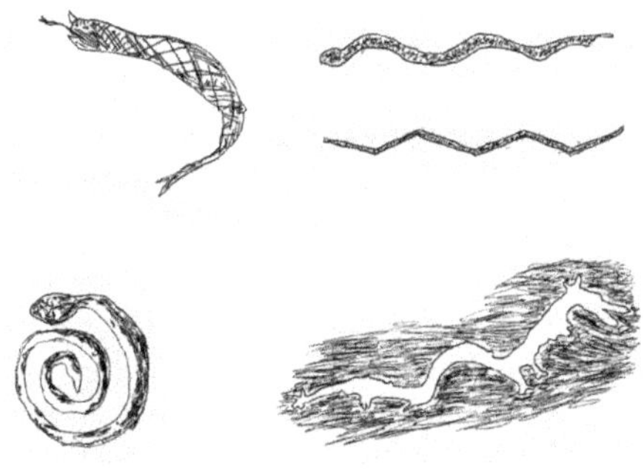

Figure 2.18. Native Amerindian rock art from New England. Images, redrawn from illustrations in Lenik (2010), consist of incised serpents on flat stones from Bridgewater, Massachusetts; a coiled serpent carved onto a stone pendant from Narragansett Bay, Rhode Island; an incised serpent on another flat stone from Bridgewater, this time displaying an extended tongue; and a rubbing of a serpent from a petroglyph site in Maine that may depict the mythological sea dragon which battled with the god Glooscap in the nearby Kennebec River.

> and body in his folds. The husband watched it all. He now understood that, the venom of the serpent having entered the woman, she had saved her life by transferring it to others, who died.

Such stories form part a long tradition of erotic tales that would be told by men in the smokehouse (Schwarz 1976), and are prescient, along with the Japanese tradition of *shokushu goukan*, to what today is the genre of low-brow fiction known as 'tentacle porn.'

Early sightings of New England sea serpents were by no means exclusive to Cape Ann. A few of these are worth recounting, such as one from Captain George Little (*in* Oudemans 1892):

> In May, 1780, I was lying in Round Pond, in Broad Bay [Maine], in a public armed ship. At sunrise, I discovered a large Serpent, or monster, coming down the Bay, on the surface of the water. The cutter was manned and armed. I went myself in the boat, and proceeded after the Serpent. When within a hundred feet, the mariners were ordered to fire on him, but before they could be made ready, the Serpent dove. He was not less than 45 to 50 feet in length; the largest diameter of his body, I should judge, 15 inches; his head nearly the size of that of a man, which he carried four or five feet above the water. He wore every appearance of a common black snake.

Drawing parallels between the likenesses of an alleged sea serpent and that of a black snake (*Coluber constrictor*), a common resident of coastal New England, lies, as we will see (Chapter 3), at the core of the humiliation heaped upon the Linnaean Society for their embarrassing blunder pertaining to the Gloucester events of 1817.

Another account is described in *Lives of Distinguished American Naval Officers* (1846) by noted writer James Fenimore Cooper, when he discusses the exploits of Edward Preble during the Revolutionary War:

> The *Proctor* was lying in one of the bays on the eastern coast [near Penobscot, Maine]... The day was clear and calm, when a large serpent was discovered outside the ship. The animal was lying on the water quite motionless. After inspecting it with the glasses for some time,...Preble [was ordered] to man and arm a large boat, and endeavor to destroy the creature; or at least go as near to it as he could...The boat thus employed pulled twelve oars, and carried a swivel [small cannon] in its bows, besides having its crew armed as boarders. Preble shoved off, and pulled directly towards the monster. As the boat neared it, the serpent raised its head about ten feet above the surface of the water, looking about it. It then began to move slowly away from the boat. Preble pushed on, his men pulling with all their force, and the animal being at no great distance, the swivel was discharged loaded with bullets. The discharge produced no other effect than to quicken the speed of the monster, which soon ran the boat out of sight...Preble stated it as his opinion, that the serpent he saw was from one hundred to one hundred and fifty feet long, and larger than a barrel.

Since this retelling by Fenimore Cooper occurred well before the dramatic 1817 events in Gloucester Harbour, he emphasizes the following: 'When it is remembered that Preble died long before the accounts of the appearance of a similar serpent, that have been promulgated in this country, were brought to light, it affords a singular confirmation of the latter.' Indeed, as will be seen, the elements of what would become the canonical description of the Gloucester creature are all seen here, thirty-eight years earlier: the animal lying motionless on the surface, its notable speed of locomotion, and its remarkable imperviousness to ordnance fire, in this case, even that from a small cannon no less.

As a final, pre-1815, example, in a letter passed on to the Secretary of the American Academy of Arts and Sciences, Reverend Abraham Cummings reported the following (*in* Oudemans 1892):

> It was some time in July 1802 that we saw the extraordinary sea monster...[near Penobscot, Maine]. I then supposed it to be a large shoal of fish with a seal at one end of it, but wondered that the seal should rise out of water so much higher then usual; but as he drew nearer to our boat, we soon discovered that this whole appearance was but one animal in the form of a serpent. I immediately perceived that his mode of swimming was exactly such as had been described to me by some of the people of Fox [i.e. North Haven and Vinalhaven] Islands, who had seen an animal of this kind before, which must confirm the veracity of their report. For this creature had not the horizontal, but an ascending and descending serpentine motion. This renders it highly probable that he never moves on land to any considerable distance and that the water is his proper element. His head was rather larger than that of a horse, but formed like that of a serpent. His body we judged was more than sixty feet in length. His head and as much of his body as we could discover was all of a blue colour except a black circle round his eye. His motion was at first but moderate, but when he left us and proceeded towards the ocean, he moved with the greatest rapidity.

Once again, we have reportage of the great speed of the creature which, in this case, and of upmost significance, is described as resulting from vertical undulations of the twenty metre-long body described as being 'formed like that of a serpent' with neither lateral appendages nor a caudal tail. These are traits that will be reiterated in the subsequent Gloucester sightings. My use of the word 'significance' stems from the simple fact of the physical impossibility of achieving 'the greatest rapidity' through nothing more than 'an ascending and descending serpentine motion' of a wiggling body having no fish or whale-like bi-lobate tail. Clearly there is something going on in these sightings that defies the inherent laws of straightforward biomechanics. The only possible solution is, as will be seen, one that departs from a strictly biological explanation for what was observed.

Chapter 3.

CULTURAL HISTORY OF THE GLOUCESTER SEA SERPENT

3.1 HAPPENINGS

The sightings of the Gloucester Sea Serpent have been described in narrative form previously, including most accessibly in popular books by O'Neill (1999) and Soini (2010). Because of this, and the fact that the entirety of the individual accounts, in the eyewitnesses' own words, are provided in Chapter 4, the present overview will be brief.

What would become a rash of sightings began in 1815 with an account from Plymouth, Massachusetts, 70 kilometres south of Gloucester.

There were no reported sightings from 1816, the 'year without summer.'

For the auspicious year of 1817, Heuvelmans (1968) reports that although 'two women saw a sea-monster like a huge serpent come into the harbor of Cape Ann which lies north of Gloucester roads,' on August 6th, neither this, nor the confirming report of the captain of a ship entering the harbour at around the same time, drew any public attention. The situation changed dramatically four days later when four additional town residents spotted the creature. When the creature was observed again on August 12th and 13th by a respected resident sea captain, the town began to take notice and wonder if something truly unusual was taking place in their maritime front parlour (i.e. the sightings of the previous eyewitnesses: he who was 'from away' and therefore not a trusted local, and those of the female persuasion, who like all of their gender, were believed to be prone to overwrought imaginations, had lacked credibility).

Eyes of many residents now began to turn toward the harbour in anticipation of spotting the monstrous visitor. And then the big debut of the sea serpent came on August 14th. Despite it being a busy Thursday in the bustling seaport, some observers had purposely turned up, in festive spirit, spy-glasses in hand, to catch a glimpse of the enigmatic animal. They were not to be disappointed. The Serpent was seen by dozens of people, including one who took a shot at it. Eighteen eyewitnesses would later be interviewed, either by Justice of the Peace Lonson Nash for the report by the Linnaean Society, or by sea captain Charles L. Sargent for his own parallel investigation. The phenomenon of sea serpent tourism now began in earnest, more than a century before and nearly half a world away from, the hysteria of the monster sightings that would bring global fame to a certain lake in northern Scotland.

The creature continued to linger in Gloucester Harbour, with multiple sightings on the 17th and 18th, at which time the Linnaean Society of New England had a meeting to begin their investigation of the Sea Serpent. Repeated sightings, on almost a daily basis, continued on to the end of the month and into the autumn. Most were made from locals on shore or from those purposely heading out into the harbour in their fishing dories; others were made by the crews of visiting ships.

News of the Serpent began to extend well beyond its frequented Cape Ann environs. Press reporters flooded into Gloucester and soon broadsheets were published based on interviews with an increasing number of eyewitnesses. At the end of August, the amateur naturalist and former general David Humphreys arrived, and although he did not himself see the Sea Serpent, he dutifully sought out and recorded statements from eyewitnesses, which he worked into a series of letters sent during the month of September to Sir Joseph Banks, the president of the Royal Philosophical Society in England. The last sighting of the year in Massachusetts took place on September 10th, near the town of Marblehead, about halfway between Cape Ann and Boston. With one notable exception, all was quiet until three weeks later when what is likely to have been the same creature was reported in Long Island Sound, New York, where, until mid-November, he was seen a handful of times.

That notable exception was the discovery, on September 27th, in a small cove on Cape Ann, of what was reported to be a baby sea serpent, as will be elaborated upon subsequently.

The following summer, the Sea Serpent returned to its old haunts. On May 10th, 1818, at an unspecified location north of Cape Ann a creature of matching description to that of the previous year was fired upon by the crew of ship on their way to Boston. By June 21st, the creature was back at Cape Ann, where the crew of a packet observed it to be engaged in combat with a humpback whale.

During July and early August, the Sea Serpent was once again observed in Gloucester Harbour by hundreds of onlookers. During this time, a shift occurred from simply admiring the creature 'playing' in the water, to actively pursuing it for the reward money offered. The harbour became filled with a flotilla of dories filled with eager hunters who, much to their collective chagrin, were unable to kill the creature despite striking it repeatedly with musket balls and even in one case having their harpoons actually bounce off the animal. Finally, on September 6th, Captain Rich captured what he purported to be the Sea Serpent on the north coast of Cape Ann and put it on display in Boston. The public, however, immediately became incensed when they found out that what Rich insisted on being the animal responsible for the sea serpent sightings, was nothing more exotic than a large tuna. This perplexing incident will be examined subsequently.

Despite this assertion, the Sea Serpent, or another creature just like it, returned the next year, 1819. After several Cape Ann sightings in June, the creature headed 30 kilometres south to Boston Harbour to set up a new residence off the Nahant peninsula for the month of August. There, as had occurred previously in Gloucester, the presence of the Sea Serpent generated a viewing frenzy with hundreds of people walking the shores or plying the waters, spyglasses and/or rifles in hand, eager for an encounter with the mysterious denizen of the deep. By the end of August, the creature had returned to make one more appearance in Gloucester Harbour and again in Boston Harbour before disappearing for the rest of the year.

Sightings in subsequent years were infrequent. This was either the result of observer habituation and a consequent un-newsworthiness of the events in terms of their reportage, or an actual reduction in the creature's presence. Six sightings were recorded in the years 1820, 1821 and 1822 in Boston Harbour, although more may well have been made, given that a local entrepreneur was confident enough to build a hotel to capitalize on the craze (Figure 3.1). Four more sightings occurred in various locations along the Massachusetts coast in 1823, including one in which the creature was reported to have been impervious to being fired upon by musket balls. After another sighting north of Cape Ann in 1824, the famous Gloucester Sea Serpent largely sinks into the mists of time, to be replaced by a diverse bevy of sea monsters recorded in other regions of the northwest Atlantic as well as the rest of the world. With one notable exception, however.

Figure 3.1. Nineteenth-century illustration of hotel at Nahant created, in part, to capitalize on tourists interested in catching a glimpse of the famous Sea Serpent in Boston Harbour (Paterson and Seaburg 1991).

Drawing the parallel to the well-known nursery-rhyme, Kierman (1950) recounts how the sea serpent came back, if not like the cat, the very next day, then sixty-nine years later. For in 1886, Cape Ann was again visited by a mysterious sea creature resembling its predecessor from 1817. A flurry of sightings occurred. Then, in another case of 'déjà vu all over again,' in 1934 there was an investigation undertaken by amateur naturalists to obtain testimonies from all surviving eyewitnesses to the 1886 events. And in a grand gesture somewhat anachronistic in the spirit of the Victorian Age, a public meeting was held to present and discuss all the evidence for the sea serpent (Figure 3.2).

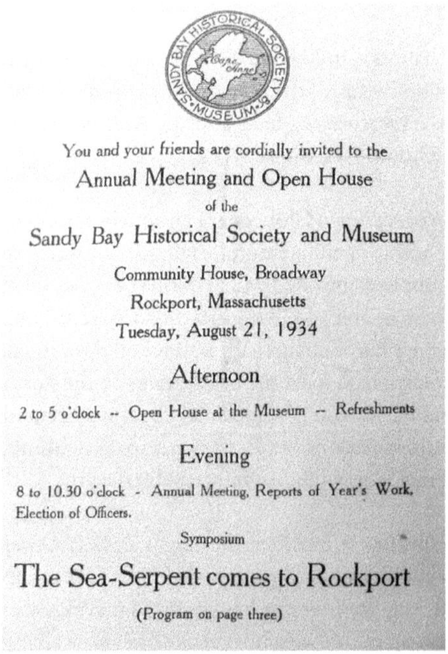

SANDY BAY HISTORICAL SOCIETY · MUSEUM · Cape Ann

You and your friends are cordially invited to the

Annual Meeting and Open House

of the

Sandy Bay Historical Society and Museum

Community House, Broadway

Rockport, Massachusetts

Tuesday, August 21, 1934

Afternoon

2 to 5 o'clock -- Open House at the Museum -- Refreshments

Evening

8 to 10.30 o'clock - Annual Meeting, Reports of Year's Work, Election of Officers.

Symposium

The Sea-Serpent comes to Rockport

(Program on page three)

Figure 3.2. Program from 1934 about an evening of lectures in Rockport, Cape Ann concerning the 1886 return of the/a sea serpent to the area (Woodbury scrapbook, Cape Ann Museum Archives).

3.2 INVESTIGATIONS

The barrage of sightings coming out of Gloucester during August 1817, of a serpent-like creature of 'immense size' that was 'jointed or like a number of buoys or casks following each other in a line' and which possessed the ability 'to move with wonderful rapidity,' motivated the Linnaean Society of New England to hold an impromptu meeting on the 18th of the month in Boston. Leading figures of the Society, the Honorable John Davis, judge, Jacob Bigelow, physician, and Francis C. Gray, attorney, created a committee to 'collect evidence which may exist respecting a remarkable animal, denominated a *Sea Serpent*, reported to have recently been seen in or near the harbor of *Cape Ann*,' which would later be gathered into a scholarly report. To facilitate this, they created a suite of questions to be asked of eyewitnesses (Table 3.1). This is of remarkable importance as it is the first, and indeed will be the *only*, attempt to ever undertake a specific investigation of a purported sea serpent soon after the initial sightings had taken place.

Demonstrating a notable naivety in terms of a misplaced confidence in the merits of a legal investigation over a scientific one, a move that would later raise the dander of learned Victorian natural historians (Regal 2012; see Chapter 1), the Committee wrote to the Honorable Lonson Nash of Gloucester, requesting him to 'examine upon oath some of the inhabitants of that town with regard to the appearance of this animal, to make the examination as early as possible, to request the persons examined not to communicate to each other the substance of their respective statements, until they were all committed to writing; to have these statements sighed and certified in due form, and sent to us.' Judge Nash wasted no time in going about his task.

Table 3.1. Questions posed to eyewitnesses sighting the Gloucester Sea Serpent for the investigation: Report of a Committee of the Linnaean Society of New England relative to a Large Marine Animal, Supposed to be a Serpent, Seen near Cape Ann, Massachusetts, in August, 1817 *(Davis* et al. *1817), reprinted in Oudemans (1892), O'Neill (1999), and elsewhere.*

1.	When did you first see this animal?
2.	How often and how long at a time?
3.	At what times of day?
4.	At what distance?
5.	How near the shore?
6.	What was its general appearance?
7.	Was it in motion or at rest?
8.	How fast did it move, and in what direction?
9.	What parts of it were above the water and how high?
10.	Did it appear jointed or only serpentine?
11.	If serpentine, were its sinuosities vertical or horizontal?
12.	How many distinct portions were out of the water at one time?
13.	What were its colour, length and thickness?
14.	Did it appear smooth or rough?
15.	What were the size and shape of its head, and the head ears, horns, or other appendages?
16.	Describe its eyes and mouth.
17.	Had it gills or breathing holes, and where?
18.	Had it fins or legs, and where?
19.	Had it a mane or hairs, and where?
20.	How did its tail terminate?
21.	Did it utter a sound?
22.	Did it appear to pursue, avoid or notice any thing?
23.	Did you see more than one?
24.	How many persons saw it?
25.	State any other remarkable fact.

Nine days later Nash wrote back to the Linnaean Society, stating that he had 'cheerfully complied with your request' and passed on eight 'duly certified' depositions which he had obtained about the 'strange marine animal.' He stated that the deponents, whom he identified as being 'men of fair and unblemished characters,' were interrogated separately and in ignorance of others so testifying. Of their accounts, he stated, though 'they differ in some few particulars, still, for the most part, they agree.' He did, however, offer his opinion, based on his own personal observation of the Sea Serpent, about several 'mistaken' attributes reported by one of the eyewitnesses, due to these being different from the others. Nash stated that although all questions were put to the witnesses, he included only the positive responses. And, in a move that is certainly open to criticism by today's standards, Nash obviously exerted a heavy editorial hand when transcribing the polished version he submitted, as demonstrated by the marked similarity of wording among some of the depositions (which will be evident in the next chapter).

The Linnaean Society obtained an additional three depositions in Boston, with all eleven being read during a meeting on September 1st. The following day, the Committee wrote back in gratitude to Mr. Nash, praising him for his accomplishments that 'will go far in giving some precise and accurate conceptions on a subject, peculiarly exposed to exaggeration and mistake,' and offering assurance that the evidence will subsequently be published. The letter went on to request three further bits of investigation from Nash: (a) that he verify a report that the Committee had heard about an unknown woman who had seen the animal 'lying dormant very near the shore;' (2) that he contact a gentleman who it was said had a drawing 'supposed to be a pretty good representation of the animal' which, if it proved so verifiable, they would like to obtain permission to have it published in the final report; and finally (3) more details about his own particular sighting of the creature. Nash dutifully provided information in response to the first and third requests, but mentioned that the gentleman in question who had made the drawing was presently in Boston.

Simultaneously, the Society, displaying their approach to the investigation as a judicial rather than a scientific inquiry, wrote to ask a Justice of the Peace in Plymouth to question under oath and obtain the written account of a 'respectable man' whom, it was generally known, had seen an animal of like description to the one in Gloucester several years before. On October 2nd, the Honorable Samuel Davis wrote back apologizing for the delay as the eyewitness in question had been engaged in out-of-town business. This second Judge Davis included the account and stated that the observer was 'impartial and unbiased,' not having read any of the Gloucester evidence, and that he was an experienced mariner who had 'frequently seen whales, and almost every species of fish.' Davis carefully noted that all the questions were asked of the Plymouth eyewitness and that the Committee should interpret negative replies as not being indicative that those attributes did not exist, but rather that they were simply not seen. Also of note, Davis wrote, is that the animal had been observed by others.

The final document was entitled *Report of a Committee of the Linnaean Society of New England relative to a Large Marine Animal, Supposed to be a Serpent, Seen near Cape Ann, Massachusetts, in August, 1817* (Figure 3.3). In addition to the previous testimonies, the *Report* also included a letter from Reverend William Jenks, of Bath, Maine, in which he recounts three additional sightings of sea serpents in Maine in the years 1809 and 1811, as well as a lengthy excerpt from Bishop Pontoppidan's natural history book describing sightings of Norwegian sea serpents from the previous century. The legalistically-inclined Committee thought that, in terms of proving the existence of a sea serpent, it was very much a situation of 'case closed':

> We have seen and heard sundry other statements of various authority relating to similar animals, said to have been seen at sea by different persons, but do not insert them in our report, because we consider the foregoing testimony sufficient to place the existence of the animal beyond a doubt; and because they do not appear so minute and so well authenticated as the preceding documents.

And if that were not enough, the *Report* concluded with the best evidence yet: nothing less than a detailed description of the specimen of the baby sea serpent that had been discovered on September 27th. This will be discussed in the next section.

Reeling from the humiliation heaped upon its conclusions in the *Report* (see below), the Linnaean Society remained mum in 1818. Into the void, Gloucester resident and merchant service sea captain, Charles Lenox Sargent, took up the baton and began to interview his neighbours about their own sightings of the Sea Serpent, many of which having had occurred the previous year. His handwritten notebook, intended possibly for later expansion and publication but made impossible by his premature death, is entitled *The Sea Serpent, Evidence About It*, is in the Cape Ann Museum Archives (Figure 3.4). The fifteen-page pamphlet contains a series of short testimonies from his acquaintances (Figure 3.5) that are presented in a warm, chatty, first-person manner distinct from the more terse and legalistic tone of the Linnaean Society *Report*. Of interest is that Sargent did not consider women to be inferior in their reliability as eyewitnesses.

The earliest accounts are three that date from August 10th, 1817, another dozen from the big debut four days later, and several more from thereafter. Sargent's conclusions were modest: 'The foregoing in addition to the Account taken last year by the Linnaean Society one would suppose enough to satisfy any reasonable person of him having been seen by numerous respectable witnesses, a strange fish of the Serpent kind from 50 to 120 feet long...' Because of this evidence, Sargent concluded his brief, independent investigation by stating his belief that the creature that had been recently caught by Captain Rich (as detailed in Chapter 4) could not possibly have been the Sea Serpent observed by so many.

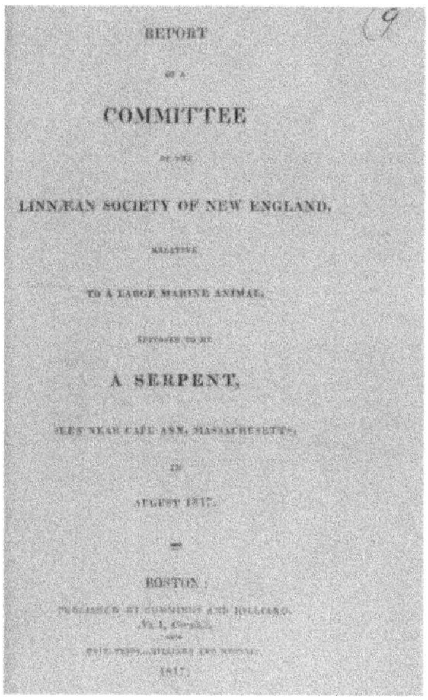

REPORT

COMMITTEE

LINNÆAN SOCIETY OF NEW ENGLAND,

TO A LARGE MARINE ANIMAL,

A SERPENT,

SEEN NEAR CAPE ANN, MASSACHUSETTS,

AUGUST 1817.

BOSTON:

1817.

Figure 3.3. One of the few remaining existing copies of the privately published 1817 report by the Linnaean Society of New England (Davis et al. *1817) detailing their investigations of the Gloucester Sea Serpent (Cape Ann Museum Archives).*

Figure 3.4. The sole copy of Charles L. Sargent's 1818 compilation of eyewitness accounts of the Gloucester Sea Serpent (Cape Ann Museum Archives).

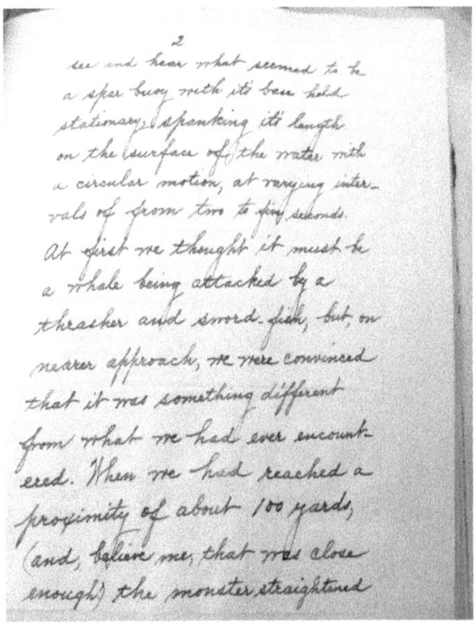

2
see and hear what seemed to be
a spar buoy with its base held
stationary, & spanking its length
on the surface of the water with
a circular motion, at varying inter-
vals of from two to five seconds.
At first we thought it must be
a whale being attacked by a
thrasher and sword-fish, but, on
nearer approach, we were convinced
that it was something different
from what we had ever encount-
ered. When we had reached a
proximity of about 100 yards,
(and, believe me, that was close
enough) the monster straightened

Figure 3.5. A representative page of eyewitness testimony derived from Sargent's interviews of his neighbours. The text is in his own hand (Cape Ann Museum Archives).

David Humphreys was a businessman, a former civil servant and military officer, a prolific writer and commentator on social issues, and a keen amateur naturalist. He travelled to Gloucester in late summer 1817 to obtain first-hand knowledge of the sea serpent phenomenon. And although he did not personally view the creature, he was nevertheless able to provide an objective sense of verisimilitude through his independent inquires and appraisals. These were expressed in a series of five letters, written from early September to early October, to the Right Honorable Sir Joseph Banks, President of the Royal Society, who had been the naturalist aboard Captain Cook's voyage to Australia. The letters are notable for their quality of writing, clarity of observational detail, and unwillingness to jump to unfounded conclusions, traits, it must be said, absent from both the Linnaean Society *Report* and Sargent's private notes. As noted by other historians of the Gloucester phenomenon of 1817 (Brown 1990; Burns 2014), Humphreys states that the creature displayed serpent-*like* attributes, but is careful to stop short of concluding that it actually *was* a giant serpent.

3.3 HUMILIATION

The last quarter of the Linnaean Society's Report (Davis *et al.* 1817) is entitled 'Account of the *Soliophis atlanticus*,' and contains details from the examination and dissection of the specimen that had been found in Loblolly Cove on Cape Ann, and which had been subsequently exhibited in Boston as being the 'progeny of the great serpent.'

The external appearance (Figure 3.6) is described as being different from any other class of serpent known to members of the Committee, largely in consequence of the 'row of protuberances along the back, apparently formed by undulations of the spine.' The overall length of the animal was about one metre, with a maximum circumference of slightly more than five centimetres. Details about the shape of the head, nature of the dentition, eyes, body scales, and colouration were provided. The committee concluded that, despite the absence of a flattened tail for propulsion, the animal is 'probably amphibious.' They went on to mention the notable vertical flexibility of the animal's spine from which they supposed that it swam by 'bending in a vertical direction', much like a leech.

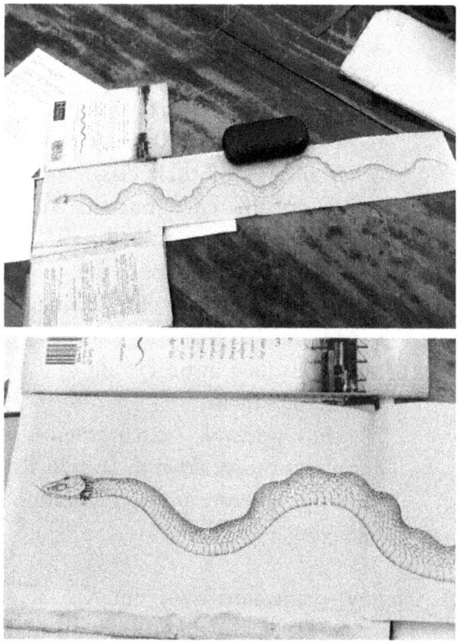

*Figure 3.6. Diagram of the deformed snake designated (*Scoliophis atlanticus*) and considered by the Linnaean Society in their* Report *(Davis* et al. *1817) to be the progeny of the Sea Serpent observed within the harbour.*

With regard to the internal structure (Figure 3.7), Davis *et al.* (1817) again reiterated that the curving undulations of the spine were singular, with 'each curve consist[ing] of about nine vertebrae.' Much ink was spilt on more detailed descriptions of the anatomy of the spinal column in comparison to those of known species of snakes. Facets of the animal's musculature and viscera were commented upon, including mention that stomach contents revealed the baby specimen to have consumed another species of serpent. An attempt was made to postulate the biomechanics of the animal's locomotion based on interpretations of how the dorsal muscles attached to the vertebrae and how the latter were articulated.

Finally, in the best tradition of the eminent naturalist from whom their organization took its name, the Committee christened the animal *Scoliophis atlanticus* (Atlantic Humped Snake). The genus name is derived from the similar root for the disease of scoliosis, which is characterized by abnormal curvature of the spine. This nomenclature will be particularly ironic given the humiliation that would soon follow.

Further entries in the *Report* were from the individuals whom had discovered, killed, and collected the specimen. Of note was that the location of capture was near a brackish pond periodically connected to the sea about twenty metres away. This led to the suggestion that this juvenile life stage might have been anadromous. Lonson Nash interviewed the father and son whom had caught the animal and generated the following bit of purple prose about their encounter in the moments preceding the serpent being skewered and beaten to death:

> He lay extended by some whortleberry bushes, when first discovered. The boy called to his father...to run, for he had never seen so strange a snake. Mr. Colby ran, with a pitchfork in his hand, and with his fork, confined the serpent against some loose rocks; when the serpent exhibited the most violent temper. Twice he bit himself, held on, and shook...as one dog shakes another, when fighting. His tail seemed likewise a weapon of defense; for he struck the head of it against the fork handle several times.

The *Report* closed by tying the two putative serpents, the one swimming in the harbour and the other crawling upon the ground, together:

> The appearance at nearly the same time and place of two creatures, agreeing with each other in certain important and conspicuous particulars, disagreeing in the most remarkable of these particulars with other animals of their class; and between whom no difference, but that of size, has been discovered, must naturally lead to a conjecture that they are of the same species. The evidences in favor of this supposition are so considerable, that your Committee would not be justified in neglecting to direct your attention to them.
>
> The principle circumstances which encourage the idea that these animals possess the same structure and configuration are the following...

A lengthy discussion then follows about the protuberances seen above the water of the adult in the harbour and those observed in the dissected specimen. Similarities are drawn between the two creatures' heads, eyes, and colour, including inferences made about the mode of locomotion:

> The structure of Scoliophis [sic] is well suited to a residence in the water, being capable of various and complicated motions. It bends horizontally, as did the great Serpent, in the act of turning; it bends vertically, as that animal was supposed to do. In the act of swimming; and it might assume any compound and intermediate motion, which should be most effective in propelling it through the water.

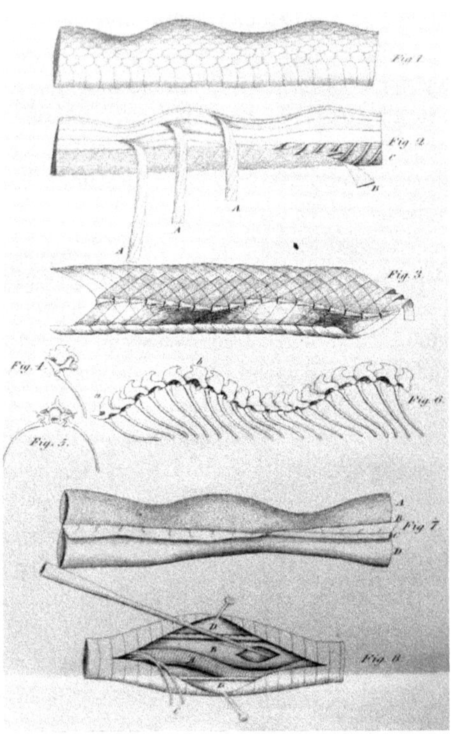

Figure 3.7. Anatomical illustrations of the so-called baby sea serpent by the Linnaean Society (Davis et al. *1817), which in reality was a deformed common black snake (*Coluber constrictor*).*

Davis *et al.* (1817) state that in 'nearly all the circumstances with regard to the appearance of the great Serpent, stated by the deponents, as facts, agree with the structure of the Scoliophis [sic].' From which they make the jump to conclude that if the two are so similar then it certainly stands to reason that 'one is the progeny of the other.' This is followed by some life history suppositions such as that the baby 'might have remained in the place where it was hatched, or might occasionally have resorted to the shore from the water. We should not expect it to venture far from the shore, until its increased size should afford it some security from becoming a prey to larger animals of the ocean.' The amateur naturalists note that other snakes deposit their eggs in the sand in spring or the summer, whereupon they 'are hatched by the heat of the sun in less than a month.' They then use this as a reason to explain why the large sea serpents 'described in our accounts and depositions, were seen near the shore, and only in the month of August, excepting once at Plymouth in June.' That said, however, the thorny question that derives from the presumed oviparous nature of the amphibious sea serpent is completely sidestepped:

i.e. why, if the animal was amphibious, no one, with the sole exception of the recent mention by one individual, had ever seen such an enormous beast emerging onto a beach to lay its eggs?

The *Report* concludes by transferring the Linnaean name of the baby specimen to that of the adult seen in the harbour:

> On the whole, as these two animals agree in so many conspicuous, important and peculiar characters, and as no material difference between them has yet been clearly pointed out, expecting that of size; the Society will probably feel justified in considering them individuals of the same species, and entitled to the same name, until a more close examination of the great Serpent shall have disclosed some difference of structure, important enough to constitute a specific distinction. JOHN DAVIS, JACOB BIGELOW, FRANCIS C. GRAY.

The race was on to identify, classify, and catalogue all life on Earth, and in this the Linnaean Society of New England would become a leader, its ticket to fame being the first scientific description of the Sea Serpent. By so doing, they would solve one of the most compelling zoological puzzles of all time while proving once and for all that the United States of America was home to truly gigantic animals. One can almost imagine the cry: 'Take that, you Old World, European [read, French] bigots.'

Alas, it was not to be.

Publication of the *Report* and popularization of its findings led to a new pastime in which to engage ardent sea serpent observers now that the adult creature appeared to have departed from the harbour. Soon, many were scouring the beaches of Cape Ann for sea serpent eggs or baby monsters. Somehow, the glaring fact that the putative baby sea serpent closely resembled the common black snake *Coluber constrictor*, ubiquitous to the Northeastern Seaboard, was conveniently ignored by both the Committee and the general public as all were swept up in a contagious spirit of enthusiasm.

The Committee was soon to become the object of local, regional, and eventually, once word spread abroad, international, ridicule. The *Report* was reviewed in the Parisian *Journal de Physique, de Chimie et d'Histoire Naturelle* by the noted zoologist and anatomist Henri Ducrotay de Blainville, who specialized in reptiles and who would later coin the term 'palaeontology.' His opinion was that there was, as Heuvelmans (1968) phrases it, 'no good reason to connect this huge animal [observed in the harbour] with the little specimen from Loblolly Cove.' The French naturalist concluded (*in* Oudemans 1892 and *in* Heuvelmans 1968):

> If we were now to scrutinize the existence of the Great Sea-Serpent, we must confess that it would be difficult to deny the appearance of an animal of very great length, very slender, and swimming with rapidity, in the sea near Cape Ann, but that it is a true snake, this is doubtful; that it is of the same genus as the *Scoliophis*, this assertion is still more doubtful, and finally that it is of the same species, reduces the number of probabilities which become totally null, if one believes that such an immense animal, as that which is observed in the sea has gone ashore to lay its eggs!

It must be remembered that the Linnaean Society of New England was largely a group of amateur naturalists, and that of the three members of the Committee charged to investigate the Gloucester Sea Serpent only one, Bigelow, had what might be called professional experience in biology, which in his case pertained to botany and human anatomy. When other professional biologists, having considerable experience in zoological anatomy, examined the carcass of the humped-back 'baby' sea serpent, they reached a completely different and much more mundane conclusion. This group was composed of William Peck, who was a professor of natural history at Harvard University with a specialization in entomology, Samuel Mitchell, who was a physician and polymath who had taught natural history at Columbia College (now University), and Alexandre Lesueur, who was a Humboldtian explorer and naturalist with a specialization in ichthyology. Their conclusion was that the odd looking specimen was nothing more than a common black snake with a spinal column that had become deformed through disease or injury, and that it most certainly had nothing whatsoever to do with whatever it was – and they did not discount the possibility that 'it' might indeed be a sea serpent – seen swimming around Gloucester Harbour.

The hitherto believing public were much taken back by this revelation of professional incompetence on the part of the Linnaean Society. As Gould (1930) comments, 'it is scarcely surprising that those competent to form an opinion should have concluded that persons capable of so egregious a blunder in a simple point of identification were not the safest guides to accurate information regarding a strange and huge creature of unknown species which they had not personally examined.' Heuvelmans (1968) sums up the fallout similarly: 'The sceptics and mockers had a field-day. The whole inquiry, though begun with such strict care, was blown away in a gale of laugher. People felt, not unreasonably, that experts who could make so grotesque a blunder could not tell the true from the false and had been taken in by hoaxers or madmen. The public swore – a little prematurely – that they would not be taken in again.' The caveat 'a little prematurely' refers to the issue of the capture of the purported Sea Serpent that occurred the following year, as will be discussed in Chapter 4.

And so, what we have in this early debate, is the exact reverse of the 'Us' (i.e. observers of, and believers in, anomalies) against 'Them' (i.e. ivory tower experts and skeptics) disjunction that lies at the psychological core of modern cryptozoology (Dendle 2006). For here, in 1817, it is the public who are the skeptics, and it is some of the academics (like Bigelow) who are the gullible believers. On a further note, what is a little sad, is that the whole charade of so-called experts misidentifying some mundane animal as the juvenile form of a sea serpent is something that continues today (Woodley *et al.* 2012). How little we have learned, it seems.

Throughout the nineteenth century, naturalists continued to express opinions about the baby sea serpent of Cape Ann. A letter from Lesueur to de Bainville was published in 1819 in the German journal, *Isis*, edited by noted naturalist Lorenz Oken, in which Lesueur expresses his criticism of the inaccurate drawing of the vertebral column made by the Committee, noting it to be of poor quality rather than implying any deliberate fraud. Four years later, in the German journal *Notizen* (*in* Oudemans 1892), Robert Froriep added his opinion on Lesueur's revelation: 'As long as the Linnaean Society, to prove their explanation, cannot depose an accurate observation or a dissection, we may be allowed to entertain modest doubt about their explanation.' For Oudemans (1892), the naivety of the Linnaean Society was glaring:

> It is quite astonishing that scientific men could come to the conclusion that the large animal, that gave rise to 51 accounts which the Committee could have gathered up to

> their days, was a full grown individual of the species they called *Scoliophis atlanticus*! If they had collected all these accounts, if they had seriously compared them, they would have come most probably to the conclusion that they did not know preciously what it was, but that it could never be a snake.
>
> Also from another point of view it is hard to explain that the Committee believed the sea-serpent to be of the same species as the little *Scoiophis*. Three persons mentioned the tongue, which was not bifid, while the tongue of *Scoliophis* is so! And the most accurate testimonies agree that the skin was smooth and had *no* scales!.

The inevitable result of this taxonomic side-show was to cast doubt and dispersion upon the sightings of the creature in the harbour. Gould (1930) refers to the 'calm indifference' with which the natural scientific community regarded the whole Gloucester phenomenon. Heuvelmans (1968) goes one step further in his appraisal, in stating that 'no scientist dared to take the sea-serpent seriously' as a result of 'the appalling bloomer made by the Boston committee.' Gosse (1861), for example, states: 'Owing to the habit prevalent in the United States, of supposing that there is somewhat of wit in gross exaggerations, or hoaxing inventions, we do naturally look with a lurking suspicion on American statements, when they describe unusual or disputed phenomena.' Ouch. Cogswell (1847) expresses difficulty in divesting European minds of suspicion when it comes to regarding North American sightings of sea serpents: 'Our relative in the West having acquired nearly as much celebrity for the endowment of a grand inventive genius as his Scandinavian ally in the cause of sea-serpents. They defer indeed, in so far as the latter for the purely benevolent purpose of what is called 'hoaxing' the unwary public.' He then goes on to recount a recent example of an American pitching a fabrication about a race of aliens discovered on the moon, which fooled many people. For Cogswell, 'such proof of a disposition to practice on the public credulity, too often repeated, necessarily communicate a colouring of insincerity to all other reports of strange events emanating from the same source [i.e. America], and certainly demand the exercise of an unusual amount of circumspection.' Double ouch. Because of this widespread suspicion of American veracity, Gosse (1861) simply ignores all accounts originating from the United States. The 1842 edition of the *Encyclopedia Britannica* sums up the opinion of British experts that 'no proper proof has yet been adduced for any of these species (sea-serpents) inhabiting the 'American Ferry', as we see that world of waters now named since the steaming days of the *British Queen* and the *Great Western*.' And so what had been an attempt to finally demonstrate American scientific exceptionalism, disastrously backfired.

For those of the general public distant from Gloucester, the sea serpent began to be thought of as nothing more than a monster myth (Gould 1930):

> The story was strange and improbable; it rested on the assertions of a comparatively small number of persons living in the same town; and it provided their less-favoured neighbours with a most excellent means of poking fun at Gloucester credulity and vaunting their own superior perspicacity and common sense. In consequence, the sober tales of the Gloucester witnesses were, in no long time, distorted, parodied and (to a great extent) laughed out of court. Practical jokers sent in marvelous reports of apocryphal 'sea-serpents'; not supporting – or, indeed, designed to support – the slightest examination. The Press, with one or two honourable exceptions, followed on the same side – after its fashion.

Indeed, as will be shown later in this chapter, it took only a year until a theatrical parody would be performed spoofing the entire debacle as a bunch of Yankee humbug at best, or shenanigans at worst.

3.4 HISTORIOGRAPHY

Historians seem to have been a confused bunch with respect to how, or even if, to refer to the remarkable events of 1817, when the town of Gloucester became an issue of dockside and parlour room discussion around the world.

Two widely distributed general histories have been written about Gloucester. In 1860, native son and descendent of early settlers, John Babson, published *History of the Town of Gloucester, Cape Ann*. At over six hundred pages, the book contains copious information about the early settlers, presented in almost Old Testament detail in terms of who begat whom. All kinds of events are described, some of national significance, others concerning the mundanity of small town life in a coastal community. Chapter XXX lists as its subjects, the following: 'The Federalists lose Ascendency. The Sea-Serpent. Boys drowned. Universal Benevolent Society and its Ministers,' etc. And herein, in a dozen paragraphs, is a short distillation about the sea serpent phenomenon. Babson describes the controversy involving Captain Rich very succinctly, stating that the captured creature was 'reported extensively to be the sea-serpent itself, but finally ascertained to be nothing of greater strangeness than a very large horse-mackerel [i.e. giant Bluefin tuna].' The section concludes open-endedly, quoting the poet Schiller:

> In view of all this testimony, and particularly of that of so many intelligent and respectable citizens of Gloucester, what shall we say? – that this monster was indeed a visitor from that 'gloomy and pathless obscure,' the home of –
>
> 'Salamander, snake, dragon, – vast reptiles that dwell
> In the deep'? –
>
> Or that all the relations concerning him must take their place in the history of mass delusions?

Thirty-two years later, another Gloucester native, James Pringle, continued the story in *History of the Town and City of Gloucester, Cape Ann, Massachusetts*. Chapter VIII, entitled 'From 1800 to 1860,' contains the section 'View in 1817,' which offered promise for insights it might contain about the famous Sea Serpent, several generations on from the events of that most notable year. Here, along with the accompanying image (Figure 3.8), is the text of that section, given in its entirety to indicate the shift that seemed to have occurred in how locals had moved from pride to embarrassment in their regard for the Sea Serpent:

> A sketch of the town in 1817, was drawn by Capt. John Beach, from a point in the harbor opposite Ten Pound Island, off Pavilion Beach. Standing out prominently in the foreground was an eight-sided wooden windmill erected on the site now occupied by the Pavilion Hotel. This had been built in 1814 by Ignatius Webber, and its long arms, fitted with sails, made it a conspicuous object, and gave the town quire a foreign aspect. It was subsequently removed to a position on Fort Square, where it became a familiar landmark, being destroyed by fire several years ago. Running along Pavilion Beach from the windmill, a narrow passage only intervening, and extending nearly to the public mall, was a long rope walk which had been built by Capt. Webber and Aaron Plummer in 1803. Conspicuous in the view on this side of the fort were two large houses, one with a gambrel roof, then used as the residence of prominent people now long since given over to tenement house purposes, with no reminder of past glories. The church steeples were also prominent. The Universalist edifice, as it stands now, the old First Parish church, with its pagoda like steeple, and the old Collins school house, standing on the

Figure 3.8. Panorama of Gloucester harbour and the sea serpent by John Ritto Penniman from a drawing made by Captain John Beach. The painting was exhibited in Salem, Boston, and New York City through the fall and early winter of 1817-1818 before being sold a year later to Charles Peale who exhibited it as part of his museum (see Chapter 1) in Philadelphia. Its present location is unknown, though it has been frequently republished, initially in history books due to its realistic portrayal of the harbour (e.g. Muir 1939) and much later in books about sea serpents (e.g. Sweeney 1972). Although missed in some recent works (O'Neill 1999; Soini 2010), the painting is used as the frontispiece in Burns' (2014) thesis as well as in an article in the magazine from the Cape Ann Museum about the two-hundredth anniversary of the sightings (Halsted 2017), and, of course, as the cover image of the present book.

> site now occupied by the building of the same name, and the pride of the town, Tappan's Hotel, were among the more noticeable features. The large square and gambrel-roofed structures along Middle Street, also stood out conspicuously and altogether showed the town at its bravest from the harbor. The most prominent object on Fort Point Hill was the whitewashed ramparts of the old fortification. A considerable cluster of evergreen trees stood on what is now Commercial Street, spared monuments, even at that day, of the woodman's axe.

That's it. Absolutely nothing about the Sea Serpent. Other chapters deal with fires and murders, interesting sights, and individual buildings of note. But nothing at all about the Sea Serpent. It is as if a deliberate attempt was made to sweep the existence of the creature from the pages of history. The irony is palatable, however, when one looks more closely at the sketch itself. For there, in the foreground, discernible among the waves, and completely ignored by the historian, is a representation of the multi-humped Sea Serpent of August 1817. Indeed, as is indicated in the painting's caption, its very purpose was to give credence to the Sea Serpent through providing the most accurate rendering possible of it swimming within its environs of Gloucester Harbour.

Two recent books focus on Gloucester's relationship with the sea. One might reasonably expect such works to give considerable attention to the sea serpent. Gloucester resident Chester Brigham's 2007 *Gloucester's Bargain with the Sea: The Bountiful Maritime Culture of Cape Ann, Massachusetts* is a comprehensive and scholarly tome of sixteen thematic chapters on such subjects as early settlers, privateers, shipmasters, schooners, and even one on myths. It also spends considerable time discussing the books and paintings produced by local authors and artists. What it does not do, however, is contain a single word about the nineteenth century sea serpent. Not a word.

Mark Kurlansky is a New York Times bestselling author. Unfortunately his 2008 *The Last Fish Tale: The Fate of the Atlantic and Survival in Gloucester, America's Oldest Fishing Port and Most Original Town* is, despite its explicit title, a bit of a dog's breakfast of a book that cruises over the surface of a handful of topics interspersed between sea food recipes and vapid polemics about the state of the world's oceans. It too contains nothing about the Sea Serpent.

It seems unfathomable that what purport to be two maritime histories of Gloucester go out of their way to completely ignore one of the only two occasions when marine-based events led to the town being swarmed by hundreds of gawking tourists as well as it receiving copious international attention. The other event is of course the loss of the fishing vessel the *Andrea Gail* in the Halloween nor'easter of 1991 and the ensuing global interest and rise in tourism generated by Sebastian Junger's gripping account *The Perfect Storm*. Both Brigham (2007) and Kurlansky (2008) do devote pages to reviewing the *Andrea Gail* event. As for the Sea Serpent, however, it is as if it has been the victim of Stalinist airbrushing, or perhaps being like that embarrassing wayward uncle whose existence is only whispered about in hushed tones in one-on-one conversations away from the table at Thanksgiving or Christmas family gatherings.

It is important to remember that sightings of the creature known to posterity as the Gloucester Sea Serpent were not restricted to that eponymous town alone. Two general histories of the sea serpent's second home, the Nahant Peninsula in Boston Harbour, both mention the mysterious creature. Fred Wilson begins Chapter XI of his 1928 *Some Annals of Nahant Massachusetts* by quoting philosopher Edmund Burke: 'Superstition is the religion of the feeble minds,' before segueing into stating his belief that 'of course, there is no such animal, and never was nor ever will be such an animal' as a sea serpent. That said, he dutifully recounts the many sightings of the creature, which he believes to be a misidentified mundane animal, seen in the waters around Nahant during 1819: 'It was only a couple of big mackerel sharks, or more likely a school of porpoises, perhaps consciously arranging themselves in a single file while they disported, in order to provide entertainment for summer visitors and profits for summer seashore hotel keepers.' For Wilson, the most interesting attribute of the famous Sea Serpent remains its role in leading to the construction of the hotel (Figure 3.1) and the fostering of local tourism during the succeeding decades. He does, however, provide a glimpse into the hysteria that gripped Boston during those heady days of 1819, quoting at length reminiscences by a J.L. Homer written in 1848 about the 'advent [of the creature] at Nahant [that] produced an intense, a fearful excitement, among all classes in Boston':

> I shall never forget the excitement produced by the first accounts received in Boston of the arrival of the sea serpent in our lower harbor, nor the consequences which grew out of that excitement. The whole story, at this distant day, now almost thirty years, would be deemed incredible. Salem witchcraft was no touch to it, in one respect. There was not an old lady at the North End who did not shake in her shores, and some of the clergy and many respectable citizens, partook of this unamiable and childish feeling. It was even feared, by some of the most timid, that, if he once condescended to leave his native element...he would overtop all the houses in the city, occasionally peeping into the scuttle of some gentlemen in the vicinity of Mount Vernon Street, for the purpose of getting a tidbit, or spiriting away some Irish maiden lady, of fair proportions and good face. Depend upon it, there was a terrible excitement in Boston and vicinity when the sea serpent first visited these shores.

Although missed by other commentators and historians in consequence of their narrow focus on only the sightings particular to Cape Ann during this period, this quotation from Nahant is important for indicating the prevalent state of mind of Bay State residents with respect to the Gloucester Sea Serpent. And whereas I have not been able to find similar commentary from published historical books pertaining to the contemporaneous sightings from Long Island Sound, I expect a more diligent investigation of local histories and documents might likewise prove rewarding in this regard.

Stanley Paterson and Carl Seaburg's 1991 *Nahant on the Rocks* is also significant, unlike the aforementioned general histories of Gloucester, in devoting space to recounting the sea serpent phenomena. An overview of the most dramatic sightings is presented before, as was the case with Wilson above, the history quickly shifts to more tangible topics such as construction of the hotel and the consequent boom in tourism development on the peninsula.

Two comprehensive and well-received maritime histories deal with coastal Massachusetts. Samuel Elliot Morison's *The Maritime History of Massachusetts 1783 – 1860* has been reprinted numerous times since its original 1921 publication, and is in many respects still regarded as the seminal tome on the subject. Chapters cover important events such as the Embargo Act and War of 1812, as well as subjects such as international trade, cod fishing, and whaling. A handful of coastal locations – Salem, Newburyport and Nantucket, and Cape Cod and Cape Ann – are covered in their own specific chapters. But despite being meritorious of its own chapter, the maritime history of Cape Ann is detailed with nary any mention made of Gloucester's famous Sea Serpent. Once again, the creature is airbrushed from history.

Fast forward to 2012 and we find a very different situation, wherein W. Jeffrey Bolster, in his *The Mortal Sea: Fishing the Atlantic in the Age of Sail*, boldly entitles one of the six chapters 'The Sea Serpent and the Mackerel Jig.' Here finally, after a gap of more than a century and a half since Babson's book, we have another historian whom neither denies nor shies away from discussing the phenomenon that swept through Gloucester, Massachusetts (and much of the world) in the early nineteenth century. Indeed, Bolster notably goes as far as to state that 'No marine environmental historian worth his or her salt can afford to ignore early-nineteenth-century sea serpents.' It is therefore important to treat the Serpent as a serious historical subject worthy of study rather than to ignore or dismiss it as 'a maddening anomaly or, worse yet, as hocus-pocus' as such actions 'can compromise our understanding of changes in the sea and of why people acted as they did' (Bolster 2012).

Bolster is interested in the role played by the Sea Serpent in the context of regional knowledge of the sea and about what that meant in terms of perceptions of the Almighty's immutable laws of Nature:

> At the very least, sight of the serpent and the discussions in its wake pointed to how little was known with certainty about the world beneath the waves, despite desires to increase fishing pressure. Contemporary publications on nature, whether in Europe or in America, rarely included the oceans. Questions regarding abundance and distribution of marine organisms, even commercially valuable ones, played second fiddle to larger questions about the study of nature as a whole, and what each class of creatures revealed about the Creator's plan...Put another way, early nineteenth-century people worked within a cosmos in which nature was imagined as fixed, even if it was not entirely comprehended. The concept that nature's dispensations could fluctuate radically seemed offensive to the idea of the harmony of God's handiwork.

In this regard, the timing of the sightings in 1817 are important as they occurred at a time intermediate between the blind acceptance of the seventeenth-century when New Englanders took for granted that such creatures existed in their unexplored coastal waters, and the late nineteenth-century when their descendants began to seriously challenge assumptions about the natural world, especially a belief in fabulous beasts.

Following a brief review of Scandinavian and earlier New England sightings, Bolster makes the point that the Gloucester encounters were different in that 'none had ever prompted a protracted scientific investigation, a heated international discussion, and a systematic pursuit.' The events in Gloucester are very briefly described, using Reverend William Bentley's personal diary as a major source, in addition to a handful of abbreviated quotations from eyewitness accounts. The issue of which group, fishermen or naturalists, is ultimately to be trusted as the arbitrators of truth is touched upon. Naturalists obviously got it wrong in terms of the capture of the baby sea serpent, *Scoliophis atlanticus*, but was Captain Rich, who reputably caught the adult creature, also a victim of his own deception? Regardless, Bolster makes the point that the scientific community obviously took the serpent seriously enough to include mention of it in two reputable texts from the period: 1833's *Introduction to Geology* by Robert Blackwell, and 1839's *Reports on the Fishes, Reptiles and Birds of Massachusetts* by D. Humphreys Stoner. For the general public, however, the possibility of startling creatures of enormous size inhabiting coastal waters, though certainly deemed credible, 'remained wrapped in mystery' due to the many unanswered questions about the Gloucester creature.

Two academic historians have made the Gloucester Sea Serpent an object of their scholarship. Chandos Brown is, to my knowledge, the first such to produce a peer-reviewed journal article on the subject. Despite its title, 'A Natural History of the Gloucester Sea Serpent: Knowledge, Power, and the Culture of Science in Antebellum America,' the 1990 article is concerned with the cultural, rather than the natural, history of the creature. In particular, through becoming a global *cause célèbre*, when it seemed that American naturalists had established the existence of a new genus, an accomplishment that tilted the hat 'simultaneously to the bounty of American nature and to the genius of American science,' the episode demonstrated how the practice of proto-science at the time was an activity influenced by regional loyalty, cultural chauvinism, and misdirected national pride.

Brown (1990) ably reviews the cultural zeitgeist of the early nineteenth century, particularly pertaining to how it shaped natural history. As mentioned in Chapter 1, President Jefferson took umbrage over the French naturalist Comte de Buffon's jingoism that 'nature is less attractive, less energetic on one side of the globe [i.e. North America] than she is on the other [i.e. Europe].' In contrast, the special character of the United States, argued its naturalists, could be defined just as much by its distinctive biology as by its physical geography, both of which were more compelling and robust than either of those in the effete Old World.

The motivations of the Linnaean Society are thought to have been based on the desire for both international acclaim in addition to revenue that the pamphlet sales would bring. The latter was important, Brown notes, citing an unpublished thesis, due to a financial shortfall in 1817 that necessitated fundraising dinners. In this regard, the appearance of the Gloucester Sea Serpent could not have come at a more opportune time.

Brown delves into the nature of the eyewitness evidence, believing that Justice of Peace Nash interpreted too far the instructions of the Committee to obtain 'only the most consistent testimony.' Siding with others, such as Gould (1930), whom have questioned the independence of the eyewitnesses as reflected by the nearly identical opening statements in some of the testimonies, Brown wonders about whether or not the actions of Nash were deliberate. Certainly, in one case, Nash went out of his way to offer his own disagreement in a way to flag the nonconforming testimony of one particular individual.

It is a mistake, Brown (1990) believes, to read the letters of David Humphreys to Sir Joseph Banks simply as additional testimonies. Notwithstanding the distinctive, non-legalistic, and at times theatrical nature of the prose, the letters also include interpretations and commentary about the epistemology of natural history, particularly in the face of how the Earth's manifold bounty reflected the Creator's grand design:

> However, precise the judgment of the Linnaean Society...such creatures as the Gloucester Sea Serpent were ultimately beyond the keening of mere humanity, for they symbolized in the elusive structure all that was mortal...in our perception. The secret 'oeconomy' of nature was, after all, the living signature of God, which must remain forever beyond the grasp of science whether motivated by love of nation or by personal pride. Humphreys hesitated, even while he exhilarated in the pursuit of the serpent. By training and habit, he passed too easily into that figural space in which the hunt for the sea serpent reenacted the original encounter in the biblical Eden, and a nation might be cast out from Paradise as readily as Eve and Adam. He recalled the words of the primer, as the younger naturalists well may have forgotten: 'In Adam's fall / We sinn'd all.' This chastened skepticism stood in quiet rebuke to the emphatic claims of the Society, which represented, if anything, the contemporary excess of Enlightenment that so tormented the old Federalists of Humphrey's age and persuasion. Humphreys and the Society inhabited nearly separate worlds of understanding, sensibility, and experience.

The discovery of the putative baby sea serpent in September 1817 was so readily accepted, Brown believes, due to the disappointment of the public arising from the failures to catch the adult in the harbour. de

Blainville's biting, almost mocking, critique that followed the publication of the *Report* realized the worst fears of the Linnaean Society in terms of their theories being rejected by such a notable savant. Nevertheless, the journalists and populace of New England continued to believe in the Sea Serpent, in spite of 'Rafinesque's protests, Rich's discovery, and de Blainville's caustic attack.' This is in contrast, Brown notes, to the jeering about Northern gullibility that was a regular feature of Southern newspapers, a sentiment that would achieve its apogee in Craft's (1819) brilliant satirical play about New England boosterism (see below) from which Brown quotes extensively.

At over three hundred, double-spaced pages, Elizabeth Burns' doctoral thesis, *Monster on the Margin: The Sea Serpent Phenomenon in New England, 1817-1849*, is the lengthiest historical study of the Gloucester Sea Serpent. For Burns (2014), because imagination overtook inquiry, the ultimate legacy of the Sea Serpent is as a cultural phenomenon rather than a scientific entity. The Sea Serpent's imaginary existence was expressed in poetic, dramatic, artistic, and fictional venues. These in turn led to the development of a tourist industry which resulted in the construction of a hotel and operation of serpent-spotting cruises. Additionally, for Burns, the Sea Serpent functions as a medium, alà McLuhan, that reflects the concerns of New Englanders writing about nature. In particular, the following subjects are covered: the creature's impact in the writing of regional identity of New England; the scientific inquiry about the creature in relation to the nationalistic discourse of natural history; characterizations of the creature in the context of recent events and future fears as a reflection of the anxieties and aspirations of New Englanders; the role of the Serpent in relation to the shift in attitude from repulsion to fascination regarding the seashore; and the Sea Serpent as a presage to the upcoming age of Barnumesque showmanship and hucksterism.

Chapter 1 focuses on the newspaper reportage of the era and the state of early American natural history. The non-scientific natural history discourse about the phenomenon of the Gloucester Sea Serpent that occurred in the periodicals was contemporaneous with the development and expansion of American news journalism. The dissemination of coverage, many articles based on reiterating the previous sightings before detailing the latest observation, was impressive indeed. For example, a single sighting led to the following flurry of reportage during the summer of 1818 in attempt to satiate the growing public hunger for such information: 27 June, *American Advocate and Kennebec Advertiser*; 1 July, *Essex Register* [Salem, MA]; 2 July, *Boston Gazette* and *New-York Columbian*; 3 July, *New England Galaxy* [Boston, MA] and *New-York Daily Advertiser*; 4 July, *Columbian Register* [New Haven, CT], *Newport Mercury*, and *Portsmouth Oracle*; 7 July, *Boston Recorder*; 11 July, *Niles' Weekly Register* [Baltimore, MD]; 13 July, *City Gazette* [Charleston, SC]; and 14 July, *Vermont Gazette* [Bennington, VT].

As Burns (2014) states, the reportage 'was unique in that it began with observations, continued through collation and analysis, and ended in a literature that broadened quickly to include the objective of entertaining the reader as well as persuading the skeptic.' Most reports and comments occurred in print under the headline 'The Sea Serpent', 'as if,' she writes, 'the creature had its own department of the paper.' She notes that descriptions of the creature in the press shifted from 'an unusual fish' to 'an uncommon animal' to 'the serpent kind,' in addition to 'it' being supplanted by 'he.' The 'incongruous fiasco of the Rich Expedition,' as Burns terms it, generated a flood of contemptuous humour as well as comments on the nature of natural history investigations in general. One such included a lampoon wherein a horse mackerel (i.e. a Bluefin tuna) is turned into a monstrous sea serpent through an optical illusion produced by a magical kaleidoscope. Lewis (2011) is cited about the methodologies of early natural history when it was mainly a non-professional gentleman's pursuit, wherein the standard procedure was to use public forums such as newspapers or magazines to comment upon or to pit one theory against another. For example, Burns describes the comments by W. Palfrey, editor of the *Essex Register*, in an 1820 editorial that looked back upon the Linnaean Society's *Report*, in particular about Justice Nash's rejection of one eyewitness' claims of having seen fifty dorsal protuberances when other observers had observed far less. Palfrey writes: 'if we allow ourselves to reject evidence at our own will and pleasure, there can be nothing but the reveries of our own imagination as a result of this liberty.' He goes on to contrast the statement by a recent eyewitness who expressed his doubts about what it was he actually saw, commenting: 'if only the original witnesses had been so candid, and so cautious we believe the story could never had gained credit with the public' and brought shame upon the Linnaean Society.

In Chapter 2, Burns (2014) reviews the scientific inquiry into sea serpents by nineteenth-century naturalists. To set the stage in terms of the dubious state of understanding about the natural world at the time of the Gloucester sightings, she recounts a remarkable court case in 1818 when the evidence presented by fishermen who claimed the whale to be a fish triumphed over assertions made by zoologists that cetaceans were in fact mammals (see Burnett 2007). True understanding eventually did come about, such that by the mid-nineteenth-century, scientists had concluded that sea serpents were 'perceptual anomalies caused by the form or movement of other sea animals already known and catalogued' (I question this conclusion, however, given that lively debates about sea serpents continued well into the late Victorian period of natural science (see Lyons 2009)).

When examining the Linnaean Society's findings, Burns concludes that 'one of the chief weaknesses of the Report is its tenacious assertion that the sea serpent is a serpent, a large reptile, instead of a serpentine creature.' In contrast, Humphreys is praised for referring to the creature as possessing 'attributes of the serpent kind' without going as far as to stating that it was actually one such. As for the nearly identical phrasing in many of the testimonies collected by Justice Nash, Burns states: 'Whether or not it is Nash's translation or a feature edited in afterwards, its seems to me that the assertion are crafted to make the descriptive evidence square with the essential nature of the one piece of concrete evidence the Committee could secure – the carcass [of the purported baby serpent].' The discredited *Report* is referred to as having turned the sublime into the ridiculous, and its fallout is reviewed, in particular, the writings of Rafinesque and Mitchell (whose interpretations will subsequently be examined here in Chapter 5).

Chapter 3 and 4 of Burns' thesis discuss the literature and multi-media phenomenon (exhibits, paintings, theatrical presentations, and narratives) that were created about the Gloucester Sea Serpent in the context of: (1) the tumultuous events that shook New England in the early nineteenth century; and (2) the attitudinal changes in regards to the seacoast. Examples include the journalistic uses of the Sea Serpent as an allegory or metaphor for the uncertain political climate, and concepts of the sublime and terror in popular literature. Some of these will be discussed below.

The concluding sentence to the thesis aptly sums up Burns' (2014) impressive scholarship on the subject: 'The Sea Serpent on the printed page did not verify his existence, but it did make him real.'

The Gloucester Sea Serpent has enjoyed a rich afterlife as a perennial staple in popular writing. Following the initial flurry of press reportage, journalists and writers have returned to the creature again and again, reiterating the events each time the sighting of another mysterious denizen of the deep becomes newsworthy. In so doing, the public's continual appetite for such information is satisfied...until the next sighting occurs, and stories about 'His Snakeship' are dusted off once more, sometimes to give the impression of historical weight to the latest appearance of the mystery animal (Dexter 1986; Drake 1884; Erickson 1956; Geller 1970; Kierman 1950; Moore 1977). Recently the trend has been for articles to be more notional in tone (Anonymous 2014; Fama 2012; Fritts 2017; Greener 2010; Halsted 2017; Jylkka 2017; Shattuck 2013).

As would be expected, the Gloucester Sea Serpent has repeatedly been the subject, to varying degrees of scrutiny or fantasy, in cryptozoology books. In their comprehensive overviews, for example, Oudemans (1892) devotes more than fifty pages to the sightings, whereas Heuvelmans (1968), who believes that 'the mass of reports from the east coast of the United States agree so well as to be almost monotonous,' gives twenty-five pages to the creature. The Gloucester Sea Serpent is also one of seven notable creatures to be given their own chapter in Gould's (1930) scholarly investigation of the subject, leading him to profess that 'there can be no doubt at all that the thing seen, whatever its nature, was a most unusual – a unique – sight.' In contrast, several recent cryptozoological works attempting to formulate a taxonomic system of classification for all sea monsters, give the Gloucester creature short shrift (Coleman and Huyghe 2003; Woodley 2008). Interpretive theories about the nature of the beast posited by cryptozoologists will be reviewed in Chapter 5.

Two contemporary trade books focus on the creature that visited Gloucester and other locations around 1817. In J.P. O'Neill's comprehensive *The Great New England Sea Serpent: An Account of Unknown Creatures Sighted by Many Respectable Persons between 1638 and the Present Day*, two chapters are devoted to the subject. O'Neill does thorough, but not exhaustive, research in assembling eyewitness accounts, drawing from the standard sources of Davis *et al.* (1817), Sargent (1818), and Humphreys (1817), but also, importantly, from more elusive reportage in regional news broadsheets of the day, many obtained from George Woodbury's folder in the Cape Ann Museum Archives. The accounts of about thirty eyewitnesses are provided in sufficient detail to necessitate the quotations being off-set from the body of the text. Other accounts are presented in abbreviated form or occasionally paraphrased, all glued together with her straight-forward, unsensational, and factual prose. In addition to the sightings, O'Neill does an admirable job in reviewing both controversial captures of the Loblolly Cove 'baby serpent' as well as Captain Rich's purported adult animal.

It is only in the last chapter where O'Neill sheds her objectively as a capable amateur historian. Here, after summarizing varying theories advanced by notable cryptozoologists over the years, she engages in a bit of arm-waving that interweaves the history of palaeontology, discovery of live coelacanths, possibility of plesiosaurs being sea monsters, state of the North Atlantic fisheries, with assumptions about the feeding ecology of the New England Sea Serpent. All this is toward imagining the importance of the Sea Serpent as an environmental sentinel whose 'disappearance having been an early warning of the dangers represented to marine life by a lack of responsible stewardship of ocean resources.' This leads O'Neill to reveal her cryptozoological beliefs when she concludes pessimistically that 'sadly, the almost total absence of recent sightings of sea serpents in New England waters may mean that they are gone, or nearly gone, for good.'

Gloucester's Sea Serpent, by Wayne Soini (2010), is the only book devoted exclusively to the eponymous creature. Soini, who was born in the town, recounts a childhood visit to the Gloucester Museum when he first saw a copy of the famous painting of the sea serpent as a Chinese dragon (see next section) that is used as the cover image for his book. Soini is a practicing lawyer, and approaches the subject through that lens, being more interested in the various people involved in the acquisition of the testimonies than in the actual nature of the creature itself. A short book of only 126 pages, it begins with an epigraph of that oft-quoted Hamlet statement to Horatio about remaining open-minded. The first four chapters provide interesting biographical details about two of the major players in the investigation by the Linnaean Committee: Justice Lonson Nash and Judge John Davis. Further background information is given about Federalist politics, the Embargo Act and War of 1812, the 1815 eruption of Tambora and the following year's weather, and several earlier sightings of the New England Sea Serpent. Eleven subsequent chapters, some of only two pages in length, focus on each day of the sightings during August 1817. Both Sargent's (1818) manuscript and Davis *et al.*'s (1817) report are used as sources.

The dramatic events of that summer are told by Soini in a quasi-novelized format filled with presumed intentions and imagined actions, such as, for example, the following description:

> Nash likely personally selected a courier. To ensure speedy, correct delivery, Nash would have tagged someone going to Boston whom he knew and trusted. He would have hand carried his packet to someone's office, the stage depot, a home, stable or another rendezvous point in Gloucester to clinch an agreement about bringing the dozen or twenty folded or rolled-and-ribbon-tied pages to Boston. One readily imagines a compact, folded or sealed packet, possibly bound with string or maybe wrapped in a cloth as a buffer against dust, rain or mud in transit, for the irreplaceable and important documents.
>
> Imagine Judge Davis, off the bench, his black robe on a hanger, seated at his chambers desk in shirtsleeves and the windows open for a little air when his clerk tapped on the doorway to announce that a messenger from Gloucester had arrived with a packet of papers.
>
> 'Bring him in, by all means!' or similar words would have come from Davis. His ambitions to have a bundle of sworn, signed statements imminently to be realized, he would have been confident now of a book and of immediate controversy and lasting fame. Now New Englanders – New England Federalists at that and Judge Davis most visibly included – would make the mark that would put the United States on the map of scientific progress. How do you like that, Mr Jefferson?

Only six eyewitness quotations are given in sufficient length to warrant them being set aside in the text. Instead, Soini paraphrases the accounts to which he adds his own statements, such as, for example, about 'The Big Show' on August 14th that 'still stands as the greatest day of sea serpent sightings on earth;' as well as: 'no timid sea serpent but a party animal,' and 'the world's most gregarious sea serpent ever, one that seemed to thrive on attention,' when describing the behaviour of the creature, referred to flippantly as 'Gloucester's pin-wheeling pet.'

Soini, like O'Neill, adopts the vernacular of 1817 in referring to the creature with the pronoun 'he' rather than as an 'it.' To Soini, the summer's events can be distilled to being a puzzle (what exactly was the creature?), a project (being given a Latinized name by the Linnaean Committee), and a flop (the ridicule brought to the Committee by the Loblolly Cove snake fiasco). The Introduction ends with the teaser that the mystery of the Gloucester Sea Serpent might finally be solved in the subsequent pages:

> Did he really visit Gloucester harbor in August 1817, a sea serpent that swam circles around sailboats in the sunshine, his chocolate-brown body incredibly long and so marvelously flexible that he could come and go at the same time; a creature that sunbathed on the beach on Ten Pound Island but, in Gloucester harbor, zipped past

> awestruck onlookers doing four-minute miles, with his head flat at the top and pointed like a serpent with bulges at the sides and a tongue that he could flip up and drop back at whim; a sea serpent that, when paired with her putative baby in Judge Davis's book, became incredible; a sea serpent prevented by that little snake from being accepted and recognized until – just now?

But despite that well written and accurately distilled description of the creature and its behaviour, Soini offers no final hypothesis, concluding that 'at the bottom, the happy visitor's extended story in Gloucester...remains an unfathomable mystery.' And although he approaches the events as a legal case framed from his own experience of weighing witness testimonies, no opinions are professed as to the veracity of the evidence nor any interpretation offered as to creature's identity.

What Soini (2010) is most interested in are the motivations behind Judge Davis undertaking the investigation in the first place, and why he would gamble the Linnaean Society's reputation on the dubious credibility of the Sea Serpent. Four possibilities are suggested. The first is the fact that Davis grew up in the coastal community of Plymouth, Massachusetts, and as an impressionable young lad, he would have been familiar with and enraptured by dockside tales told by sailors of mysterious creatures spotted at sea. The second possibility concerns his defence of America in the ongoing cultural war with France, particularly in relation to the assertion made by Buffon and others that Europe had a more spectacular fauna than the United States. Thirdly, Davis was a known polymath and simply had an inherent intellectual curiosity about the world around him. And finally, as a committed Federalist, Judge Davis may have had a grudge against Jefferson, who was not only President of the country and responsible for the costly Embargo Act, but also the president of the American Philosophical Society through his capacity as being the nation's best known amateur scientist. The latter hypothesis is elaborated upon:

> Davis – tenured for life in an office sometimes called 'the last one-man fiefdom in Western civilization' and enshrined in a courtroom that put judges on a raised platform, above the bar to whom they are literally the law – in his most unpopular decision, upheld Jefferson's Embargo Act, a law that crippled New England economically. It was a law that the seaport boy on the Federal bench surely and personally thought unwise and no doubt a decision over which he may have brooded, hoping one day to even scores with Jefferson. The sea serpent, in Davis's hands, was a weapon of revenge. Not the bone collectors to the South, with Jefferson among them, but Judge Davis and his small team of Northerners (and Federalists) would load, aim, fire and finally win the cultural war for the United States of America.

No doubt, all of these reasonable possibilities came together in a perfect storm that was responsible, in Soini's assessment, for Judge Davis' Committee dispensing with their hitherto demonstrated scientific discipline in a rush to obtain a specimen of the Sea Serpent in the form of the Loblolly Cove snake. The reactions and interpretations of the natural history community in regard to the Gloucester Sea Serpent phenomenon will be reviewed and discussed in Chapter 5.

3.5 IMAGINATION

Where reality fell far short in terms of definitively capturing, and science, in defensively identifying, the Gloucester Sea Serpent, the void was filled by the active imagination of those in the arts and letters community. Burns' (2014) most able scholarship provides detailed analysis and cultural commentary on these offerings, which, in consequence, will be only briefly noted herein.

Right from the very start, the Gloucester Sea Serpent formed the inspiration for a series of poems published in regional newspapers and periodicals. Some of these share in common a desire to establish the maritime majesty of the serpent. In 1817, 'The Sea Monster!' referred to the 'mighty mogul of the sea;' in 1819, 'A Valedictory Poem. Address to the Sea Serpent' used the phrase 'your snakeship' in reference to the creature; in 1845, 'Ode to the Sea Serpent' called it the 'Great monarch of the water!;' and in the 1818 effort 'The Sea Serpent,' penned under the nom de plume 'Leviathan,' the creature is described as asserting his greatness and undisputed sovereignty over the seas. As Burns (2014) shows, in these poems, as well as others, such as 'Lines to the Sea Serpent' (1836) and the annual end-of-year offerings for 1817 and 1818 in the *Salem Gazette,* the Sea Serpent provided a vehicle in which to comment about many pressing political issues of the time, such as immigration, Manifest Destiny, American imperialism, and the marginalization of slaves.

Other poems of the times eschew political witticism for simple humour, as for example one called 'The Sea Serpent' by the nom de plume 'Snapshots,' that I found undated in George Woodbury's scrapbook in the Cape Ann Museum archives:

I'm the animal hold, whose deeds are told
  From Bath to Brighton gay;
The marvelous lengths of my tail unfold
  From Deal to Boston Bay.
You may safely bet I will soon be met
  At harbor and cove and bight;
I'm the family pet of Mine Host, et cet.,
  The beau of the season's height...

At ship and shore I bellow and roar,
  And shake my barnacled head,
With nostrils fringed and teeth galore,
  And mane of a fiery red.
Or peacefully I glide in a ripple wide,
  With fins on my eyebrows grim –
Anatomies all are quire defied,
  When it's I who's in the swim.

There are some, indeed, who are fully
  Agreed
(A slur that I cannot help);
I'm only a spar or a drift of weed,
  A whale or a mile of kelp.
But, fact or fake, I'm a poor old snake
  Of a new passe design,
Who once a year must an outings take
  For the men who are paid by the line.

Playwrights also partook in the cultural phenomenon of the Gloucester Sea Serpent (Burns 2014). In 1819, a pantomime – a British form of low-brow comedic theatre still popular among young audiences around Christmas time – was performed in England. *The Sea Serpent! Or, Harlequin Yankee* told the exaggerated tale of courtship amongst Native Americans in which the Sea Serpent plays a dramatic role. This was followed two years later, in Boston, with the staging of another 'panto,' *The Sea Serpent, or, Harlequin in Gloucester*. In 1833, J. Planché inserted the Sea Serpent – actually a hybrid half man-half beast – into Greek mythos in his elaborately entitled *The Deep, Deep Sea, or, Perseus and Andromeda; and Original, Mythological, Aquatic, Equestrian Burletta IN ONE ACT*. But the most interesting thespian creation, and one oddly not examined by Burns (2014), was *The Sea Serpent, or Gloucester Hoax: A Dramatic jeu d'esprit in Three Acts*, by Harvard alumnus William Crafts (Figure 3.9).

Staged in his native city of Charlestown, South Carolina in May 1819, at a time when New England naturalists had become the butt of worldwide ridicule following the true revelation about their discovery of the putative baby sea serpent as well as the letdown resulting from the capture of the putative adult in the form of a tuna, Craft's play is a biting satire and parody of Yankee conceit in the face of the ensuing farce that the investigation had become.

'Tis so with all New-England's gallant race,
Eager to learn, investigate, and trace
The world around them – knowledge is their aim,
Hence is their happiness, and hence their fame.

After all manner of descriptions from eyewitnesses, Justice Portly, a thinly-disguised Lonson Nash, seeing a way to keep business coming to Gloucester, states:

A finer job I never could desire.
For sceptic appetites I'll make a feast,
One thousand affidavits at the least...
Let a new sign upon my door appear:
Snake! Snake! Snake affidavits, taken here!

Later, the Mayor expresses his enthusiasm for the benefits accruing from the convenient invention of the Sea Serpent:

This unexpected visitor, I guess,
Will our good citizens of Glo'ster bless:
Our inns are full, the world is on the road,
And every Hackman has a double load.

To which, the Justice concurs:

Aye, we shall turn him to a good account;
Of wealth and wonder the exhaustless fount;
Let Salem boast her museum, and her witches,
Her statues Newb'ry, Marblehead her riches –
We from them all the shining now will take,
The snake and Glo'ster, Glo'ster and the snake!

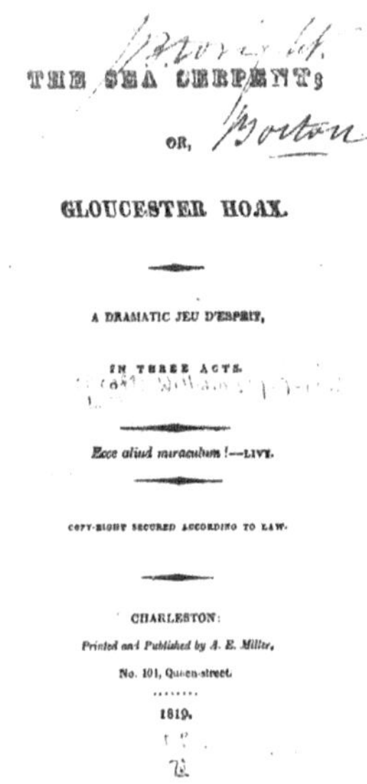
THE SEA SERPENT;
OR,
GLOUCESTER HOAX.

A DRAMATIC JEU D'ESPRIT,
IN THREE ACTS.

Ecce aliud miraculum!—LIVY.

COPY-RIGHT SECURED ACCORDING TO LAW.

CHARLESTON:
Printed and Published by A. E. Miller,
No. 101, Queen-street.

1819.

Figure 3.9. Cover of pamphlet of the bitingly satirical play about the Gloucester Sea Serpent phenomenon by William Craft (1819).

Soon hunting parties are after the creature, and scientists begin to debate its nature, before producing a shed snake-skin as evidence:

> The world's turned topsy, turvy, and mankind
> In quest of monsters, monstrously inclined,
> Rack their invention, prodigies to find
> When found what are they? Like a maniac's dream,
> Exactly the reverse of what they seem:
> An hour's amusement, or a week of wonder,
> An infant scare-crow, or a learned blunder!

And then offering the apt and cautionary statement that:

> Where there are blockheads ready to believe,
> There will be knaves as eager to deceive.

while continuing on with a story about 'victims of false credulity' whom were fooled by a shaved bear dressed in man's clothes and passed off as a foreign Native.

Meanwhile, back at the Mayor's office, an artist is hired to paint a picture of the creature of exaggerated size so as to increase the degree of wonder. And the hunt commences:

> Now, know ye all, if any shall ensnare
> The aforesaid monster, and shall bring him here,
> So that our fisheries shall thrive once more,
> And cod and mackerel shall frequent our shore,
> Five hundred dollars shall reward his toil;
> And all, and singular the serpent's spoil:
> Moreover, he shall have New-England's thanks,
> And honored mention to Sir Joseph Banks;
> His vessel, too, for such marked resolution,
> Shall moor, 'long side the frigate Constitution.

And so the play ends with the capture of a horse mackerel (Bluefin tuna) and with that, the hero, Timothy, wins the bounty and is able to marry his true love, and thus Justice Portly can provide over the wedding service and in so doing maintain his livelihood. The closing words of the morality play cast scorn on all:

> How will philosophy endure the shame?

And:

> So fares it always with the advent'rous race,
> Who banish nature – and give monsters place;
> Imagine miracles without their need,
> Look wise in fancy – but are dupes indeed.

The Gloucester Sea Serpent has been fictionalized a handful of times (Burns 2014). Several of these, often published in regional magazines or newspapers, are in what latter would become known as the genre of speculative or science fiction. Published in 1817, the short story 'Aug. 1, 2217' describes the perfectly tame serpent of the future, who makes his annual visit to Gloucester where he floats placidly, allowing scientists to measure his length and calculate his growth rate over the past four centuries. The 1823 short story 'Boston. September 11, 1855* Sea Serpent Caught,' depicts an idealized version of Captain Rich's quest, in this case by aid of a specially designed steamboat. Two other offerings are more in the thriller genre. The 1818 story 'Anaconda of the Ocean' portrays a schooner, ironically named *The Sea Serpent*, that is attacked by the eponymous beast off the coast of Cape Cod. Eugene Batchelders' 1849 *Romance of the Sea Serpent, or Icthyosaurus* is an extended work of prose in which the monster leaves the ocean to run amok, Godzilla-like, amongst the hoary halls of Harvard. Interestingly, Reynolds (1988) believed that such 'Dark Adventure' tales of mythical sea monsters served as an inspiration for Melville's classic *Moby-Dick*, published in 1851. Another short story, 1833's 'The Sea Serpents and the Steam Boat,' capitalizes on the developing tourism trade of sea serpent sighting cruises along the New England coast which were designed along the lines of, and conducted in the same locations as, today's whale watching tours. And in Nathaniel Hawthorne's 1841 story 'Egotism, or the Bosom Serpent,' the Serpent, according to Burns (2014), becomes internalized as the expression of destructive jealousy and as a symbol of monstrous arrogance.

The children's picture book, *The Serpent Came to Gloucester*, by M.T. Anderson (2005), illustrated by Bagram Ibatoulline (Figure 3.10), proves that the mystique of the Sea Serpent continues to inspire imagination today. The Author's Note section at the end of the book provides an historical synopsis of the 'true story' and cites a number of the key references to demonstrate the author has done his research on the subject. The structure of the story is told from the point of view of an old man reminiscing to his grandchild about events from his youth, when on a sunny day, while hanging clothes with his mother, the Serpent came to town: 'It came from the sea, from the lonely sea, It came from the glittering sea.' The young boy runs through the town yelling to all to get out their spyglasses and come down to the harbour to observe the beast. There they see it playing 'As it swam in the sea, in the lovely sea. As it swam in that ancient, old sea.' Boston zoologists arrive to document the beast, and soon thereafter it disappears for the winter, leaving the townsfolk sad. The next year, men, armed with rifles, knives, nets, and harpoons arrive 'singing killing songs' and begin the hunt. Striking the beast and being dragged in a Nantucket sleigh-ride, they reel in a large horse mackerel which they claim to be the serpent. After the hunters retreat, the boy rows out to where the real sea serpents are still at 'play in that lonely sea.' The story concludes with the old man nostalgic for the days before steamships ploughed the waters and forests were felled, a time when serpents roamed the now depauperate seas: 'And I fear that its shall not come again, my love, From the ancient and wrinkled old sea.' The accompanying illustrations throughout are beautifully rendered, the beast depicted as an

Figure 3.10. Cover of a wonderfully worded and beautifully illustrated children's book about the Gloucester Sea Serpent by M.T. Anderson and Bagram Ibatoulline. (2003).

enormous, multi-humped creature with a characteristic serpentine head. Of particular interest, is a small engraving shown, in the background, hung on the wall of the home of the old man, in the penultimate illustration in the book, and also redrawn to decorate the front and back inside covers. For this is one of the most iconic of all the historic representations of the Gloucester Sea Serpent (see below).

Figure 3.11. The iconic illustration of the Gloucester Sea Serpent from a pamphlet (Anonymous 1817) which has been republished many times, including as a nineteenth-century print displayed in the Cape Ann Museum Archives.

The 1817 pamphlet, *A Particular Account. Monstrous Sea-Serpent, the Largest Ever Seen in America, Which Has Lately Made Its Appearance in Gloucester Harbor, Cape Ann, (Mass.) And has been seen by hundreds of respectable Citizens, With an Exact Representation of that Wonderful Monster*, contained the illustration shown in the recent children's book (Figure 3.11). The artist based his version of 'the wonderful monster' on the compilation of eyewitness accounts contained therein, and this image has frequently been published in the years since, as for example, on the cover of O'Neill's (1999) book (Figure 3.12), in addition to being reworked as a T-shirt design (see Appendix One). Unfortunately, it bears little resemblance to a careful reading of the delivered testimonies. The scene portrayed is of Gloucester Harbour, a few waterfront buildings discernable in the background, and three sailing crafts and two rowing dories on the water, in between which is shown the Sea Serpent. The creature's head is raised at least seven metres in the air, atop a curved neck, and it sports an extended tongue. The body, of an estimated size relative to the surrounding boats and people, of more than thirty metres, is composed of four coils of front-to-back diminishing size. It is difficult to tell whether or not the Serpent is scaled.

The most famous engraving of the Gloucester Sea Serpent (Figure 3.13), also from 1817, is, despite its original caption of being 'taken from life,' even more fantastical, and though not representative of any of the eyewitness accounts, has nevertheless been widely republished, including, most notably, in a colorized version, on the cover of Soini's (2010) book (Figure 3.12), in addition to being used in the Cape Ann Museum's advertising in 2017 (see Appendix One). The depicted scene shows a distant view of the Gloucester waterfront nestled in rolling hills, atop several of which are a lighthouse and a couple

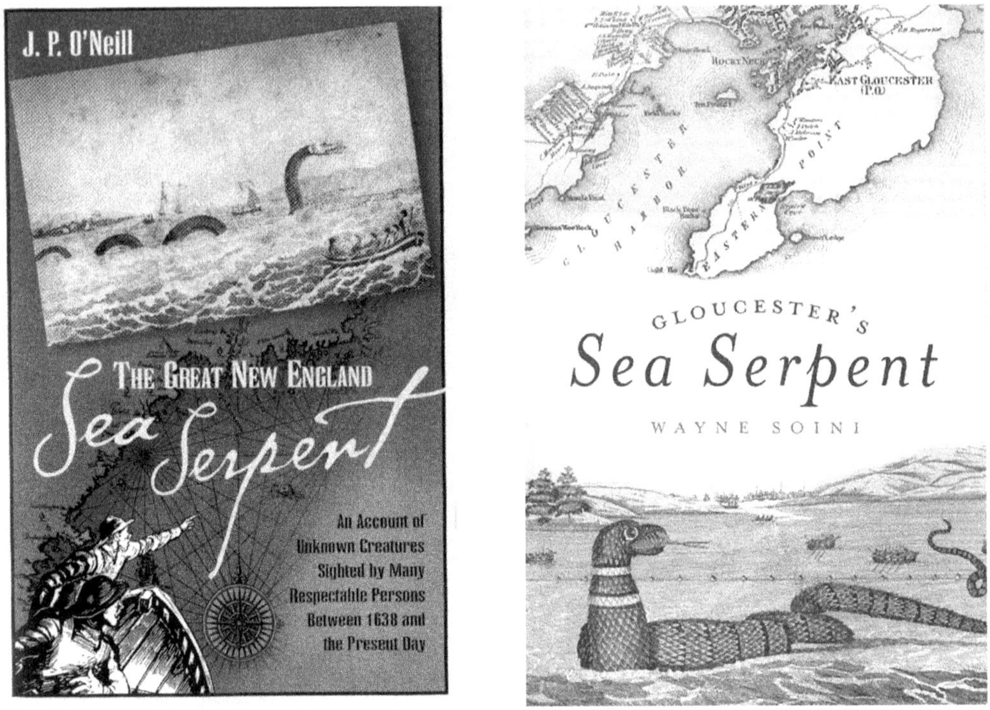

Figure 3.12. Covers of two popular books (O'Neill 1999; Soini 2010) that make use of the iconic image (Figure 3.11) and the Chinese dragon image (Figure 3.13) of the Gloucester Sea Serpent, neither of which match the bulk of eyewitness descriptions of the creature seen in the harbour.

of pine trees. In the harbour, half a dozen rowboats are manned by four to nine people. Separating this from the Sea Serpent in the foreground is what looks like a long line of buoys of a fishing net that extends from border to border. The Sea Serpent bears more in common to classic images of Chinese dragons than it does anything ever seen in Gloucester Harbour. The head is held rigidly erect, and with its large eye, hooked beak, and feathery scales, it gives the impression of a giant aquatic parrot...until one notices the mouth full of fangs and a long, serpent-like extended tongue. Sporting a handsome collar of two white bands, the rest of the body is shown as likewise being scaled and floating on the surface, with a long, curly-cue tail that is being waved about in the air much like that of an annoyed cat. Whereas in the previous illustration, the Serpent had been shown gazing blindly ahead, oblivious of the surrounding boats, here 'His Snakeship' is starring back at the puny boats, perhaps sticking his tongue out, either to detect their delectability or to mock their pursuit of him.

A broadside summarizing various sea serpent sightings (Bowen 1817) contained another artistic representation of considerable license. The scene depicted is of the Sea Serpent being pursued and shot at by individuals in a small boat (Figure 3.14). Once again, the creature's head is held high in the air with a forked tongue emerging from a maw of sharp teeth. The animal's body is shown, not as vertical coils as in the earlier iconic drawing, but rather as side-to-side flexures, all lying atop the waves. The body is segmented like an earthworm's, and is mottled with white spots.

Figure 3.13. The most famous and, despite its title ('Sea Serpent engraved from a drawing taken from life as appeared in Gloucester Harbour, August 23, 1817'), most inaccurate, illustration of the Gloucester Sea Serpent. Widely republished, including being used for advertising and selling gift-shop items by the Cape Ann Museum in 2017 (see Appendix One), it remains the stuff of cryptozoological fantasy.

Heuvelmans (1968) presents two dramatic illustrations from 1819 of ships locked in battle with the Great Sea Serpent off the coast of Massachusetts and in Long Island Sound (Figure 3.15). Given that these are not linked to any specific account, it seems they are very much in the journalistic spirit of frightening people, a tradition that harkens back to the days of the 'here are dragons' period of cartography. And they have their successors. In 1910, a hoax photographer created an image of the monster off the coast of Ipswich, just north of Cape Ann, devouring a small child (Figure 3.16). And Schmidt's sketch in Moore (1977) is, like many recent renditions, equally fantastic (Figure 3.17). In this case, the raised head of the serpent resembles an angry goose, with either water – or is that saliva? – dripping from its beak, while the body (not shown) sports a long, crested dorsal fin much like that of an eel or giant oarfish.

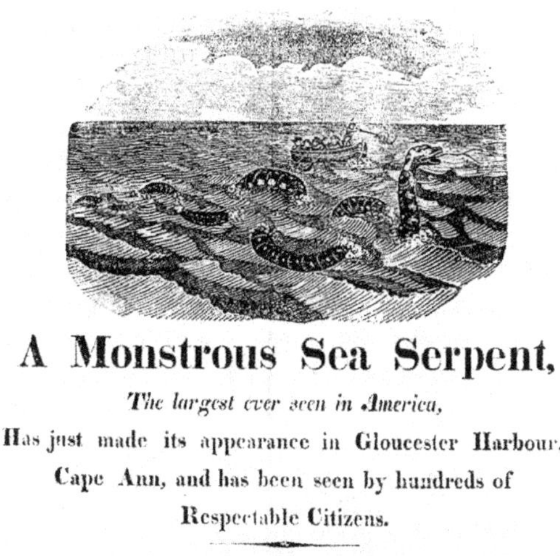

Figure 3.14. Illustration of the imagined Gloucester Sea Serpent used atop Bowen's (1817) broadside republication of Anonymous (1817).

Another illustration in a magazine in a display case in the Cape Ann Museum reworks the mythologized monster of Gloucester (Figure 3.18). Later representations of regionally observed sea serpents followed the same artistic pattern. Also inside the Cape Ann Museum display case (and therefore presumably bearing some relationship to Gloucester) is portrayal of an 1848 encounter of a ship with a giant serpent (Figure 3.19A). The 1951 edition of Henry David Thoreau's *Cape Cod* contains a depiction of a sea serpent (Figure 3.19B; see also O'Neill 1999), as described in his journal entry for June 14, 1857 (Thoreau 1962). Also in the Cape

Figure 3.15. Two 1819 drawings in Heuvelmans (1968) of sea serpents attacking and sinking vessels off Massachusetts and in Long Island Sound.

Ann Museum Archives, the Woodbury scrapbook contains a drawing from an eyewitness of the sea serpent that returned to Cape Ann in 1886, which once again shows a row of humps elevated above the water surface (Figure 3.19C; first published in O'Neill 1999, which also shows another depiction of the same visitation). An illustration of the 'marine monster' seen off Nahant in 1902 (Figure 3.19D) is notable for portraying what the eyewitnesses 'thought they saw' – a sinuous multi-humped animal with a large head and blowing air through its nostrils, compared to 'what they probably saw' – just a whale. So by this time, at the beginning of the new century and with a new awareness of what is fantasy and what is reality, a mundane explanation is offered for a sea serpent sighting. At this time as well, the portrayal of what can be regarded as the archetypal form of the sea serpent begins to assume a role as a symbol of lampoon (Figure 3.19E).

During the media phenomenon that the Gloucester Sea Serpent became, magic lantern shows and exhibitions were popular ways to increase exposure to the mysterious creature for residents of distant locations such as Boston, Salem, New York City, and Philadelphia. A characteristic of the exhibitions were paintings of the Sea Serpent and its surroundings,

Figure 3.16. Hoax photograph from 1910 by Ipswich resident George Dexter of a sea serpent off Cape Ann seen devouring a child (www.historicipswich.org 03/02/2014; curtesy of Gordon Harris of Historic Ipswich).

often displayed with accompanying objects in the form of curiosities or oddities of the natural world, including, in at least one instance, the soon-to-be-refuted infant sea serpent (Burns 2014). One painting, now lost, that received considerable attention was 'The Great Sea Serpent as seen in Gloucester Harbor on 14 August 1817' (sometimes entitled 'View of Gloucester Harbour with the Sea Serpent of 1817'), by John Penniman, that was claimed to be based on sketches made by Captain John Beach, an avid promoter of the sea serpent. The panorama (Figure 3.8 and cover) is intriguing given its juxtaposition of an accurately rendered Gloucester Harbour and town in the background (so realistic, in fact, that it has since been recognized in historical works about the architecture and planning of the town) in juxtaposition with the fantastic Sea Serpent swimming in the foreground. Unlike the previously mentioned offerings of a more exaggerated and garish kind, the Sea Serpent shown here appears much more subtle, barely discernable amidst the choppy waves. The creature is depicted more in tune with the eyewitness accounts, the head and the six humps – not coils – are only half a metre or so above the water surface. A typical serpent's head with flicking tongue is shown once again, and the neck and trunk of the body are scaled, traits that were indeed observed, but only by a small fraction of all the eyewitnesses.

Figure 3.17. A modern rendition of the Gloucester Sea Serpent as a scary monster (Moore 1977).

Figure 3.18. Another, undated, reimagining of the Gloucester Sea Serpent from an illustration displayed in the Cape Ann Museum Archives.

Figure 3.19. Additional interpretations of sea serpents from the region: (A) an 1848 encounter of a sea serpent by the ship Daphne; *(B) a sea serpent seen off Cape Cod in 1857; (C) a sea serpent seen off Cape Ann in 1886 as drawn by an eyewitness (Cape Ann Museum Archives); (D) a 1902 sighting off Nahant of a whale possibly misidentified as a sea serpent, and (E) a 1905 cartoon proposing a tongue-in-cheek answer for a report from Lynn of a lighthouse keeper whose chickens had mysteriously disappeared (*The Daily Evening Item *and* The Daily Evening News *from the Cape Ann Museum Archives, and republished in O'Neill 1999).*

Soini (2010) rightly comments that such imaginative products were created for the specific purpose of satisfying consumer desires, and were displayed for an admission fee, often accompanied by appropriately Barnumesque prose, such as the following example:

> The Sea Serpent
> A Monster of Uncommon Size
>
> Who has paid a long visit to our eastern coast and has excited the admiration of scientific men, and the western world in general, has been accurately painted by an artist of the first talents, under the direction of a gentleman, whose genius and minute observation of the monster entitle him to the confidence of the public! And now exhibiting at UNION HOTEL, 68 William Street on upward of three hundred square feet of canvas
> PRICE OF ADMISSION, 25 CENTS.
>
> The Painting of this wonderful animal, which is now exhibited at Washington Hall, is, in many respects, one of the most interesting displays ever presented to the public. It covers a canvas about 35 feet by 20, representing a beautiful view of a bay opening to the ocean, with boats, vessels, and hills, in the distance, and the Serpent stretching his enormous and formidable length across the front. The painting is well executed, and every object in it – and the land, waves, vessels, clouds, sky, light and shade, so justly arranged as to please the taste of general beholders. When we consider that the principal design is to give the public a correct idea of a terrible animal which is now known to exist, as it were, in the neighbourhood of our city, which appalls the courage and baffles the skill of everyone attempting its approach, we feel fully warranted in asserting, that it is a spectacle so magnificent, so sublime, as to be surpassed only by the awful monster which it represents. The interest of the spectator is increased by the consideration that there is no fancy, no fiction, and no poetry in the view – but that it is only a PENCH. DISPLAY of a monster whose force is irresistible, whose movement is swift as the wind, and the terror of whose eye can no more be painted than the strength, the swiftness, or the noise of his movements over the ocean waves.
>
> The views are taken from the representations of a celebrated naturalist whose enterprising vigilance has given him an opportunity, six different times, to approach so near the animal as to see his eyes, teeth, tongue, and the colour of his head and neck, distinctly. We may therefore look on the painting as a fact, and contemplate the animal, with his lofty head erect in the air, and his long and spiral volumes dashing over the waves as a moving miracle, bearing down stubborn incredulity and oppressing the beholder with the weight of the most 'TERRIBLE' and the most 'SUBLIME.' In all the arts there is nothing like the painting – for the academicians never knew of any thing in nature like the SEA SERPENT; and every AMERICAN should be proud that we are the first to give this interesting subject of the pencil a SHAPE AND FORM.

Finally, it must be noted that in contrast to these fantastical renderings by artists, sketches that were made in direct association with particular eyewitness accounts (see Chapter 4), sometimes by the observers themselves, are much truer to the composite description of the sighted creature. This is particularly the case in relation to the number of protuberances seen, which were often considerably more than the three or four coils generally shown in the artist's renderings shown above.

Chapter 4.

NATURAL HISTORY OF THE GLOUCESTER SEA SERPENT

4.1 'SO INTERESTING A LINK IN THE CHAIN OF ANIMATED BEINGS'

4.1.1 Eyewitness accounts

Unless otherwise referenced, accounts are from the *Report of a Committee of the Linnaean Society of New England Relative to a Large Marine Animal, Supposed to be a Serpent, Seen near Cape Ann, Massachusetts, in August 1817* (Davis *et al.* 1817), many of which were reprinted, in part, in Oudemans (1892). If known, original publication sources, usually local newspapers, are often mentioned. Whenever a more accessible, modern source reprints the information, that is mentioned. Space constraints necessitate that entries be abbreviated. Most of what I have omitted is either of a background or contextual nature about the sighting particulars (e.g. observation distance, weather conditions, sighting made from on shore or in a boat, number of witnesses, etc.) or material that is repetitive to previous accounts. The emphasis below is on presenting information concerning the physical and behavioural attributes of the purported Sea Serpent that are novel with respect to the temporal sequence in which the accounts are presented.

The Davis *et al.* (1817) report is readily available on-line for further consultation. Unless specifically mentioned otherwise, the sightings occurred in Gloucester Harbour or near Cape Ann. The two sectional epigraphs are from Harvard professors William Peck ('So interesting a link...') and Jacob Bigelow ('The most interesting problem...'), both made in reference to the Gloucester Sea Serpent. The hand-drawn map of Gloucester Harbour (Figure 4.1) is by Wayne Soini, author of *Gloucester's Sea Serpent*, which was handed out during his bicentenary tour of the sighting locations in August 1817 (see Appendix One) and kindly made available for purposes of the present book.

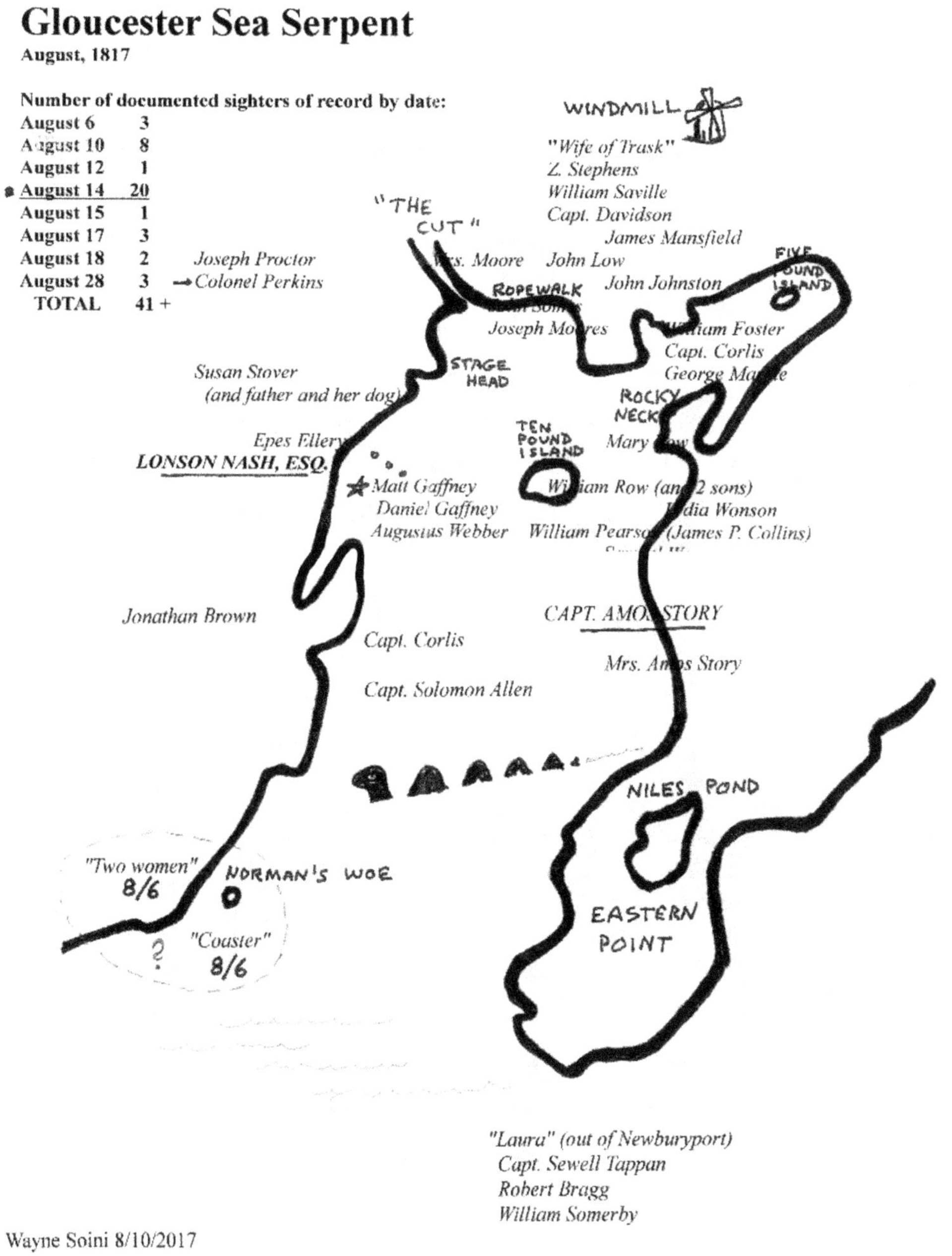

Figure 4.1. Map by Wayne Soini, author of Gloucester's Sea Serpent *(2010), showing locations of sightings during August 1817.*

1815

Account 1: 20 June, 1815, Plymouth, MA:

> I, Elkanah Finney of Plymouth…Mariner, testify and say: That about the twentieth of June, A.D. 1815, being at work near my house, which is situated near the sea-shore in Plymouth at a place called Warren's cove, where the beach joins the main land…I looked towards the cove, where I saw something which appeared to the naked eye to be a drift sea-weed. I then viewed it through a perspective glass, and was in a moment satisfied that it was some aquatic animal, with the form, motion, and appearance of which I had hitherto been unacquainted. It was about a quarter of a mile from the shore, and was moving with great rapidity…The animal went about a quarter of a mile to the northward; then turned about, and while turning, displayed a greater length than I before seen; I supposed at length a hundred feet. It then came towards me, in a southerly direction, very rapidly, until he was I line with me, when he stopped, and lay entirely still on the surface of the water. I then had a good view of him through my glass, at the distance of a quarter of a mile. His appearance in this situation was like a string of buoys. I saw perhaps thirty or forty of those protuberances or bunches, which were about the size of a barrel. The head appeared to be about six or eight feet long, and where it was connected with the body was a little larger than the body…I could not discern any mouth…While he lay in this situation he appeared to be about a hundred or a hundred and twenty feet long. The body appeared to be of a uniform size. I saw no part of the animal which I supposed to be a tail…I could not discover any eyes, mane, gills, or breathing holes. I did not see any fins or legs. The animal did not utter any sound, and did not appear to notice any thing. It remained still and motionless for five minutes or more…The next morning I rose very early to discover him…He often disappeared, and was gone for five or ten minutes under the water. I thought he was diving or fishing for his food…His quickest motion was very rapid: I should suppose at the rate of fifteen or twenty miles an hour. Mackerel, menhaden, herring, and other bait fish abound in the cove where the animal was seen.

Account 2: 20 June 1815, Plymouth, MA, many people, comments by Justice of the Peace Samuel Davis:

> 'Certain house carpenters, who were at work on a building near the spot [i.e. Captain Finney's house], also saw it; as well as many others – these persons dwell with emphasis on the long and distinct wake made in the water by the passage of the fish.'

1817

Account 3: 6 August, two unnamed women, in Heuvelmans (1968):

> 'Two women saw a monster like a huge serpent come into the harbor of Cape Ann which lies north of Gloucester roads.'

Account 4: Early August, a captain of a ship from Maine that put into Gloucester Harbour 'on account of the weather,' recorded by David Humphreys (1817) in a September 2nd letter to Sir Joseph Banks:

> [The captain reported] that he and his crew, in entering the harbor, had been alarmed and frightened at the sight of a terrible looking creature, resembling a Serpent, close along-side the vessel, with its head at, or in the chains of the bow, and its tail extended beyond the stern. According to the rate of tonnage mentioned, the vessel must have been sixty feet long, and the monster of course still larger. So extraordinary and improbable a story, told by a person who was unknown, appearing to gain no credit: he, in disgust, quitted his hearers and sailed as soon as the wind permitted.

Account 5: 10 August, Susan Stover, Reports gathered by Charles Lenox Sargent in his 1818 notebook, *The Sea Serpent: Evidence About It*:

> In Company with her father near the shore by the house saw the serpent very near, his head very plain of the shape of a dog'; he was so near the shore her father took off his hat that he should not frighten him; cannot say how long he was; saw him turn in the water and the parts pop in the opposite direction.

Account 6: 10 August, Lydia Wonson, from Sargent (1818):

> [At a distance of less than a quarter of a mile, the creature] appeared about 60 or 70 feet on the surface of the water; with his head out of the water; looked like the buoys of a seine; saw him draw himself up into a coil and extend himself again, and this with a good spyglass; nearly ½ an hour he was plainly and fairly in view; his head out of the water about as large as a horse's head.

Account 7: 10 August, William Row, from Sargent (1818):

> Standing near the cove...of Rocky Neck [and] saw the serpent and two sharks come into the cove; was within [a] stone's throw of him. Certainly saw 100 feet born on the water. Seemed to be in pursuit of bait as he moved rapidly and frequently put his head out of water as if to swallow his game; his head was as large as a horse's...Furthermore, [he] states that his sons took one of the sharks that accompanied him into the harbor.

Account 8: 10 August, Amos Story, mariner:

> I saw a strange marine animal, that I believe to be a serpent...It was between the hours of twelve and one o'clock when I first saw him, and he continued in sight for an hour and a half. I was setting on the shore, and was about twenty rods from him when he was the nearest to me. His head appeared shaped much like the head of a sea turtle, and he carried his head from ten to twelve inches above the surface of the water. His head at that a distance appeared larger than the head of any dog that I ever saw. From the back part of his head to the next part of him that was visible, I should judge to be three or four feet. He moved very rapidly through the water, I should say a mile in two, or at most,

three minutes. I saw no bunches on his back. On this day, I did not see more than ten or twelve feet of his body.

Account 9: 12,13,14 August, based on three separate sightings, Solomon Allen, shipmaster:

I have seen a strange marine animal, that I believe to be a sea-serpent, in the harbour in said Gloucester. I should judge him to be between eighty and ninety feet in length, and about the size of a half barrel, apparently having joints from his head to his tail. I was about one hundred and fifty yards from him, when I judged him to be the size of a half barrel. His head formed something like the head of the rattle snake, but nearly as large as the head of a horse. When he moved on the surface of the water, his motion was slow, at times playing about in circles, and sometimes moving nearly straight forward. When he disappeared, he sunk apparently directly down, and would next appear at two hundred yards from where he disappeared, in two minutes. His colour was a dark brown, and I did not discover any spots upon him.

Solomon Allen, in answer to a series of questions posed by Lonson Nash:

Q. What parts of it were above the surface of the water, and how high?
A. Its joints or bunches, appeared about eight or ten inches above the surface of the water...
Q. How many distinct portions of it were out of the water, at one time?
A. I should say fifty distinct portions.
Q. Did it appear smooth or rough?
A. It appeared rough and scaly.
Q. Had it ears, horns, or any other appendages?
A. I perceived none.
Q. How die its tail terminate?
A. He seemed to taper towards (what I thought) his tail, though I had no distinct view of his tail...
Q. How many persons saw it?
A. Twenty or thirty persons were in view of me.
Q. Did he open his mouth when you saw him, and if so, how wide?
A. Yes, when I looked at him from the shore with a glass, at about tow hundred yards distance, his mouth appeared to be open about ten inches. I had no glass, when I saw him from the boat.
Q. Did he carry his head above the surface of the water?
A. Yes, at times, about two feet, then again he would carry the top of his head just on the surface of the water.
Q. Did he turn short and quick, and what was the form of the curve that he made?
A. He turned short and quick, and the first part of the curve that he made in turning resembled the link of a chain; but when his head came parallel with his tail, his head and appeared near together.

Account 10: 14 August, Epes Ellery, shipmaster:

> I saw a sea animal that I thought to be a serpent, in the harbour in said Gloucester. I was on an eminence, near low water mark, and about thirty feet above the level of the sea, when I saw him. I should judge that he was about one hundred and fifty fathoms from me. I saw the upper part of his head, and I should say about forty feet of the animal. He appeared to me to have joints, about the size of a two-gallon keg. I was looking at him with a spy-glass, when I saw him open his mouth, and his mouth appeared like that of a serpent; the top of his head appeared flat. His motion when he turned was quick, but I will not express my opinion of his velocity. The first part of the curve that he made in turning was of the form of a staple, and as he approached towards his tail, he came near his body with his head, and then ran parallel with his tail, and his head and tail then appeared together.

Epes Ellery, in answer to a series of questions posed by Lonson Nash:

Q. What parts of it were above the surface of the water, and how high?

A. I did not count the number of bunches, but they appeared about six inches above the surface of the water.

Q. Were the sinuosities vertical or horizontal?

A: Vertical.

Q. Did it appear to pursue, avoid. Or notice any thing?

A. It did not appear to avoid any thing. He appeared to be amusing himself.

Account 11: 14 August, William H. Foster, merchant:

> When I first discovered him, his head was above the surface of the water, perhaps ten inches, and he made but little progress through the water. He was apparently shaded with light colours. He afterwards went in different directions, leaving on the surface of the water, marks like tho made by skating on ice. Then he would move in a straight line west, and would almost in an instant, change his course to the east, bringing his head, as near as I could judge, to where his tail was; or in fact, to the extreme hinder part visible, raising himself as he turned, six or eight inches out of the water, and shewing a body at least forty feet in length.

Account 12: 14 August, Matthew Gaffney, ship carpenter:

> I was in a boat, and was within thirty feet of him. His head appeared full as large as a four-gallon keg, his body as large as a barrel, and the length I saw, I should judge forty feet at least. The top of his head was of a dark colour, and the under part of his heed appeared nearly white, as did also several feet of his belly, that I saw. I supposed and do believe that the whole of his belly was nearly white. I fired at him, when he was nearest to me. I had a good gun, and took good aim. I aimed at his head, and I think I must have hit him. He turned toward us immediately after I fired, and I thought he was coming at us; but he sank down and went directly under our boat...He did not turn down like a fish, but appeared to settle directly down, like a rock...His motion was vertical, like a caterpillar.

Mathew Gaffney in answer to a series of questions posed by Lonson Nash:

Q. How fast did it move?

A. I should say he moved at the rate of a mile in two, or at most, three minutes...

Q. Does he turn quick and short, and if so, what is the form of path that he makes, in turning?

A. He turns quick and short, and the first part of the curve that he makes in turning, is in the form of the staple; but his head seems to approach rapidly towards his body, his head and tail moving in opposite directions, and when his head and tail came parallel, they appear almost to touch each other.

Q. Did he appear more shy, after you had fired at him?

A: He did not; but continued to play as before.

Account 13: 14 August, Lonson Nash, letter to the Linnaean Society:

> I judged him to be about two hundred and fifty yards from me. At that distance I judged him (in the largest part) about the size of a half barrel, gradually tapering towards the two extremes. Twice I saw him with a glass...When he moved on the surface of the water, the track of his rear was visible, for at least half a mile. His velocity, when moving on the surface of the water, I judged was at the rate of a mile in about four minutes. When immersed in the water, his speed was greater, moving, I would say, at the rate of a mile in two or at most three minutes. When moving under water, you could often trace him by the motion of the water, on the surface, and from this circumstance I conclude he did not swim deep. He apparently went as straight through the water, as you could draw a line. When he changed his course, he diminished his velocity but little – the two extremes that were visible appeared rapidly moving in opposite directions, and when they came parallel, they appeared not more than a yard apart...I should not be surprised, if he should be found one hundred feet long...If I saw his head, I could not distinguish it from his body...I never saw more than seven or eight distinct portions of him above the water, at any one time, and he appeared rough...When he disappeared, he apparently sunk directly down like a rock.

And, in a further correspondence, Nash states:

> I think Mr. Allen is likewise mistaken as to the distinct portions of the animal that were visible at one time. I saw, at no time, more than eight distinct portions, though more have been visible; still I cannot believe that fifty distinct portions were seen at one time. I believe the animals to be straight, and that the apparent bunches were caused by his vertical motion.

Account 14: 14 August, Jonathan Brown, from Sargent (1818), in Soini (2010):

> '[He] saw 60 or 80 black substance[s] on the water between the beach and Stage Point; many people with glasses all around the shore looking at him, saw him move in a semi-circular form, repeatedly.'

Account 15: 14 August, William Row, second sighting, from Sargent (1818):

> '[He] was plain in sight all day and frequently at rest on the water, and his humps plainly to be distinguished.'

Account 16: 14 August, Mrs. Moore, from Sargent (1818):

> Without any previous knowledge of him being a serpent in our water and [she] saw the creature...and was so alarmed the trembled like a leaf and stood some time in agitation to view. That he was very near and could not have been less than 100 feet plainly visible... [and] saw him turn...in quick motion.

Account 17: 14 August, Mrs. Trask, from Sargent (1818), in Soini (2010):

> 'With a good glass she saw the serpent on the water; seemed like gallon kegs tied together; [and] saw him suddenly disappear and came up again at a great distance.'

Account 18: 14 August, Z. Stevens, from Sargent (1818), in Soini (2010):

> '[I saw him] lying on the surface of the water perfectly still, the humps plainly to be seen extending about 50 feet; his head and tail was not visible: he had a good glass and is confident what he saw was of the serpent kind.'

Account 19: 14 August, Captain Corlis, from Sargent (1818):

> In a boat between the fort and Ten Pound Island, he plainly saw the serpent lying still on the surface at the distance of about 30 feet and apprehensive of being too near him he immediately rowed from him; his move alarmed the serpent and he made off with great speed. Is certain he saw 40 feet of him out of [the] water or rather lying on the surface. Is certain he could not mistake him for any [other] than a strange fish of the serpent kind.

Account 20: 14 August, Mary Row, from Sargent (1818):

> She saw the Serpent between Rocky Neck and Ten Pound Island from her dwelling house. He was in rapid motion and turned suddenly his head going one way and his tail the other. Saw his head out of [the] water very plain [and it] seemed as large as a horse's head; feels sure she saw 100 feet plainly on the surface of the water.

Account 21: 14 August, Captain Davidson, from Sargent (1818), in Soini (2010):

> 'He was standing near the windmill and saw the creature with a good glass 60 or 70 feet plain out of water...Is certain what he saw was a live substance on the surface and not the wake of any fish.'

Account 22: 14 August, John Lowe, from Sargent (1818), in Soini (2010):

> As he was standing on the windmill point he saw the Serpent between the beach and stage point...He...could plainly distinguish hi humps as before described, that he lay perfectly still in the water for some time and, as he rose and lowered in the water, he was more distinctly seen at some times than others...[He] is accustomed to seeing objects on the water and could not mistake it for other than a fish of the serpent kind from 80 to 100 feet.

Account 23: 14 August [I concur with Soini (2010) and Burns (2014) that this was misdated in Sargent's pamphlet as occurring in the following year], Joseph Proctor, from Sargent (1818), in Soini (2010):

> 'Standing with other [he] saw the serpent from 60 to 80 feet as plain out of the water as a rock; saw him still and in motion; saw him repeatedly at different times with the same appearance of humps and at times rapidly in motion.'

Account 24: 14 August, John Somes, from Sargent (1818), in Soini (2010):

> [From his rope walk on the beach, he saw the creature] pass repeatedly across the harbor and at times he was perfectly still on the water: saw his humps plain but did not count them, saw 60 to 70 feet of him on the surface of the water; saw his head and neck very plain out of water; and they had every appearance of a Serpent: says he could not mistake in describing which he saw.

Account 25: 14 August, Joseph Moores, from Sargent (1818), in Soini (2010):

> [He saw] a black substance on the surface of the water about 60 feet long...[and] he put his head near his tail and could plainly see one part of him move one way and the other part in a contrary direction and a continuity of the parts from one end to the other.

Account 26: 14 August, Samuel Wonson, from Sargent (1818), in Soini (2010):

> [I] saw the serpent between the house and Ten Pound Island, came with great rapidity from Norman 's Woe and 'brot to' abreast of me perfectly still at about 100 fathoms distance could distinctly see 100 feet out [of the] water on the surface [which was] so near [that] his children were alarmed and ran into the house.

Account 27: 15 August, James Mansfield, merchant:

> I saw a strange creature of enormous length, resembling a serpent...I should say he was from forty to sixty feet in length, extended on the surface of the water, with his head above the water about a foot. He remained in this position but a short time, and then started off very quick, with much greater velocity than I have seen him move at any other time. I saw bunches on his back about a foot in height, when he lay extended on

> the water. His colour appeared to me black, or very dark. I should say, he moved a mile in five or six minutes.

Later, in answer to a series of questions posed by Lonson Nash, Mansfield states that the creature's 'head appeared to be about the size of the crown of a hat…I saw no ears, horns, or other appendages,' and that its sinuosities were vertical.

Account 28: 15 August, Joseph Lee, south of Cape Ann, published in 1875 in the *Gloucester Daily Telegraph*, from O'Neill (1999):

> I was on bord of schooner *Hazard* laing to anchor off little island between the town of Manchester and Cape Ann. I judge I was abought one hundred and fifty yards from him. But to have a better view of him (Figure 4.2) I went into the boat and went as near him as I though it safe. I was within twenty yards of him. I thought it not safe to go any further. I then turned back on bord of the schooner again. I saw him in the above fashen. He was about one hundred feet in length as nie as I could judge…I saw him twenty minnets.

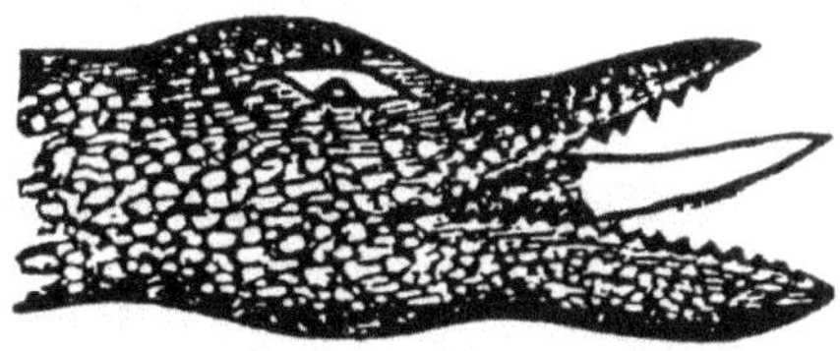

Figure 4.2. Drawing of the head of the unidentified marine object (UMO) observed by Joseph Lee just south of Cape Ann in August 1817.

Account 29: 17 August, William Saville, from Sargent (1818):

> 'He was standing near the windmill when something appeared above the water from 40 to 50 feet long in distinct bunches but soon after being alarmed by the noise…he sank in the water and he saw him no more.'

Account 30: 17 August, William H. Foster's second sighting:

> I again saw him. He came into the harbor, occasionally exhibiting parts of his body, which appeared like rings or bunches. As he drew near, and when opposite to me, there rose from his head or the most forward part of him, a prong or spear, about twelve inches in height, and six inches in circumference at the bottom, and running to a small point.

W. H. Foster, in answer to a series of questions posed by Lonson Nash:

> Q. Might not the prong or spear that you saw, have been the tongue of the serpent?
> A. I thought not; as I saw the prong before I saw his head.
> Q. At what distance was [sic] you, when you saw the spear of the serpent?
> A. I should judge forty rods; I had a spy-glass when I saw the prong or spear.
> Q. Did he appear jointed, or only serpentine?
> A.He appeared jointed.
> Q: Were its sinuosities vertical or horizontal?
> A: Vertical.
> Q: What was its colour?

A. It appeared brown…
Q. What was the size and shape of his head?
A. At the distance where I was, his head appeared as large as a man's head; but I cannot describe its shape…
Q. How fast did it move?
A: At the rate of a mile in a minute, at times, I have no doubt.

Account 31: August 17, John Johnston:

> I saw a strange marine animal, that I believe to be a serpent, lying extended on the surface of the water. His length appeared to be fifty feet at least, and he appeared straight, exhibiting no protuberances….We immediately rowed from him, and at first concluded to pass by his tail; but feared we might strike it with the boat, concluded to pass around hi shad, which we did, by altering our course. He remained in the same position, till we lost sight of him. We approached so near to him that I believe I could have reached him with my oar.

Account 32: before 18 August, Mr. Mansfield, recounted in a 1820 letter from Colonel T. H. Perkins, published in the *Boston Daily Advertiser* in 1848 at the time of the *Daedalus* debate, from Oudemans (1892):

> Whilst at Cape Ann I talked with many persons who had seen the serpent, and among others with a person of the name of Mansfield, one of the most respectable inhabitants of the town. His account to me was, that a few days before, as he was taking a ride with his wife in a chair, the road taking them close to a bank which overlooks the harbor (and is nearly a perpendicular precipice), he saw an uncommon appearance, which induced him to descend from the carriage, when he saw the sea-serpent. In which until then he had been an unbeliever. The animal was stretched out, partly over the white sandy beach, which had four or five feet of water upon it, and lay partly over the channel. He desired his wife to get out of the carriage, which she did. He said he had made up his mind as to the length of the snake, but wished the opinion of his wife on the same subject…She answered that she could not undertake to say how many feet in length he was, but that she thought him as long as the wharf behind their house, an object with which she had been familiar. Mr. Mansfield said he was of the same opinion. The wharf is one hundred feet in length. It is to be observed that the person above spoken of had been such an unbeliever in the existence of this monster, that he had not given himself the trouble to go from his house to the harbor when the report was first made of such an animal being there.

Account 33: 18 August, letter of Mr. S. Perkins to Mr. E. Everett, from Oudemans (1892):

> The captain and the crews report that, off Cape Ann Harbour, they saw a Sea Monster of the Snake kind, lying on the water, of immense length. That the crew were so much alarmed, that they got away as soon as they could, and that they estimated it at 100 feet long.

Account 34: 18 August, William B. Pearson, merchant:

> The serpent passed out under the stern of our boat...then he stood in towards us again, and crossed our bow. We immediately exclaimed: 'here is the snake!' From what I saw of him, I should say that he was nothing short of seventy feet in length. I distinctly saw bunches on his back, and once he raised his head out of the water. The top of his head appeared flat, and was raised seven or eight inches above the surface of the water... His colour was dark brown...His motion was vertical. I have seen him move with great velocity, I should say at the rate of a mile in three minutes, and perhaps faster. His size I judged to be about the size of a half barrel...He turned very short, and appeared as limber and active as an eel, when compared to his size. The form of the curve when he turned in the water, resembled a staple; his had seemed to approach towards his body for some feet; then his head and tail appeared moving rapidly, in different directions, and when his head and tail were on parallel lines, they appeared not more than two or three yards apart.

W.B. Pearson, in answer to a series of questions posed by Lonson Nash:

> Q. How many distinct portions of it were out of water at one time?
> A. Ten or twelve distinct portions.
> Q. Can you describe his eyes and mouth?
> A. I thought and believe, that I saw his eye at one time, and it was dark and sharp.
> Q. How did its tail terminate?
> A. I had not a distinct view of his tail; I saw no bunches towards, what I thought, the end of his tail, and I believe there were none. From where I judged his navel might be, to the end of his tail, there were no bunches visible.

Account 35: 18 August, Colonel T. H. Perkins, in an 1820 letter published in the *Boston Daily Advertiser* in 1848 at the time of the *Daedalus* debate, reprinted in the *Zoologist* of 1849, from Oudemans (1892):

> I observed an agitation in the water at the entrance of the harbor, like that which follows a small vessel going five or six miles an hour through the water...As nearly as I could judge, there was visible at a time about forty feet of his body. It was not, to be sure, a continuity of body, as the form from head to tail (except as the apparent bunches appeared as he moved through the water) was seen only at three or four asunder. It was very evident, however, that his length must be much greater than what appeared, as, in his movement, he left a considerable wake in his rear...The head was flat in the water...I was struck with an appearance in the front part of the head like a single horn, about nine inches to a foot in length, and in the form of a marlinspike.

In a letter dated 20 August, 1817, the Colonel's brother, S.G. Perkins, wrote to an acquaintance in Paris, summarizing the events of the last several weeks in Gloucester, in particular, the comments of his brother:

> He is represented to be from 50 to 100 feet long, of the size of a barrel about the body, which his formed into parallel rings, which – when he is on the top of the water – are so prominent, that they resembled buoys attached to each other. Its motions, when in

> pursuit of its prey, are very rapid, and create a wake like a small vessel passing thro' the water...It comes into the harbor daily, in pursuit of the herrings, which resort here in great quantities.

And in 1850, in *A Second Visit to the United States of North America*, the famous geologist Sir Charles Lyell, recounts an interview he had with Colonel Perkins:

> 'He counted fourteen projections, six feet apart, on the back, which he imagined to be vertical flexures of the body when in motion; but he also saw the body bent horizontally into the figure of the letter S.'

Accounts 36, 37 and 38: August 1817, pamphlet published by the Brattleboro Book Store with the banner heading 'A Monstrous Sea Serpent, the largest ever seen in America, Has just made its appearance in Gloucester Harbour, Cape Ann, and has been seen by hundreds of Respectable Citizens.' Accompanied by an illustration (Figure 3.11), the publication compiled the reportage form a number of local newspapers, in O'Neill 1999). Three such accounts are:

> I have just seen the fish, sporting in the water, and it shews the length of fifty feet within a quarter of a mile from the shore...The street is full of persons who are going to enjoy the sight of it. It appears in joints like wooden buoys on a net rope almost as large as a barrel. Two muskets were fired at it, and appeared to hit it on the head, but without effects...It appears like a string of gallon kegs 100 feet long...It had a head like that of a horse, and as he moved he looked like a row of casks following in a right line.
>
> He is at least eighty, and more probably, an hundred feet long, and nearly of the size of a flour barrel, at the largest place. As to the bushes, or protuberances, which have been mentioned, these are thought to be nothing more than the appearances occasioned by his manner of motion. He does not wind, laterally along, as Serpents commonly do, but his motion is undulatory, or consisting in alternate rising and depression, somewhat like the motion of the caterpillar... His head, though in its front it is circular, is not flat, like a common serpent's...but resembles that of a dog's...with apparently very thick scales, and its whole appearance very terrific...more often he moves along, with his head under water, showing the line of his back, or with his head immediately above the surface. He appears to be round, with large scales, which, when he contracts his folds, gives a rigidy appearance to his back, but when he extends himself they inclose, and do not prevent his appearing smooth...He often turns over quick, bringing his head near his tail, and putting himself in the form of a staple.
>
> His motions are serpentine, extremely varied, and exceedingly rapid. He turns himself around almost instantaneously. He sometimes darts forward, with his head out of the water, at the rate of a mile in 3 minutes, leaving a wake behind of a half a mile in length...He appears to be full of joints and resembles a string of buoys on a net rope, as is set in the water to catch herring. Others describe him as like a string of water casks. His back is black. Various attempts have been made, without success, to take him. Four boats went out on Thursday, filled with adventurous sailors and experienced gunners, armed with muskets, harpoons, etc. Three muskets were discharged at him from a distance of 30 feet; two balls were thought to hit his head, but without effect. He immediately after plunged into the water, and disappeared for a short time, after which he moved off to the outer harbor, and was seen no more that night...It is conjectured that he has resorted to

this harbor for the purpose of preying upon a very numerous shoal of herring, which has appeared lately here. If he is instrumental, as is supposed, in driving these herring into the harbor, he has rendered an essential service to the town...They...are of the opinion that he is cased in shell.

Account 39: 22 August, letter from Justice of the Peace Nash to the Linnaean Society Committee:

I have seen and conversed with the woman, who was said to have seen the serpent dormant on the rocks, near the water...but she can give no material evidence. She says that she saw something, resembling a large log of wood, on the rocks, on the extreme eastern point of Ten Pound Island (a small island in our harbor), resting partly on the rocks, and partly in the water. The distance was about half a mile. She took a glass, looked at the object and saw it move, and when she looked for the object again, it had disappeared.

Account 40: 23 August, Amos Story, second sighting:

He then lay perfectly still, extended on the water, and I should judge that I saw fifty feet of him at least...I continued looking at him about half an hour, and he remained still and in the same position, until I was called away. Neither his head nor tail were visible. His colour appeared to be a dark brown, and when the sun shone upon him, the reflection was very bright. I thought his body was about the size of a man's body.

Accounts 41, 42 and 43: 28 August, respectively from Sewell Toppan, schooner master, and Robert Bragg and William Somerby, mariners:

At about two miles, or two and half miles east of the eastern point of Cape Ann being becalmed, I heard one of my men call to the man at the helm, 'what is this coming towards us'...I then got on top of the deck load, at which time I saw a singular kind of animal or fish, which I had never before seen, passing by our quarter, at a distance of forty feet...I saw a part of the animal or fish ten or fifteen feet from the head downwards including the head; the head appeared to be the size of a ten-gallon keg, and six inches above the surface of the water. It was of a dark colour...The motion of his head was sideways and quite moderate; the motion of the body, up and down. I have seen whales very often; his motion was much more rapid than whales or any other fish I have ever seen; he left a very long wake behind him; he did not appear to alter his course in consequence of being so near the vessel...I have been to sea many years, and never saw any fish that had the least resemblance to this animal. Judging from what I saw out of the water, I should judge the body was about the size of a half barrel in circumference.

He left a very long wake behind him...He had a head like a serpent, rather larger than his body and rather blunt; did not see his eyes; when astern of the vessel about 30 feet, he threw out his tongue about two feet in length; the end of it appeared to me to resemble a fisherman's harpoon; he raised his tongue several times perpendicularly, or nearly so, and let it fall again...I think he moved at the rate of 12 or 14 miles an hour; he was of a dark chocolate colour...and was about two and a half feet in circumference...From what appeared out of the water I should suppose he was two and a half feet in circumference;

> he made no noise; his back and body appeared smooth; s small bunch on each side of his head, just above the eyes; he did not appear to be at all disturbed by the vessel...His motion was much swifter than any whale that I have ever see, and I have observed many – did not observe any teeth; his motion was very steady, a little up and down.
>
> His head was like a serpent's tapering off to a point. He threw out his tongue a number of times, extended about two feet from his jaws – the end of it resembled a harpoon – he threw out his tongue backwards several times over his head – and let it fall again – I saw one of his eyes as he passed; it appeared very bright, and about the size of the eye of an ox. The colour of all that appeared was very dark, almost black. He did not appear to take any notice of the vessel, and made no noise. There appeared a bunch above the eye – Should judge him to be about two and a half feet in circumference. Have often seen whales at sea. The motion of this animal was much swifter than that of any whale. The motion of the body was rising and falling as he advanced, the head vibrating from side to side. The colour of his tongue was a light brown.

Account 44: August, William Pierson, from Humphreys (1817) in a September 5th letter to Sir Joseph Banks:

> Once he saw the monster approach the boat. Placing himself in the bow, *harpoon in hand*, he collected all his force to throw it. Arrived within twenty feet, the tremendous creature sheared off, plunged suddenly ten or twelve feet beneath the surface, passed close along-side, and rose about two hundred yards from the stern.

Account 45: 7 September, a captain, from Humphreys (1817) in a September 5th letter:

> He was met abruptly by a boat employed in fishing off Salem harbor. The crew, unprovided with any instrument of annoyance or defense, found themselves, at once, almost in contact with, and as it were in the very jaws, of this *hideous monster.*
>
> For several hours he followed, and at a number of times...was not more than ten feet from the boat. Apprehensive of his sinking them in the deep, or devouring them alive, they must have been more than human, not to have experienced indescribable horror and dismay. Death inevitably, in his worst form, seemed to stare them in the face.
>
> A situation more terrific can hardly be conceived by the most fertile imagination...
>
> In following them so closely, and for so long a time, he seems to have been only intent on the multitudes of fish which fled before him; which he is known to pursue with eagerness, and to seize such of the fugitives as fall in his way for *sustenance*...Happily...he has never showed any indications of a disposition to chase or attack any moving object or living creature, except for the before-mentioned purpose.

Account 46: September, two captains, from Humphreys (1817) in a September 13th letter:

> The captains of two vessels report, that they have seen him since he left Cape-Ann, at two different times; once, laying stretched out straight on the surface of the water like a spar: at another time, coiled in serpentine folds, in the manner of a common snake. At neither of these times did he appear frightened, or in any degree disturbed by their sudden and near approach.

Account 47: 10 September, halfway between Cape Ann and Boston, from O'Neill (1999):

> 'The creature was reportedly seen...'coiled up on the surface of the water reposing after a hearty breakfast of herring."

On September 27th two boys and their father killed a small snake on the shore of a Cape Ann cove, which would later be concluded by the Linnaean Society to be a baby sea serpent. This was detailed in the preceding chapter.

Account 48: 30 September, Long Island Sound, William Bentley, entry from his diary, from Bolster (2012) and Burns (2014):

> 'We have seen the Serpent...His body when out of the water looks like the buoys of a net, or a row of kegs, or a row of large casks.'

Account 49: 3 October, Long Island Sound, James Guion, recounted in a letter from Thomas Herttell, from Oudemans (1892):

> 'He judged his speed to be little or no less than a mile a minute. He describes the irregularity and unevenness of his back, about fifty feet of which appeared above the surface of the water.'

Account 50, 5 October, Long Island Sound, Thomas Herttell, from Oudemans (1892):

> Viewing it [from his home in Long Island Sound] with my glass convinced me it was a large living animal, – His back, forty to fifty feet of which were seen above the surface of the water, appeared to be irregular, uneven, and deeply indented. I did at this time remark that his head was more elevated above the water than the ridges or humps on his back...His body more depressed below and his head more elevated above the water, going with increased velocity...creating a swell before him not unlike that made by a boat towed rapidly at the stern of the vessel. From the time I first saw him till I lost sight of him perhaps could not have exceeded ten minutes, in which short time he had gone probably no less than six or seven miles...That it was a sea serpent of great bulk, to me is certain. – That it is what is usually called a Sea-Serpent, and the same which appeared in Gloucester harbor, is only probable.

Accounts 51 and 52: 8 November, Long Island Sound, Fitch Welch and Eli Green, and Mark Baldwin, from Humphreys (1817) in a November 14th letter:

> About half a mile from the shore, they discovered a thing on the water, which at first view resembled a ship's buoy, but on approaching it within five or six rods, they found it to be a live animal of a dark brown colour, a flat head with a bunch or crest upon it, his eyes about the size of the eyes of a horse, with a white ring around each of them, about three inches in diameter.
>
> About 7 or 8 feet only of his body was visible, but the part seen was much larger than his head, being about the size of a barrel, and tapering off to the extremity of the head.

As the boat approached him, he raised his head 3 or 4 feet above the water, moved it backwards and forwards, and fiercely eyes the men. The water was 16 or 18 feet deep, and from the direction of the neck, and that part of his body which was seen, the beholders supposed he was lying on the bottom.

He was also seen by a Mr. Mark Baldwin about the same time, who was on the shore, and says he appeared to him like a man standing in the water.

1818

Account 53: 10 May, Joseph Woodward, captain, reported in several newspapers, from Heuvelmans (1968):

I, the undersigned, Joseph Woodward...being on route from Penobscot to Hingham... and being about 10 leagues from the coast, perceived last Sunday, at two P.M. something on the surface of the water which seemed to me to be the size of a large boat. Supposing it might be part of the wreck of a ship, I approached it; but when I was within a few fathoms of it, it appeared, to my great surprise and that of my whole crew, that it was a monstrous serpent. When I approached nearer, it coiled itself up, instantly uncoiling again, and withdrew with extreme rapidity. On my approaching again, it coiled itself a second time, and placed itself at the distance of 60 feet at most from the bow of the ship.

I had one of my guns loaded with a cannon ball and musket balls. I fired it at the head of the monster; my crew and myself distinctly heard the ball and bullets strike against his body, from which they rebounded, as if they had struck against a rock. The serpent shook his head and tail in an extraordinary manner, and advanced toward the ship with open jaws...All the crew were seized with terror, and we thought only of getting away. He almost touched the vessel; and had not I tacked as I did, he would have certainly have come on board. He dived; but in a moment we saw him appear again, with his head on one side of the vessel, and his tail on the other, as if he was going to lift us up and upset us. However, we did not feel any shock. He remained five hours near us, only going backward and forward.

The fears with which he at first inspired us having subsided, we were able to examine him attentively. I estimate that his length is at least twice that of my schooner, that is to say, 130 feet; his head is full 12 or 14; the diameter of the body below the neck is not less than six feet; the size of the head is in proportion to that of the body. He is of blackish colour; his ear-holes are about 12 feet from the extremity of his head. IN short, the whole has a terrible look.

When he coils himself up, he places his tail in such a manner, that it aids him in darting forward with great force: he moves in all directions with the greatest facility and astonishing rapidity.

Account 54: 21 June, off Cape Ann, Shubael West, mariner, from Oudemans (1892):

'[The] master of the Packet Delia, describes it [the sea serpent] as seen...engaged with a whale', which O'Neill (1999) describes as 'battling a large humpback whale.'

Account 55: July, Joseph Moores, from Sargent (1818):

> He saw the same creature or to appearance the same...at about 100 yards which near his house he took a compass and took the bearing of his head and tail reversing his station he took another set of bearings and by this he calculated him to be 120 feet long; is certain that he could not have moved any during the operation as he had a mark from his head of the point of land in view; had he have moved he would have [?] the point in which he did not.

The pamphlet contains a rough sketch of Mr. Moore's trigonometric methodology, which would later be elaborated on when printed in a compilation of sea serpent reports in the *Boston Gazette* (Figure 4.3).

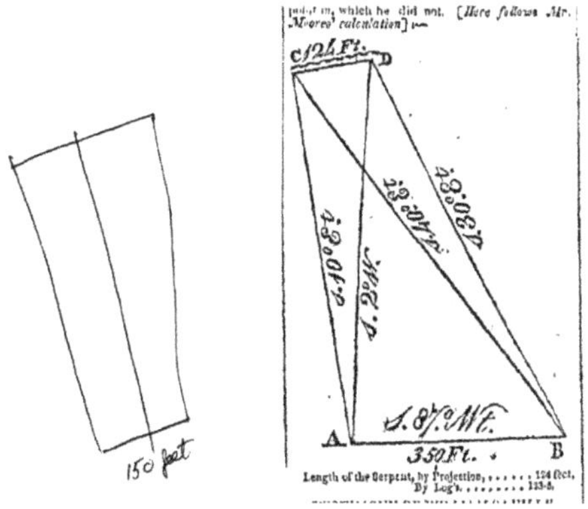

Figure 4.3. Eyewitness Joseph Moore's trigonometric method used to estimate the size of the UMO observed in Gloucester Harbour in August 1817. Left illustration: redrawing of a crude sketch from Sargent's 1818 pamphlet (Cape Ann Museum Archives); right illustration: version of the drawing from the October 22nd edition of the Boston Gazette *(reprinted in Burns 2014).*

Accounts 56-61: summarized newspaper reports, from Dexter (1986):

> In July of 1818 the sea-serpent reappeared in Gloucester Harbor. On 25 July, William Sargent counted 32 humps [Account [56]. Four days later some fishermen pursued it with muskets, but without success [57]. The following day Captain Webber and several whalers got within a few yards, but their harpoon bounced off the animal [58]. On 12 August, Timothy Hodgkins, a local sailor returning from Newburyport, observed the sea-serpent in Annisquam Harbor [Cape Ann; 59]. He estimated the length to be 120 feet and counted 20 humps five feet apart. The body was equal to a 70-gallon cask with a head as big as a barrel. Four days later many persons saw it near Annisquam Lighthouse where it was pursued by a fleet of whaleboats [60].

These last sightings were also covered in local newspapers at the time. Captain Webber, observing from the shore [61], reported 'that the mother was 100 feet long, the infants were each 50 feet long' (Burns 2014). William Sargent [56] believed that there were in fact four serpents, a mother and 'three young ones, and they make great destruction among the bait...the fish spring above the water to escape their enemies, but in vain, for the Serpents pass and repass, devouring as they go' (Burns 2014). Furthermore, as the mother serpent 'passed under his boat' a head and protuberances were seen. Heuvelmans (1968) adds for Account 57 that the monster 'received, with apparent unconcern, seven or eight salvos [of musket-fire] and went on playing on the surface.'; and for Account 59, he notes that Hodgkins had first thought that the serpent's humps to have been a school of pilot whales, before discovering otherwise:

> His head was elevated from three to five feet; the distance was about six feet from his neck to the first bunch; we counted twenty bunches, and we supposed them on an average about five feet apart, and his whole length could not be less than one hundred and twenty feet...His head was of a dark brown color, formed like a seal's, and shined in

glossy appearance...His body was of the size of a sixty or eighty-gallon cask, his head as large as a barrel...There was nothing that appeared like fins or gills. We did not discern a tail. There was a quick vibration of the parts we saw, which probably was his mode of swimming.

Account 62: 18 August, W. Sergent, in the *Niagara Patriot*, from Lyons (2009):

Mr. W. Sergent [I am assuming that this is a different person than William Sargent, noted above] along with two young men were fishing and observed what they thought was a hake approaching. It came within a few feet of Mr. Sergent's boat with its head about a foot above the water, and passed under the boat with a gliding motion, 'apparently without any muscular exertion.' Mr. Sergent described the head as resembling: 'that of a toad, having projections which he calls smellers, like the hake, on each side. On top of the head were three spears or horns, the middle one the largest, and all three at that time laying flat on the head, The body appeared about the bigness of a lime cask. The color was dark brown...The scales about the size of the crown of his hat, and the barnacle on the body about four inches.' He estimated the creature to be about seventy to one hundred feed long. He attempted to obtain a barnacle from the serpent's back, but when he struck the water with an oar, 'the Serpent sunk beneath the surface without any struggle and disappeared.'

Account 63, 1 September, Captain Smith, in the *Niagara Patriot*, from Lyons (2009):

'Two weeks after Mr. Sergent's account was printed, a Captain Smith also reported sighting what he thought was the serpent. But later in the day, he saw nearby at two different times what he unequivocally identified as a whale, and concluded that he had been mistaken in claiming he had seen the serpent.'

Account 64: 11 September, Andrews Norton, letter to a friend, from Dexter (1986):

Described the Cape Ann sea-serpent as 120 feet long and said the Gloucester Board of Health had sent two boats into Gloucester Harbour to intercept the sea-serpent.

In early September the famous sea serpent was indeed reported to have been caught by Captain Rich and put on display, as detailed later in this chapter.

1819

Accounts 65 and 66: 6 June, on a sloop approaching Cape Ann, Hawkins Wheeler and Gersham Bennett, mariners, from Oudemans (1892):

His motion was at a rate of about four miles an hour...I had a fair and distinct view of the creature, and from his appearance am satisfied that it was of the serpent kind. The creature was entirely black; the head, which perfectly resembled a snake's; was elevated from four to seven feet above the surface of the water, and his back appeared to be composed of bunches or humps, apparently about as large as, or a little larger than a half barrel; I think I saw as many as ten or twelve...The tail was not visible, but from the head to the last hump that could be seen, was, I should judge, 50 feet. The first view I had of him appeared like a string of empty barrels tied together, rising over what little swell of the sea there was. What motion I could discern in the body of the animal was undulatory, but he evidently moved his tail under water, and the ripples produced by it indicated a sweeping motion, making a wake as large as that made by the sloop.

I there saw a serpent of an enormous size and uncommon appearance, upon the water; his head was about the length of the anchor stock above the surface of the water, viz. about seven feet...The colour of the animal throughout, as far as could be seen, was black, and the surface appeared to be smooth, without scales – his head was about as long as a horse's and was a proper snake's head – there was a degree of flatness, with a slight hollow on the top of his head...he was moving in the same direction with the sloop, and about as fast. The back was composed of bunches about the size of a flour barrel, which were apparently about three feet apart – they appeared to be fixed, but might be occasioned by the motion of the animal, and looked like a string of casks or barrels tied together – the tail was not visible, but the wake of his tail which he evidently moved under water, showed a horizontal or sweeping motion, producing a wake as large as the vessel made. He turned his head two or three times slowly round towards and from the vessel, as if taking a view of some object on board.

Account 67: August, Swampscott (Boston Harbour), Mr. Buffum, from Henry David Thoreau's *Journal*, 14 January, 1858:

[The serpent] was to be seen off and on for some time. There were many people on the beach the first time, in carriages partly in the water, and the serpent came so near that they, thinking that he might come ashore, involuntarily turned their horses to the shore as with a general consent, and this movement caused him to shear off also. The road from Boston was lined with people directly, coming to see the monster...Buffum says he has seen him twenty times, once alone, from the rocks at Little Nahant, when he passed along close to the shore just beneath the surface, and within fifty or sixty feet of him, so that he could have touched him with a very long pole, if he had dared to.

Account 68: 11 August, Nahant (Boston Harbour), Mr. Smith and many others, in an August 16th letter from James Prince, Marshal of the District, to Judge Davis, from Oudemans (1892):

Intending to pass two or three days at Nahant, with my family, we left Boston early... [and] on passing the halfway house, on the Salem turn pike, Mr. Smith informed us the sea serpent had been seen the evening before at Nahant-beach, and that a vast number of people from Lynn, had gone to the beach that morning, in hopes of being gratified with a sight of him: This was confirmed at the hotel.

Account 69: 13 August, James Prince and several hundred others, Nahant (Boston Harbour), from Oudemans (1892), in Wilson (1928):

> I presume I may have seen what is generally thought to be the sea-serpent...On our arrival on the beach, we associated with a considerable number of persons, on foot and in chaises, and very soon an animal of the fish-kind made his appearance. His head appeared about three feet out of water; I counted thirteen bunches on his back: my family thought there were fifteen – he passed three times at a moderate rate across the bay, but so fleet as to occasion a foam in the water – and my family and myself, who were in carriage, judged he was fifty feet in length, and, at the extend, not more than sixty; whether, however, the wake might not add to the appearance of his length; or whether the undulation of the water, or his peculiar manner of propelling himself, might not cause the appearance of protuberances, I leave for your better judgment...
>
> As he swam up the bay, we and the other spectators moved on, and kept abreast of him; he occasionally withdrew himself under water, and the idea occurred to me that this occasionally raising his head above the level of the water, was to take breath, as the time he kept under water was of an average about eight minutes; after being accustomed to view him, we became more composed; and made the annexed figure of his outlines (Figure 4.4; of interest is the illustration labelled as 'James Prince's sea-serpent, 1819' from Heuvelmans (1968) which shows only 6 coils of a serpent)...
>
> He did not turn without occupying some space, and taking into view the time and space which he found necessary for his ease and accommodation, I adopted it as a criterium to form some judgment of his length – I had seven distinct views of him from the long beach so called, and at some of them the animal was not more than a hundred yards distance. After being on the long beach about an hour, the animal disappeared...I concluded he had left in consequence of the number of boats in the offing in pursuit of him – the noise of oars must have disturbed him, as he appeared to us to be a harmless timid animal...

Figure 4.4. Drawings of eyewitness James Prince of the UMO he observed off Nahant Peninsula (Boston Harbour) in August 1819. The second illustration is another nineteenth-century illustration of the same sighting (Heuvelmans 1968).

> [While on another nearby beach] we had more than a dozen different views of him, and each similar to the other; one however so near, that the coachman exclaimed: 'Oh, see his glistening eye'. Thinking I might form some calculation of his length by the time and distance of each turn; and taking an angle with my two hands of the length he exhibited, that is to say, from his head to his last protuberance, and applying the same angle to other objects, I feel satisfied of the correctness of my decision that he is sixty feet long, unless the ripple of his wake deceived me – nor my dear sir, do I undertake to say he was of the snake or eel kind, though that was the general impression on my family, the spectators and myself. Certainly it is a very strange animal.

Account 70: 19 August, off Nahant Peninsula (Boston Harbour), Samuel Cabot and many others, from Oudemans (1892), in Wilson (1928):

> As my curiosity was directed towards the boats to ascertain the course they were taking, my attention was suddenly arrested by an object emerging from the water at a distance of about one hundred and fifty yards, which gave my mind at the first glance the idea of a horse's head. As my eye ranged along I perceived at a short distance eight or ten regular bunches or protuberances, and at a short interval three or four more. I was now satisfied the *Sea-Serpent* was before me, and after the first moment of excitement produced by the unexpected sight of so strange a monster taxed myself to investigate his appearance as accurately as I could...
>
> The Head [was] serpent shaped, it was elevated about two feet from the water, and he depressed it gradually, to within six or eight inches as he moved along. I could see under his chin, which appeared hollow underneath or to curve downward. His motion was at that time very slow along the Beach, inclining towards the shore; he at first moved his head from side to side as if to look about him. I did not see his eyes...His bunches appeared to be not altogether uniform in size, and as he moved along some appeared to be depressed, and others brought above the surface, though I could not perceive any motion in them. My next purpose was to ascertain his length. For this purpose I directed my eye to several whale boats at about the same distance, one of which was beyond him, and by comparing the relative length, I calculated that the distance from the animal's head to the last protuberance I had noticed, would be equal to about five of those boats. I felt persuaded by this examination that he could not be less than eighty feet long...
>
> After I had viewed him thus attentively for about four or five minutes, he sunk gradually into the water and disappeared; he afterwards again made his appearance for a moment at a short distance...
>
> I had another view of him, for a longer time, but at a greater distance. At this time, he moved more rapidly, causing a white foam under the chin, and a long wake, and his protuberances had a more uniform appearance. At this time he must have been seen by two to three hundred persons on the beach and on the heights each side, some of whom were very favorable situated to observe him.

Account 71: August, Nahant (Boston Harbour), Dr Hamilton, in Jardine's *Naturalist's Library*, vol. xxv, from Gould (1886):

> Dr Hamilton states that an animal of similar appearance [i.e. to that seen previously in Gloucester] was seen again, in August 1819, off Nahant, Boston, and remained in the neighbourhood for some weeks. Two hundred persons witnessed it, thirteen folds were counted, and the head, which was serpent-shaped, was elevated two feet above the surface. Its eye was remarkably brilliant and glistening. The water was smooth, and the weather calm and serene. When it disappeared, its motion was undulatory, making curves perpendicular to the surface of the water, and giving the appearance of a long moving string of corks.

The UMO was seen regularly by hundreds of people throughout the month of August. A long shed was actually built next to Faneuil Hall in downtown Boston in anticipation of housing the body of the creature whose capture was thought to be imminent.

Account 72: 1820 letter of Colonel T.H. Perkins describing the Aug 1819 sightings by his acquaintances, published in the *Boston Daily Advertiser* in 1848, at the time of the *Daedalus* debate, from Oudemans (1892):

> The apparent bunches on his back they consider as arising from the construction of his body, and that the movement was vertical not horizontal. At one time his head was about two or three feed above water, but soon disappeared to the level of the sea...They saw him turn and bring his body into a letter S, the head being at right angles with the tail. From fifteen to twenty three bunches, or apparent bunches, were counted by the different persons who saw him, and his size round they thought to be that of a common firkin or half barrel. No one thought they saw the whole length of the body at a time, the tail seeming always to be considerably under water. The greatest length given to him was one hundred feet and no one who had a good sight of him thought him less than eighty feet in length.

Accounts 73, 74 and 75: early August, Nahant (Boston Harbour), Nathan Chase, in an 1881 letter to J.B. Holder which appeared in an article in *Century Magazine* of 1892, from Gould (1930) and O'Neill (1999), along with the original drawing (Figure 4.5):

> He then passed along about one hundred feet from where I stood with his head two feet out of the water. His speed was about that of an ordinary steamer. What I saw of his length was about sixty feet. It was difficult to count the humps, or undulations on his back, as they did not all appear at once. This accounts in part for the varied descriptions given by other parties. His appearance on the surface was occasional and but for a short time. The

Figure 4.5. Illustration accompanying account of eyewitness Nathan Chase who observed the UMO off Nahant in August 1819.

> colour of his skin was dark, differing but little from that of the water, or the back of a common fish...I saw the creature just as truly, though not as clearly, as I ever saw anything. I have no doubt that the uncommon, strange rover, which has been seen by hundreds of men and boys, is a form of snake, Plesiosaurus or some such form of marine animal.

Benjamin Newhall:

> As he approached the shore...he raised his head apparently about six feet, and moved very rapidly. I could see white spray on each side of his neck, as he plunged through the water. He came so near as to startle many of the spectators, and then suddenly retreated. As he turned short, the snake-like form became apparent, the body bending like an eel. I could plainly see what appeared a succession of humps upon his back.

John Marston:

> While walking over Nahant Beach in common with many others who had been aroused by the excitement, I saw in the water, within two or three hundred yards of the shore, a singular-looking fish in the form of a serpent. His head was out of the water, and he remained in view about twenty minutes...I would say that the creature was at least eighty feet in length. I saw the entire body, not his wake. It would rise in the water with an undulating motion, and then all his body would sink except his head. This would be repeated. The sea was quite calm at the time. I have been constantly engaged in fishing since my youth, but never saw anything like this before.

Account 76: 26 August, in Gloucester Harbour, Cheever Felch, ship chaplain, in *American Journal of Sciences*, from Oudemans (1892):

> [The] Commander of the Schooner, seeing some appearance on the water, said – 'there is your sea-serpent', meaning it is a laugh on me, for believing in its existence; but it proved to be no joke. The animal was then between thirty and forty yards distance from us...He soon sun; but not so deep but we could trace his course. He rose again within twenty yards distance of us, and lay sometime on the water...He continued some length of time, playing...From my knowledge of aquatic animals, and habits, and intimacy with marine appearances, I could not be deceived...His head is about three feet in circumference, flat and much smaller than his body. We did not see his tail; but from the end of the head to the farthest protuberance, was not far from one hundred feet. I speak with a degree of certainty, from being much accustomed to measure and estimate distances and length. I counted fourteen bunches on his back, the first one, say ten or twelve feet from his head, and the others about seven feet apart. They decreased in size towards the tail. These bunches were sometimes counted with, and sometimes without a glass...His motion was sometimes very rapid, and at other times he lay nearly still. He turned slowly, and took up considerable room in doing it. He sometimes darted under water, with the greatest velocity, as if seizing prey. The protuberances were not from his motion, as they were the same whether in slow or rapid movement...I should say he must be one hundred and thirty feet in length, allowing for his tail. That there is an aquatic animal in the form of a snake, is not to be doubted...The sketch or picture of Marshal Prince is perfectly correct. I could not, with my own pencil, give a more correct likeness.

Account 77: September, letter sent to *Philosophical Magazine*, from Oudemans (1892):

> 'The great Sea-Snake has been seen again towards the middle of September, in the bay of Massachusetts, and three yellow collars observed on his neck, which led some to believe it might be another individual and species.'

1820

Account 78: August, Nahant (Boston Harbour), Colonel Perkins, from Oudemans (1892):

> The creature was sighted 'several times' from the piazza of Perkins' house.

Account 79: 5 August, Swampscott (Boston Harbour), Mr's Reynolds, Lewis, King and Ingalls, from Oudemans (1892):

> I discovered...an animal different from any that I had ever seen before; he was lying on the surface of the water, which wsa at that time very smooth, and appeared to be about 50 or 60 feet long...He had a head about two feet long, and shaped somewhat like an egg, which he carried out of the water when he was moving. There were several protuberances on his back, the highest point of which appeared to be seven or eight inches above the level of the water. He was perfectly black.

Other details form these accounts mention that the UMO was 60 feet long and had 23 protuberances.

1821

Account 80: Summer, Nahant (Boston Harbour), from Heuvelmans (1968):

> A group of witnesses reported seeing a creature with five distinct humps about five feet apart.

Accounts 81 and 82: Sometime in the summer, Nantucket Island, MA, Francis Joy; and 2 August, Portsmouth Harbour, NH, Samuel Duncan, whaler, from O'Neill (1999):

> Reference to the sightings of 'an unknown creature.'

Account 83: 12 August, Nahant (Boston Harbour), Captain Beal and Mr. Sampson, mariners, and Mr. Hagins, from Oudemans (1892):

> His motion was slow and apparently playful, with his head raised from the water about three feet. The circumference of the animal was about the size of a common barrel; his head shaped like that of a horse and the protuberances on his back were about six feet apart...[The] Serpent kept in distinct view more than 30 minutes; his length appeared about 60 feet, but having no glass on board it could not be as ascertained with certainty.

1822

Account 84: Summer, Nahant (Boston Harbour), Perkins family, from O'Neill (1999):

> Seen from the family stone cottage.

Account 85: Summer, Nahant (Boston Harbour), many people, from Heuvelmans (1968):

> The beast was seen almost every day at Nahant in the summer.

1823

Account 86: 12 July, Lynn (Boston Harbour), Francis Johnson, fisherman, from O'Neill (1999):

> I heard a noise in the water, and saw, about four rods distant, something resembling the head of a fish or serpent, elevated about two feet above the surface, followed by seven or eight bunches, the first about six feet from the head, all were about six feet apart, and raised about six inches above the water. It then [moved] eastwardly, at the rate of five miles an hour, with an undulating motion, like that of a caterpillar; its color was dark like that of a porpoise. I firmly believe what I saw to be the animal hitherto described as the 'sea serpent'.

Accounts 87 and 88: 6, 7 August, Squam Bar and Sandy Bay, MA (Cape Ann), from O'Neill (1999):

> 'As many as fifty witnesses gave the usual description and reported firing upon it, but the balls did not seem to penetrate.'

Account 89: 11 August, Plum Island (17 kilometres north of Cape Ann), Ruggles family, from O'Neill (1999):

> [The family] were enjoying a swim on Plum Island beach in Newburyport when they were startled by the appearance of a huge animal. The creature came up a few yards from the shore, where just a few moments previously they'd been bathing. They then watched it for more than half an hour as it moved up the shore in the direction of Shad Cove, where they lost sight of it.

1824

Account 90: summer, Plum Island, MA, Ruggles family, from Oudemans (1892) and Heuvelmans (1968):

> Seen by 'the whole family' at a distance of 'about 100 feet.' 'The head was two feet long and of a brown colour. Mr. R. could distinctly observe the teeth in the mouth when

opened. He could not discern the tail, but several times, about thirty feet behind the head, he observed parts of the animal in an undulating motion.'

4.1.2 Anecdotal analyses

Given the absence of a 'smoking gun' in the form of bodies or conclusive photographs, the study of sea serpents is based on descriptions pertaining to sightings. As such, *sea-serpentology* is a form of maritime history, *not* of marine science. The debate regards just how much credence should be placed on the reliability of such eyewitness accounts. Cryptozoologists are very testy about this, believing that scientists, those, in their minds, self-imagined high priests of truth and wisdom, devalue all anecdotal evidence, and only operate under the confines of controlled experiments, which consequently thwarts genius and leads to entrenched protectionism of the status quo. This of course is nonsense, as even the most cursory reading of the literature on the history of science will clearly demonstrate. But the issue does remain, as to how scholars should regard the information contained within anecdotes which, by their very definition, are essentially personal stories and thus susceptible to all the vagaries of human perception and memory. Courtroom cases abound in which different people can view the same thing differently, and many famous instances exist wherein time can influence recollection (Loftus and Palmer 1996; Wright and Loftus 2008). Unusual objects or events, be they small-scale slights of magic (Wiseman and Lamont 1996) or purported world-transforming miracles (Ehrman 2017), generate conflicting accounts. Some scholars therefore argue that as anecdotes of unusual phenomena cannot be objectively analysed, they should not be part of the scientific process (Shermer 1997, 2003). However, there are numerous statistical approaches that can and are routinely applied to subjective data which are in common use in the social sciences. Furthermore, Paxton (2009) believes that when anecdotes about a particular natural phenomena, such as sea serpent sightings, are statistically analysed *in toto*, they can indeed be regarded as being data '*as long as it is realized that the raw data are reports of sightings subject to a variety of biases, and not necessarily accurately recalled sightings of animals known or unknown. Thus the conclusions should be constrained appropriately*' (his use of italics for emphasis). It is important to keep this important caveat in mind for the rest of this book.

Throughout the preceding historical overview, I have referred to the Gloucester Sea Serpent by such generic terms as 'creature' or 'serpent' in order to be consistent with the vernacular that has developed over time. I have, however, purposely, avoided using the pronoun 'he' rather than the more nondescript 'it.' A few others, when they have referred to specific cryptids, have cautiously used such appellations as 'X' (Gould 1934) or 'N' (Campbell 1997). The question of how to refer to the Gloucester Sea Serpent is not a moot point, as words that environmental scientists and scholars use do matter (France 1989, 1993, 2001a; Regier and France 1990). The same pertains to investigations of anomalous phenomena. 'Flying saucers' *per se* do not exist, whereas UFOs, or unidentified flying objects, most certainly abound, many being explained through weather-related causality; in which case the word 'flying' seems out of place. Champagne (2001) and Paxton (2009) both undertook statistical analyses of what they refer to as 'unidentified marine animals.' This is certainly an improvement to avoid falling into an *a priori* cryptozoological hole by default through using the name 'sea serpents' or 'lake monsters' as have many others (e.g. Coleman and Huyghe 2003; Harrison 2001). But given that there is a high likelihood that many sightings of aquatic cryptids derive from non-zoological phenomena, Champagne and Paxton's designation is also too limiting a term. As such, henceforth I will go one step further and alter the word 'animals' to 'objects' to be more inclusive. And so, as I have done elsewhere (France 2016a,b, 2017, 2018), what we are concerned with are 'unidentified marine objects,' or *UMOs*.

Focusing on the nearest reported distance (NRD) of observers to UMOs, as Paxton (2009) did, makes sense based on ecological sampling theory. For if UMOs represent misidentified animals, then the greater the distance between the observer and the UMO, the more likely it is that a mistake in identification could occur. Also, given that the detection of smaller animals is more uncertain compared to larger ones in any line-transect survey, the possibility exists for a positive relationship to be found between observation distance and estimated animal, or in this case, UMO, size. Other questions examined by Paxton included whether differences in NRD existed between shore- and boat-based eyewitness accounts, or were influenced by whether or not witnesses were identified or remained anonymous, as well as the length of time elapsed between when the observation occurred and when it was reported. What is particularly interesting is that Paxton, in his survey of the complete corpus of world-wide sightings, purposely excluded all distance data from the 1817 Linnaean *Report* about the Gloucester sightings as they were, due to being specifically collected as part of the Society's investigation, deemed to be too accurate for his purpose of investigating potential influences on observation bias.

The results found by Paxton were somewhat surprising. Contrary to what might be expected, most sightings were at close distances of less than 200 metres. Indeed, more than half of all sightings were within 100 metres of the observed UMO. This suggests that 'misidentification of inanimate objects or known animal species due to great distance was unlikely.' In other words, people were just as likely to posit a UMO as being a 'sea serpent' if it was observed close to or far away from them. Also, no relation was found between reported observation distance and reported length of the UMO. Imagined sea serpents were therefore estimated to be of 'monstrous' proportions irrespective of their proximity. There were no differences in the distances and lengths of shore- or boat-based sightings. This dispels the supposition that all UMOs are large animals endemic to deeper waters whom are incapable of inhabiting shallow regions of the coastal littoral. (An earlier analysis by Champagne (2001) of some of the same as well as additional data had found that most UMO sightings occurred near continental margins, though of course this might be a confounded result due to their simply being more people to observe animals in this area compared to the open sea.) A subsequent secondary-analysis of the same data (Paxton and Shine 2016) revealed there to be no evidence that consecutive sightings of the same UMO by the same observer (i.e. *intra*-variability) had a higher consistency in physical descriptions, such as estimated lengths, then did simultaneous sightings of the same UMO by different observers (*inter*-variability).

The question about the reliability of eyewitnesses has always been at the centre of debates concerning the possible existence of sea serpents, as, for example, the opinions of Sir Richard Owen about the *Daedalus* and other famous Victorian sightings (France 2017; Regal 2012), as discussed in Chapter 1. Oudemans (1892) begins his seminal *The Great Sea-Serpent* with a chapter on cheats and hoaxes into which he places dozens of accounts to preclude them form further consideration. Heuvelmans (1968), in turn, in *In the Wake of Sea-Serpents*, the book that can be said to have done more than any other to popularize cryptozoology, reduced the number of what he considered to be credible sightings from 587 to 358 'when the hoaxes, mistakes and vague reports have been eliminated.' In both cases, the egregious subjectivity displayed in considering whether a particular account either is or is not deemed trustworthy, makes one question the entire legitimacy of the exercise (Magin 1996), and forces one to conclude that these two compilers displayed considerable confirmation bias in their attempts to force all 'reliable' observations to fit into their own particular, preconceived theories. Due to these shortcomings, and because false identification may be the result of ambient conditions (illumination, etc.), emotions (fear, etc.), duration of the observation, and/or lack of qualification (education, experience, etc.), Champagne (2001) developed an objective way to evaluate the credibility of UMO sightings (Table 4.1). He considered a summed score of five, of a possible total of 11 points, to be the minimum standard for reliability of any particular sighting of an UMO. Of the 1247 observations that he examined from the entire corpus of UMO sightings, only 351, or 28%, met his screening criteria for reliability.

Paxton (2009), though he does not cite Champagne's work, concludes with a cautionary statement concerning the nature of conclusions drawn from eyewitness anecdotes due to inherent vagaries in the chain progressing from the presence of a phenomenon, through the act of its observation, to the memory and then recording of the event. That caveat acknowledged and understood, I do concur with the over-arching thesis that the careful examination and statistical analysis of anecdotal data from eyewitness might be useful in trying to understand historical sightings of purported sea serpents. As such, I undertook similar analyses of the Gloucester UMO data. Comparisons to other data sets are facilitated by the statistical testing of differences between proportional-frequency distributions, a simple and standard procedure often employed in exploratory data analysis, which I have successfully

Table 4.1. Criteria used by Champagne (2001) to establish a rating system of eyewitness credibility with respect to UMO sightings (one point awarded for each criterion except # 8 which received two).

1.	Observation reviewed by competent and unbiased investigator, preferably with life science training or having experience, and found to be credible.
2.	Observation by qualified person having life science training or occupational experience (e.g. fisherman or mariner), or with above average observational skills.
3.	Observation included a detailed account with physical description and/or behavior of the UMO, with notation about observational circumstances (e.g. weather, date, duration, etc.).[1]
4.	Observer with a known reputation for veracity, or with an occupational or societal position that would be harmed if sighting was determined to be fabricated.
5.	Sighting of UMO documented by more than a single witness.[2]
6.	Observation of UMO made at a distance of less than or equal to 62 meters (200 feet).[3]
7.	Observation was for a duration of more than one minute to reduce likelihood of misidentification.
8.	Observation included tangible, physical evidence (carcass or tissue sample, etc.).
9.	Observation substantiated by photographic or sonar evidence.
10.	Observation substantiated with corroborating evidence (changed landscape, footprints, damage to objects such as boats, etc.).

[1] 'Descriptions such as 'a big, snake-like thing' or 'it looked like an upturned boat' were not considered detailed.'

[2] 'It was assumed that if more than one observer witnessed the event, the opportunity for exaggeration or fabrication would have been diminished.'

[3] Based on Champagne's belief from his personal experience that this is the maximum effective observational distance with the naked eye.

used previously in this regard to generate broad conclusions of ecological import (e.g. France 1994, 1995a,b,c, 1996a,b; France and Steedman 1996; France *et al.* 1995).

The proportional-frequency distributions of observation credibility ratings for the Gloucester UMO sightings were significantly different (X^2 test, $P<0.01$) from those found by Champagne (2001) for all UMO sightings (Figure 4.6). Indeed, 63% of the Gloucester-Massachusetts-New York (henceforth, simply 'Gloucester') sightings had ratings of 5 or higher, more than twice that recorded by Champagne for his extensive data set. This highlights the unusually credible nature of the observations made about the Gloucester UMO. It also justifies the decision of Paxton (2009), who was interested in the general question of the vagaries of observer reliability for the entire corpus of UMO sightings, in purposely excluding those reports investigated by the Linnaean Society of New England. As well, this supports the argument made by cryptozoologists, such as Bauer (2013), who was testy about summarily being dismissed for engaging in 'abominable science,' that 'Any claim for objectivity in this work [i.e. Prothero and Loxton (2015)] is undercut as the book knowingly, willfully, and deliberately fails to discuss two of the very strongest cases. One of them, the Gloucester sightings by innumerable witnesses (O'Neill 1999), is by far the best-documented evidence that sea-serpents exist.' As will be shown subsequently, if the Gloucester UMO is held up as the gold standard upon which to place faith in the existence of sea serpents, then cryptozoologists are setting themselves up to be bitterly disappointed. In point of fact, I

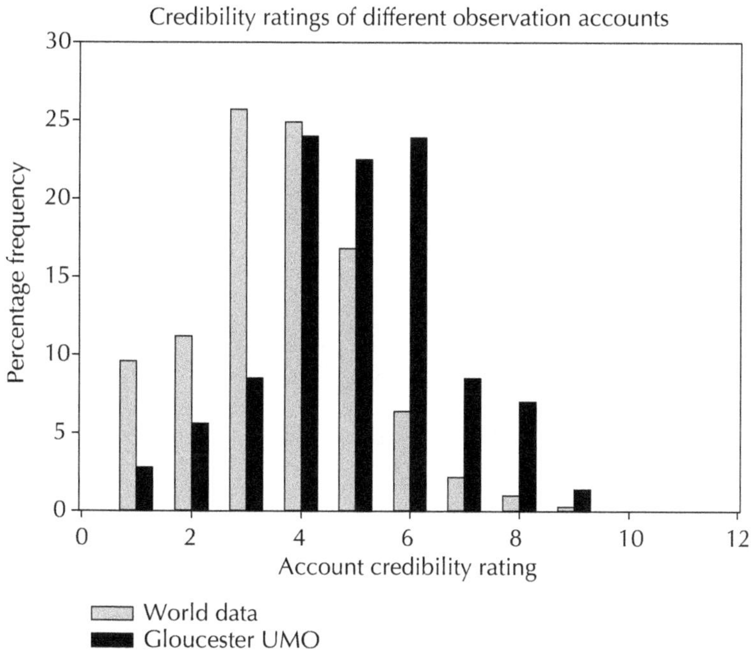

Figure 4.6. Proportional-frequency distributions of observations by credibility rating calculated using Champagne's (2001) methodology (see text). Data sources are for UMO sightings from around the world (n=1,247) (Champagne 2001) and for those from Gloucester (n=90).

will show that the Gloucester UMO does function as a gold standard, but actually to demonstrate just the opposite: an alternative, parsimonious explanation for certain sea serpent sightings.

It is worth mentioning one caveat about the reliability of the observations made of the Gloucester UMO, and that concerns the similarity of some of the testimonies collected by Justice of the Peace Nash for the Linnaean Society *Report*. This raises suspicion about the degree to which Nash might have been leading the witnesses. If so, it is worth noting that where the biases exist, it may be in the recording and reportage rather than the observations themselves. A case can be made that any weakness due to a conformity of testimonies resulting from over-zealous editing is more than balanced out by the astuteness of the Committee's questions. As well, it is important to note that the testimonies from August 1817 in the Linnaean *Report* represent less than a quarter of all those that I have collated and analysed herein.

Nearest reported observation distances for the Gloucester UMO ranged from less than a meter to 200 m (3 – 600 feet), the majority of sightings occurring between 0 and 20 m (66 feet). Paxton (2009) found that over 50% of reported sightings from boats for all UMOs were within 100 m (328 feet) of the observer. Although, as expected, some boat-based sightings of the Gloucester UMO were recorded at closer distances than those made from the shore, partition of data into these categories did not statistically influence the interpretation of subsequent analyses, and so they are grouped together. The same was true in regard to separating data based on the type of eyewitness, i.e. if they were identified as mariners or landlubbers, and whether or not a spyglass had been used in the observation. Overall, the

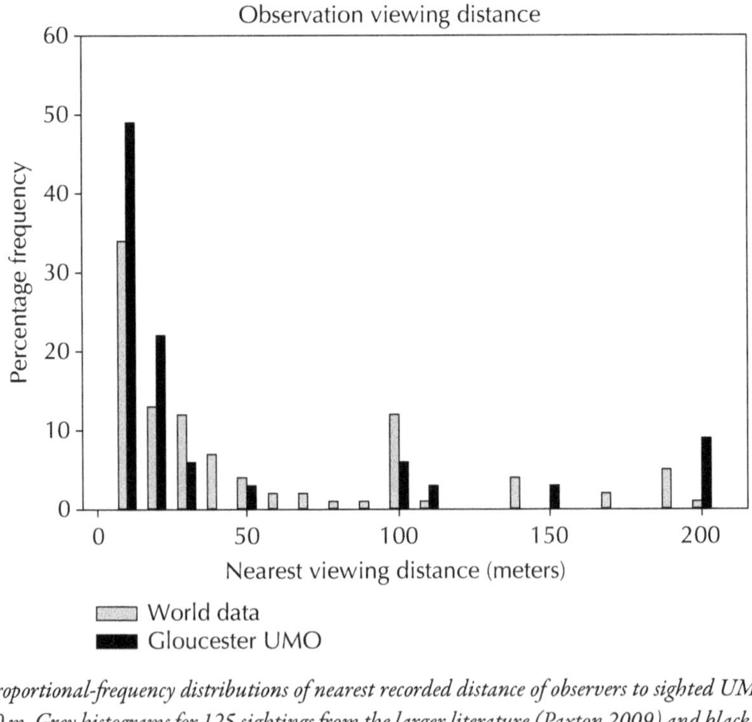

Figure 4.7. Proportional-frequency distributions of nearest recorded distance of observers to sighted UMOs within the range 0 to 200 m. Grey histograms for 125 sightings from the larger literature (Paxton 2009) and black histograms for 32 sightings for the Gloucester UMO.

Gloucester UMO was observed more frequently at closer distances than was the case for other UMO sightings (Figure 4.7). These differences were marginally significant (X^2 test, $P<0.10$), and therefore give credence to the accuracy of descriptions made about the Gloucester UMO, an issue of importance as it is upon these descriptions that I will base my own theory, to follow.

Regression analyses determined that no significant ($P>0.05$) relationships existed between reported observation distance and either the estimated reported length or number of segmented humps of the Gloucester UMO. There was also no significant ($P>0.05$) relationship found between reported UMO length and number of humps. For categorical data, t-tests showed that whether or not a head was detected, the back was likened to resembling kegs or barrels, or a marked flexibility existed in the body, were not significantly ($P>0.05$) influenced by observation distance. Also, same-time sightings by multiple eyewitnesses did not seem to be less consistent in terms of estimated UMO size then the few cases that existed of repeated sightings by the same individual (though the data for the latter were insufficient for comparative statistical analysis).

The present analyses therefore indicate that the reported observation distance has little utility for weighing the veracity of, and therefore for potentially screening, morphological or behavioural descriptions of the UMO putatively identified as the Gloucester Sea Serpent. In short, whether or not the observer was in a boat or on the shore, someone whose profession was sea- or land-based, a spyglass was or was not used in the sighting, what their observation distance was from the UMO, or whether the sighting was made by a single individual or group of people, had no bearing on the descriptions recorded. As such, all anecdotes, those recorded as legally sworn testimonies by certain personages of standing to the Justice of the Peace Lonson Nash for the official Linnaean Society *Report*, those from regular town-folk given to Charles L. Sargent for his pamphlet, those individuals interviewed by David Humphreys, or those obtained from various local newspapers, are meritorious for inclusion in subsequent interpretations of the natural history of the Gloucester Sea Serpent.

There are three descriptive attributes of the Gloucester UMO that were quantified by many of the eyewitnesses: (1) the estimated size of the presumed creature; (2) the number of irregular component body parts or 'protuberances' or humps; and (3) the estimated speed of locomotion.

Length estimates for the Gloucester UMO ranged from 10 to 120 feet (3-40 m), with the majority, 26% and 40%, respectively, being situated in two groups between 50 and 60 feet (17-20 m), and above 90 feet (30 m) (Figure 4.8), and with an overall median length of 68 feet (23 m). The only other regionally specific, purported sea serpent whose corpus of sightings have been similarly analysed, is the so-called *Cadborosaurus* ('Caddy') from the Pacific Northwest (Leblond *et al.* 2014). The frequency distributions of estimated lengths from eyewitness accounts for the two UMOs were significantly different (X^2 test, $P<0.01$), the Pacific Northwest UMO typically being smaller in size: i.e. only 9% of its reported lengths were observed to be more than 90 feet (30 m), and the median length was 34 feet (11 m).

Although more than half of the eyewitnesses of the Gloucester UMO noted the presence of a segmented body, only 18 actually counted the number of humps or protuberances. These counts ranged from 0 to 80, with the majority, 44%, being from 10 to 20 humps (Figure 4.9). For the Pacific Northwest UMO, about a third of the eyewitnesses noted the presence of coils or humps, but Leblond *et al.* (2014) do not provide data on the specific number of segments observed.

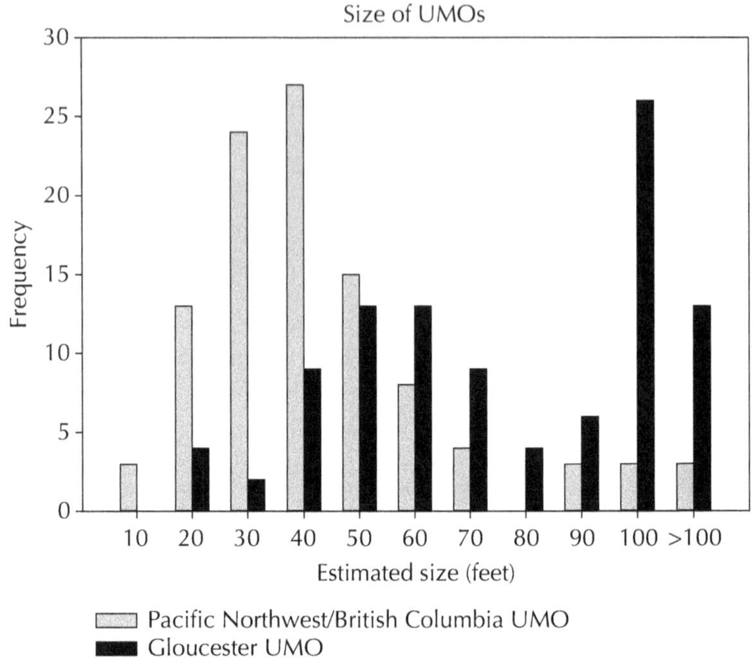

Figure 4.8. Proportional-frequency distributions of lengths observed for UMOs from the Pacific Northwest (n=79; Leblond et al. *2014) and Gloucester (n=53).*

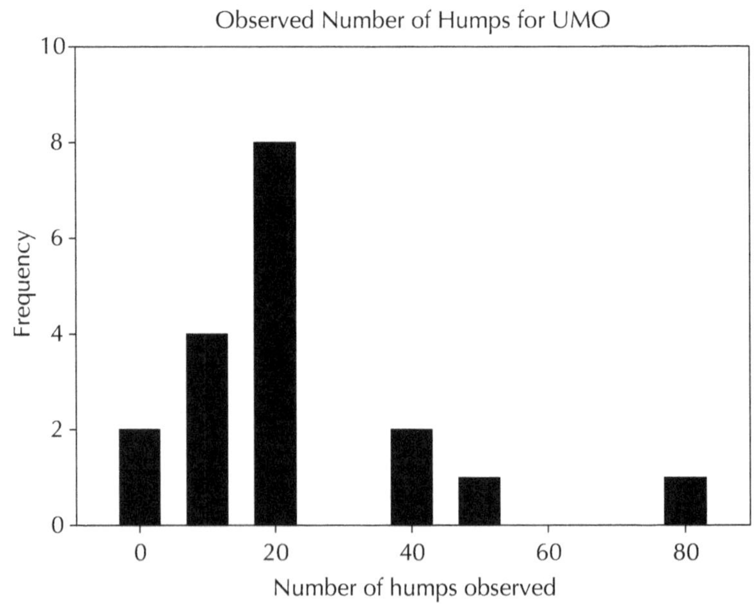

Figure 4.9. Frequency distribution of number of humps recorded by 18 eyewitnesses for the Gloucester UMO.

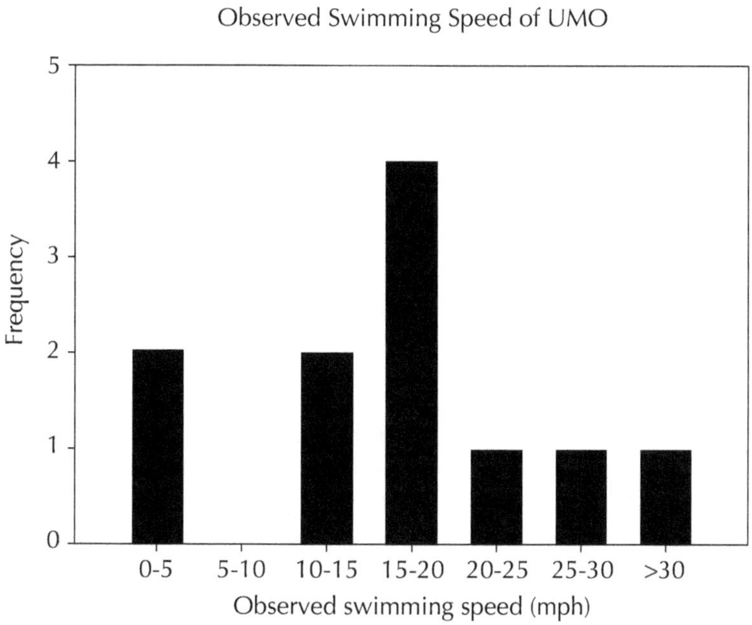

Figure 4.10. Frequency distribution of estimates of speed made by 11 eyewitnesses for the Gloucester UMO.

According to the eyewitness accounts, the Gloucester UMO, when not lying motionless on the water, displayed variable swimming speeds, ranging from 5 to 60 miles per hour (8 – 100 kmph), with most estimates being of a rate of between 10 and 20 mph (17 – 33 kmph) (Figure 4.10).

4.1.3 Assembling the composite cryptid

The Linnaean Society's *Report* (Davis *et al.* 1817) begins with the briefest of overviews about the 'repeatedly' observed UMO, offering only a few lines about its 'very singular appearance,' 'immense size,' 'wonderful rapidity' of movement, and 'jointed' shape resembling 'a number of buoys or casks following each other in a line.' In the *Report*, the absence of a comprehensive synthesis about the nature of the UMO, an inherent belief in it being a serpent, and the sporadic and incomplete collation of eyewitness reportage, has allowed for subsequent writers in the many years since to posit any number of theories as to the identity of the Gloucester Sea Serpent. These will be examined in Chapter 5.

In this section, I formulate a composite sketch of the famous Gloucester UMO based on parsing the eyewitness descriptions. Before that, however, it is useful to review how prominent natural historians at the time, as well as cryptozoologists since then, have attempted their own summaries of the famous Sea Serpent.

David Humphreys (1817) generated an 'aggregate result' of the 1817 sightings in his September 5th letter to Sir Joseph Banks:

> That his head and tail resemble those of a common snake: some, however, think the head more flat, like that of the turtle (tortoise) and adder, – that his colour is nearly black, –

> his body about the bigness of a flour-barrel – and his length from sixty to one hundred feet, or more. Some judicious people who saw him stretched out to his full length, by comparing the extent with the distance between two objects on shore, are inclined to believe the last mentioned estimate approximates nearest to the truth.
>
> For the rest, there is an extraordinary uniformity of opinion. All concur, in his having neither the gills, fins, scales, nor tail of a fish. All I believe, with one or two exceptions, represent his movement to have been like that of a caterpillar, but infinitely more rapid. They imagined, also, he moved faster *under*, than *above water*. There is, naturally, a considerable difference between the reports of those who saw him at rest, or in motion, as to his smoothness or roughness, as well as regard to the curvatures and bunches on his back. To some he appeared jointed, or like a string of kegs or buoys connected on a rope: to others, uniformly smooth and round.

The botanist William Danbdrige Peck, also regarded as America's first native-born entomologist, was, like me two centuries later, a professor of biology at Harvard University. He published his synopsis of the Linnaean Society *Report*, titled 'Some Observations on the Sea-Serpent', in the 1818 volume of the *Memoirs of the American Academy of Arts and Sciences*:

> The accounts of all these persons are very consistent; to the greater part it appeared to be straight, or without gibbosities or protuberances on the back; one person thought it had protuberances, but it seems probable that the upper flexures of its undulations, occasioned this opinion.
>
> Its velocity is variously estimated; by some it was thought to move a mile in one minute, by others in three, four, or five minutes. It has great lateral flexibility, as is shewn by its turning short and moving in an exactly contrary direction, advancing the head in a line parallel with the body; hence its undulations when under water and equally surrounded by the medium, may be either vertical or horizontal at the will of the animal. The judgment of its velocity, however, without knowing its precise distance and without instruments to observe it, is extremely liable to err.
>
> In the testimonies above referred to, the imagination seems to have had no influence, and we certainly know from them, that the existence of the animal to which they related is indisputable; we know that it moves by vertical undulations, at least while near the surface of the sea; that it is laterally as flexible as other serpents; and that its motion, at times, is very swift; but our knowledge is circumscribed by these limits. It is to be hoped, that if it again visits our shores, some successful means may be devised of taking it and presenting an opportunity of completing our knowledge of so interesting a link in the chain of animated beings.

The following year, Constantin Samuel Rafinesque published his paper, 'Dissertation on Water-Snakes, Sea-Snakes, and Sea-Serpents,' in the *Philosophical Magazine*, in which he comments upon the 'Massachusetts Sea Serpent':

> From the various and contradictory accounts given of this monster by witnesses, the following description may be collected. – It is about 100 feet long; the body is round and nearly two feet in diameter, of a dark brown, and covered with large scales in transverse rows; its head is scaly, brown mixed with white, of the size of a horse's and nearly the shape of a dog's; the mouth is large with teeth like a shark; its tail is compressed, obtuse, and shaped lie an oar. This animal came in August last into the bay of Massachusetts in

> pursuit of shoals of fishes, herrings, squids, etc., on which it feeds. Its motions are very quick: it was seen by a great many; but all attempts to catch it have failed, although 5000 dollars have been offered for its spoils.

Humphrey's and Peck's summaries seem reasonably close to the observed accounts from eyewitnesses, especially given the fact that they were largely working from the limited subset of those contained in the Linnaean Society *Report* and a few interviewed eyewitnesses. On the other hand, Rafinesque's summary contains some erroneous information that does not match reality. The difference between the syntheses is an important one in that it is an early manifestation of the remarkable divergence between the objective, academic study of UMOs with no favoured candidate established *a priori*, and the subjective cryptozoological model of attempting to force observational evidence to construe to a preconceived idea (i.e. confirmation bias) of what the creature is imagined to be. And like propagation of uncertainty or cumulative errors through successive mathematical operations, crypotozoologists commenting upon earlier cryptozoological theories that don't jive with their own, produce an increasingly unsteady house of cards. For example, Oudemans (1892), himself a believer in a single, one-size-fits-all, sea monster as a relic paleo-pinniped (seal) and not an enormous snake, is completely dismissive of Rafinesque's synopsis, stating 'We observe that Mr. Rafinesque gives here some characters to the Massachusetts's Sea-Serpent, with which we have met nowhere else, apparently only for the purpose of rendering his supposition more plausible,' before continuing with a point-by-point evisceration of Rafinesque's overview, in which he cannot resist the repeated use of exclamation points to dive the coffin nails home. These and other diverging interpretations will be examined in Chapter 5.

Rupert T. Gould was a lieutenant-commander in the Royal Navy, a radio commentator, and the author of a series of well-written and regarded books about aquatic anomalies, including the first authoritative account of the 'Loch Ness monster.' Rather than presenting a cacophony of hundreds of sea serpent sightings as did Oudemans before him, and as would Heuvelmans later, Gould, in his 1930 book *The Case for the Sea-Serpent*, zeros in on only the, in his mind, seven most convincing cases for the existence of such creatures, one being, of course, that from Gloucester. This is his factually accurate synopsis of the candidate sea serpent:

> Only one creature was seen...It was observed by many persons simultaneously, for periods ranging from a few minutes to two hours and upwards, and at distances varying from a few feet to a mile.
>
> It was seen at all times of the day, sometimes in rapid motion, sometimes at rest. When moving, it appeared to curve its back in vertical undulations; when at rest, its back seemed, at times, to be undulating, and at others smooth. It presented the appearance of an enormous serpent, of a black or dark-brown colour, its body (so far as this could be seen) having a diameter of something under three feet, tapering slightly towards the extremities. Its length was variously assessed at from seventy to one hundred and twenty feet. Its skin appeared smooth to most of the witnesses, but rough to two.
>
> The head was generally described as like a serpent's. Three witnesses deposed to seeing a long tongue projected almost vertically from the mouth. One witness only spoke of seeing the eye, bright and resembling that of an ox. No legs, fins, gills or mane were observed. There was great unanimity of opinion as to the monster's extreme lateral flexibility.

> It appeared to take little notice of surrounding objects, human beings, and even gunshots. It was not heard to utter any sounds. On two occasions it was seen, or believed to be seen, lying partly on the shore and partly in the water.

In order to counter the less-than-complete body of evidence used by Humphrey and Peck, and the overt confirmation bias of Rafinesque and Oudemans, I generated my own version describing the composite UMO of Gloucester Harbour and nearby New England-New York between 1815 and 1824, based on the prevalence of observed traits and citing the particular eyewitness accounts in the process. Creation of such a composite UMO is undertaken for purposes of distilling the anecdotal data. The question as to whether this composite UMO truly represents repeated visits by a *single* marine animal or is rather related to a *series* of different animals visiting the area over the years will be discussed in Chapter 5. Note: Due to excerpting the accounts for purposes of brevity, not all the information contained within each is presented in the texts that were shown herein; i.e. the identifications below are based on the complete, non-excerpted accounts.

The most notable anatomical feature of the sighted UMO was its considerable or monstrous length (accounts 3, 4, 44, 66, 89), being ascribed by eyewitnesses to be various sizes, as for example, slightly less than twenty feet (account 8), from forty to fifty feet (accounts 10, 11, 12, 18, 19, 27, 29, 31, 36, 40, 41, 49, 50, 65), sixty to eighty feet (accounts 6, 21, 22, 23, 24, 25, 34, 35, 69, 70, 73, 75, 79, 83), or around a hundred feet and larger (accounts 1, 7, 9, 13, 16, 20, 28, 32, 33, 37, 53, 55, 59, 61, 62, 64, 72, 76). The bulk of the UMO's length was composed of a series of irregular or uneven, loosely articulated joints, humps, or 'bunches' (accounts 9, 10, 15, 18, 22, 23, 24, 27, 29, 30, 37, 50, 59, 61, 66, 73, 74, 76, 83), sometimes identified as being of variable number, ranging from less than ten such (accounts 13, 80, 86), to around a dozen (accounts 34, 35, 65, 69, 70, 71, 76), to between twenty and fifty (accounts 1, 9, 32, 59, 72, 79), and even more (account 14). Significantly, a number of accounts specifically describe the UMO's body as resembling a string or a chain of buoys, corks, or kegs tied or roped together and floating on the surface (accounts 1, 6, 9, 17, 35, 36, 38, 48, 51, 65, 66, 71). This latter attribute is most important to my thesis to be expanded upon in Chapter 5.

The UMO was often observed to have neither a discernable tail, nor any appendages (accounts 1, 9, 18, 27, 34, 40, 59, 72, 76). The UMO's component body parts, including the premier one which was taken to be the head, were frequently likened to being of a size similar to those of a keg, cask, barrel, or firkin (a small container used for holding liquids or butter) (accounts 1, 9, 10, 12, 13, 34, 35, 37, 41, 51, 59, 62, 65, 66, 72, 83). Five witnesses describe from one to three pointed object(s) projecting from near the UMO's presumed head, which were waved about, sometimes backwards, before falling back into the water, and which resembled a prong, a spear, an harpoon, or a marlinspike (accounts 30, 35, 42, 43, 62). Again, this latter attribute is of critical importance and will be examined in detail in Chapter 5.

The most notable behavioural feature of the UMO was its rapid speed of movement (accounts 1, 7, 8, 10, 12, 13, 16, 19, 20, 23, 27, 30, 34, 35, 38, 41, 42, 43, 49, 50, 53, 65, 66, 69, 73, 74, 76, 86), which often produced a massive trailing wake (accounts 2, 13, 35, 38, 41, 42, 50, 65, 66, 69, 70, 74). Witnesses noted the remarkable flexibility of the UMO, which twisted and turned upon itself, coiling and uncoiling quickly with different portions of its body capable of moving side-by-side in diametrically opposing directions (accounts 5, 6, 9, 11, 12, 13, 20, 25, 34, 35, 37, 38, 46, 53, 72, 74). Elements comprising the elongated body of the UMO were observed to independently undulate in an up and down motion, as if floating on the surface (accounts 10, 12, 22, 30, 34, 37, 38, 41, 42, 43, 65, 70, 71, 75, 86, 90). These latter two attributes are critical to my thesis to be examined in Chapter 5.

Sometimes, the entire bulk of the UMO was seen to be pulled downward, rapidly disappearing into the depths with little or no warming, sinking as if a stone (accounts 9, 12, 13, 17, 18, 19, 44, 62, 70). At other times, the UMO was seen floating motionless upon the water, occasionally for extended periods (accounts 1, 15, 22, 24, 26, 31, 40, 46, 47, 76, 79). Intriguingly, the portion of the UMO observed above the sea surface appeared to be totally oblivious to its surroundings (accounts 10, 34, 41, 42, 43, 45, 46, 53, 62), including, most remarkably, to even being hit by musket shot, harpoons, and in one case, even a cannon (accounts 12, 36, 53, 57, 58, 87/88). Finally, the UMO was observed in shallow water very near the shore (accounts 5, 67, 89), and in a few cases, a portion of its body was actually seen stretched out upon land (accounts 32, 39). All these attributes, particularly the last two – imperviousness to injury and terrestrial sprawling – are key to my thesis developed in Chapter 5.

As will be discussed in the next chapter, there is really only a single explanation that can adequately fit all the aforementioned characteristics described in the sightings.

4.2 'THE MOST INTERESTING PROBLEM IN THE SCIENCE OF NATURAL HISTORY'

4.2.1 To catch a serpent...but then, perhaps not

Efforts were underway to capture or kill the Sea Serpent within a few weeks of it making its first appearance in Gloucester Harbour in August 1817, for all the reasons discussed previously in relation to the mandate and *modus operandi* of natural history of the time. By mid-August one newspaper reported that 'Shark hooks, variously baited, have been set afloat in the harbor' (Gould 1930). Heuvelmans (1968) described the situation thus: 'A price was put upon his head. Traps, nets and hooks were set for it, and an appeal went out to the Nantucket whalers.' And on August 16th, the *Essex Register* remarked, 'Whether he will be taken remains doubtful. The only hope seems to be from some powerful wound on the head, which he might receive from a discharge of small shot from musquetry or cannon' (*in* Moore 1977). However, the fact that the UMO seemed to be impervious to all manner of projectiles hurled at 'him,' did not bode well in this regard. Rewards were offered of between $350 and $5,000, with one rich gentlemen reported to have offered ten times the latter amount by the end of the month 'just to have in his possession the Neptunian frolicker's skin,' as Moore (1977) phrases it, before continuing:

> Gloucester by late August, 1817, found itself in a state of festive, chaotic siege. Shark hooks were baited with chunks of raw meat. Stray shots rang out in the August darkness. The Monster, for some reason, was said to have begun showing 'signs of caution.' Periods of several days elapsed between appearances. The Independent Chronicle remarked despondently, 'The chance for taking or killing this creature seems to be small.'
>
> A sharp, forlorn September wind began to ripple the codline nets in the harbor, and the Gloucester Monster was gone. The leviathan vanished from the environs of Gloucester as surely as if it had evaporated – like the June, July and August killing frosts of 1816.

These ersatz Captain Ahabs would have to wait until the following year.

As described by Colonel Perkins (*in* O'Neill 1999):

> Believing that the possession of the sea-serpent would be a fortune to those who should have him in their power, many boats were fitted out from Cape Ann and other places in the neighborhood of his haunts...Among others a Captain Rich...of Boston took command of a party, which was fitted out at some expense, and went into the bay, where they cruised along the shore two or three days without seeing the serpent.

Then, on 20 August 1818, Captain Richard Rich, eager to collect the $5,000 reward for the animal's capture or slaying, was at the front of a 'large whale boat' and accompanied by another two smaller dories, and engaged in his second day of hunting for the creature which had been seen off Squam Bar on the north side of Cape Ann. Having spotted his prey, a chase ensued for seven hours, with the boat being unable to catch up the rapid animal, until, as if oblivious to the imminent threat, the animal swam up to and dived under the boat. Then, Captain Rich, in a scene that seems out of fiction, a là fisherman Santiago or captains Ahab and Quint in their own quests, respectively, for a marlin, sperm whale, or great white shark, went in for the kill (from an unnamed newspaper, *in* Oudemans 1892):

> After several unsuccessful attempts, we have at last fastened on this strange thing called the Sea Serpent. We struck him fairly but the harpoon soon drew out. He had not been seen since, and I fear the wound will make him more cautious how he approaches these shores. Since my last, yesterday, we have been constantly in pursuit of him; by day he always keeps a proper distance from us, to prevent our striking oars. But a few hours since I thought we were sure of him, for I hove the harpoon into him as fairly as ever a whale was struck, he took from us about 20 fathoms of warp before we could wind the boat, with as much swiftness as a whale. We had a short ride when we were loose from him to our sore disappointment.

Another member of the crew, who had commanded one of the boats, recounted to his brother that the harpoon 'pierced' the creature about two feet, and 'drew the boat a considerable distance but went with such a velocity that he broke that part of the boat through which the rope passed and drew out the harpoon.' Clearly it was a very powerful creature. The crew member further stated that the creature 'has no scales on him, and no bunches on his back, but his skin is smooth, and looks similar to an eel' (to which a skeptic might ask as to whether this resembles other eyewitness descriptions of the UMO).

That brief 'Nantucket sleigh ride' must have wetted Rich's appetite, for he diligently persisted in his quest for more than three weeks, until an article in the *Boston Weekly Messenger* reported that on the 6th of September, the intrepid captain and his crew finally met with success in catching an animal, 'which, from its singular appearance in the water, they had been led to believe was the Sea Serpent so described.' The article, using language implying an unwillingness to believe in the landing, went on to state the assurance of 'some' of the crew 'who either believed or wished to believe that this was in fact the animal which has been the object of so much curiosity and speculation.' Flush with 'men against the sea' Yankee pride, wherein the monstrous leviathan of worldwide renown had been outwitted and brought to submission through the courage and ingenuity of local seamen, thousands flocked to see the specimen put on display on a Boston wharf. Imagine then the great disappointment of the public when it turned out that Captain Rich's sea serpent was in fact nothing more than a lowly horse mackerel, or a tunny; what we today would call a giant Bluefin tuna (*Thunnus thynnus*). That 'disappointment' would soon shift to anger.

Oudemans (1892) presents a letter from Andrew Norton, a Massachusetts Cantabridgian gentleman, to an acquaintance in Sweden that captures the mood of excitement at the discovery. I give the quotation in its near entirety for the additional glimpse it reveals of the ambitious and competitive spirit, as well as ego, characteristic among naturalists of the day concerning sea serpents.

> Last Friday morning upon going to breakfast at Dr. Ware's, I found there the papers of the day, in which was announced the most interesting fact, that the Sea Serpent had been taken by the expedition fitted out for that purpose...He was said to be 120 feet long, and the Board of Health had sent down two boats to stop him in the Harbour. After talking about it all breakfast time, I immediately went to Reed's stable, got a horse and chaise, put a news-paper in my pocket, rode to [Harvard University] Professor Peck's, showed him the paper, and offered to carry him into Boston, and to procure a boat to go out with him into the Harbour, that he might examine it. He was not well, and said at first that he could not go; but gradually grew warm upon the subject, and concluded at last that it would never do for him not to see it. When I had fairly got him into the chaise, his spirits rose with the exertion he had made, with the thoughts of the memoir and letters which he should write, and with the triumph which he anticipated over the Linnaean Society and their 'diseased black snake', as he contemptuously called it (meaning the small serpent, killed near the shore at Gloucester); for he pledged himself that we should find that the sea-serpent had no bunches on his back. I too anticipated with great satisfaction the honorable mention of me, which his gratitude would induce him to make in his memoir upon the subject, and expected confidently to float down to posterity behind Mr. Peck, upon this enormous animal.
>
> We entered Boston, and rode immediately to the end of Central Wharf to the store of a Mr. Rich, who had fitted out the expedition. The first person we saw was Judge Davis, whose countenance foreboded evil. [One can only wonder at the strained relationship that must have existed between the judge and the professor who had humiliated him vis-à-vis the fiasco of the baby sea serpent the preceding autumn.] His first words were to inform us that we had come in to be disappointed, for that the serpent was not taken! (I am not in the habit of using notes of admiration, but the present occasion seems to require one). The sailors, however, armed, as he said, that they had taken some most extraordinary fish of very large size, which he was going to see. I had little appetite left for seeing extraordinary fishes, but went to accompany Mr. Peck.
>
> We preceded to a wharf to the South End, and making our way through a crowd, obtained admission into the dark lower room of a store where we found a considerable number of other gentlemen waiting. After some delay the fish was dragged in from the small vessel in which it had been brought, wrapped in a sail. As soon as it was uncovered and fairly exposed to view, it was pronounced by all who knew any thing on the subject to be nothing but a Thunny, or Horse Mackerel, of a common size.
>
> We had been gradually prepared for the disappointment, so that the shock was not so great as you might suppose. The report in the morning's paper had arisen from a *mystification* performed upon the person who brought it to Boston, by the crew of the vessel engaged in the expedition. The sailors who dragged in the fish were part of this crew; and instead of their being tossed over the wharf into the water, by way of

> punishment for their imposition, and to teach them better morals, as they infallibly would have been by any mob out of Boston, there was actually a collection made to reward them for their trouble in taking the fish and bringing it to exhibit. This fact, I think, deserves to be recorded for the honor of Boston, and particularly of us gentlemen present.
>
> I have only to add that if you should learn that any one of the German literati is writing a volume upon Sea-serpents, I beg you will assure him, that we do not consider the circumstance, connected with the deception just mentioned, as affecting the evidence before obtained for their real existence.

Irrespective of the imagined opinions of European naysayers, for the general public this was just one further piece of evidence too many. It demonstrated that those of Gloucester believing in sea serpents were suffering from the same form of egregious gullibility or collective delusion that had afflicted their ancestors in nearby Salem and Danvers two centuries before with respect to witches. At least in these more sophisticated times of the early nineteenth century, no one would be hung or pressed to death, but still that did not mean that the eyewitnesses and promulgators of such fictions should be treated with anything but derision and disdain.

Alternatively, Charles Sargent captured the feelings of many of his Gloucester neighbours whom continued to believe in the existence of the Sea Serpent, in the concluding pages of his unpublished pamphlet, penned during the autumn of 1818:

> The foregoing in addition to the testimonies taken last year by the Linnaean Society one would suppose enough to satisfy any reasonable person of [?] having been seen by numerous respectable citizens, a strange fish of the Serpent kind from 50 to 120 feet long and any person having once brought themselves to believe such an animal to have been seen as herein described cannot readily admit a fish quite common and but 9 feet long to take place of the serpent in their imagination. We have no tradition of there ever being any such serpent before in our waters as the foregoing witnesses describe, but the Horse mackerel or Albicore have been repeatedly in our waters and sometimes caught as large as the one taken by Captain Rich.

Captain Rich's specimen therefore fuelled rather than resolved the underlying mystery behind the Gloucester Sea Serpent.

4.2.2 The controversy… 'But for a tuna, Richard, a tuna!'

Despite the novelty of seeing a giant Bluefin tuna at a time when they were rare enough to warrant being displayed to popular attention in museums (Hancock and Reilly 2017), the controversy began immediately. A debate raged about the validity of the catch being the self-same Sea Serpent that had been sighted by hundreds the previous year in Gloucester Harbour. Given the ensuing scandal, including the calling into question the veracity of the evidence presented by the major player in the affair, Captain Richard Rich, one wonders if any of the more erudite smiled or even chuckled at the historical parallelism in relation to the famous act of selling out by the Captain's earlier namesake, the first Baron of Leez (recently voted as the worst Briton of the Sixteenth Century by *BBC History Magazine*; i.e. remember the brilliant line and diss 'But for Wales, Richard, Wales!' in Robert Bolt's marvellous *A Man for All Seasons*). Given the long history of whales being misidentified as sea serpents (e.g. France 2016b; Galbreath 2015; Paxton *et al.* 2005), the parallelism is indeed humorous in that both Richs were certainly enriched for their efforts. History agrees that the former Richard Rich perjured himself for personal gain; the question here is whether the latter Richard Rich conflated a fish into a sea serpent for the same.

In a September 10th article in the *Boston Weekly Messenger* (*in* O'Neill 1999), editor Nathan Hale expressed his disbelief in the tuna-cum-sea serpent theory:

> The inquiry naturally arises, can this fish, or any number of them, be the monster so often described as a Sea Serpent? We answer decidedly no...The existence of some remarkable animal in our waters last summer, particularly near Cape Ann, was proved by the most satisfactory Account...The descriptions which we have had this season of the Serpent have been less consistent and satisfactory, and undoubtedly often exaggerated. But neither these exaggerated descriptions nor the error of persons who by mistake have been pursuing what had nothing of the remarkable and characteristic appearance of the Sea Serpent ought to lead us to suspect all former Account.

To support his case, the editor included a letter from a mysterious correspondent known to posterity by only by the James Bond-ish designation 'M.' Not without reason, did Heuvelmans (1968), in his brief review of the entire episode, use the words: 'But the plot soon thickened.'

> Since the great excitement occasioned last summer by the appearance on our coast of an animal supposed to be a marine serpent, public expectation has been on the watch for the return of so extraordinary a visitor. Whether the same animal has been this year applied to objects unlike the original one, is a point on which opinions differ. It is not surprising that to persons to see this aquatic monster, almost very unusual marine phenomena should be identified with the expected serpent...
>
> As we may not this season be gratified with a nearer interview with his serpentine majesty than through the person of his representative the Thunny, it may be well to examine how far the pretensions of the latter give him the right to be considered a legitimate connexion of the former.

'M.' continued, stating that tuna feed on herring and other bait fish, are found in the coastal littoral, and are known to swim in 'vast shoals with great swiftness.' However, because tuna are less than four metres in length, with a convex shape, 'it does not seem probable that a single fish of this kind could produce, by any means, appearances like those ascribed to the sea serpent...The correspondence of the two appears specious.' Next, 'M.' reviewed the observations made of the sea serpent (*in* O'Neill 1999):

> Everyone allows that the phenomenon at Cape Ann harbor last summer had a head. This was carried several feet above the water; it was seen by many, accurately described by a number, and actually fired at, from the distance of about a rod [i.e. 10 metres]. Now as feature of this kind could not well belong to a train of separate animals, unless one travelled in the air and the rest in the water. Again, the Gloucester Serpent was often seen at rest, as well as in motion, with great length of back uncovered by the water. This was repeatedly described as being 40 or 50 feet in length, not only by the spectators at Cape Ann, but by Judge Herttell and others of New York, who reported his appearance in Long Island Sound.

All of which led 'M.' to the conclusion that despite the frequent appearance of shoals of tuna in Massachusetts waters, 'it is hardly probable that in a maritime town like Gloucester, a great portion of whose inhabitants are sailors, familiar with every sea, that any common marine phenomena should have passed as new.'

In a fair minded – and one might suppose also mercantile – gesture, editor Hale arranged for Captain Rich to state his side of the story in the 17 September, 1818 issue of the *Boston Weekly Messenger* (*in* O'Neill 1999), which was preceded by his own reconciliatory comments: 'Capt. Rich with a very laudable spirit undertook the arduous and apparently hazardous enterprise of drawing the monster from his hiding place, and of solving, if possible, an important problem in natural history. If the public have been disappointed by the result of his enterprise, it is not his fault. On the contrary he ought to have the credit of the best intentions, as well as of skillful and bold exertions.'

Despite this, Hale preceded Rich's account by reprinting a synopsis of testimonies of the previous year's eyewitnesses from the Linnaean Society *Report*. If there is any doubt about whether this was done as a way to stack the deck against the hapless Rich, as well as to showing his own cards, this was removed by Hale's somewhat disingenuous statement:

> We leave it to our readers to conclude whether the above Account, made deliberately on oath by men of respectability, is utterly false and groundless. If so we should be glad to know on what grounds human Account is to be credited. We have no doubt that Capt. Rich has taken an animal which he and others believed to the Sea Serpent; but whether he has taken the animal described by the above deponents, we entreat our readers to judge for themselves.

In ignorance of the contextual framing that would precede his account published in the newspaper, Captain Rich ably described his capture of the purported Sea Serpent:

> Sufficient time having been elapsed for the public mind to cease fermenting, I now offer my statement facts, without comment leaving it for others to judge of the good or ill it may produce.
>
> Having been fully satisfied in the belief of the existence of a Sea Serpent (although I must acknowledge, I at first doubted), and feeling confident that the means were within the power of man, if rightly used, to overthrow this monster – I determined on a trial – so proceeded to Gloucester and completed my expedition. Presuming I had a formidable foe to contend with, I was well prepared for attack. While selecting my crew I took none but men of respectability and integrity, and among the whole of eleven, eight had seen the supposed Serpent. Some of these had made oath to the account already published of his existence, appearance and character. I selected such men, in order that I might not be deceived, should he make his appearance, having never seen him myself.
>
> When all was in readiness, information was brought us that the Serpent was off Squam Bar. We immediately proceeded thither, where on our arrival we were assured he had been seen, and that we might expect his appearance whenever it should be perfectly calm. The next day brought with it dead calm, with the much wished for appearance of the Serpent. My crew all agreed to a man that what we saw was the 'supposed Serpent.' Which had been seen both at that place and at Gloucester Harbour.
>
> I was perfectly satisfied, so precisely did it answer the description that had been given of him, and had I never approached nearer, I could have given Account upon oath that

> I had seen a Serpent not less than 100 feet in length. We did not keep at distance 'and wonder at what we saw.' Our object was to take it so we accordingly gave chase...
>
> After many attempts we at length fastened to him. We soon found that he possessed great strength, but he reappeared and we succeeded in taking him. We felt convinced that we had caught the creature which had caused so much wonder. I brought it to Boston, not doubting that those interested would be satisfied with what I had done...
>
> If I am asked 'how is it possible for a Fish like this to produce such a wonderful appearance', I can only answer: His peculiar movement and his velocity produced a greater deception than I had ever saw before, and that describing his body as being like kegs fastened together, struck me so forcibly, that had I not followed it up and discovered the deception, I should have given my Account to the long list already given, of the existence of a Sea Serpent on our coast.
>
> I now take my leave of the public, hoping they will do me the justice to say that I used no deception.

Rich's words will be parsed in the next chapter. Here it is sufficient to note the captain's obvious frustration at having his integrity impinged, and in some cases, even maligned. One feels for him, as it can be difficult to assume the role of an iconoclast. With his final statement, 'I used no deception,' Rich, true to his stated desire, fades from history. He was very much mistaken, however, in his opening assertion that sufficient time had elapsed 'for the public mind to cease fermenting' on the subject.

One can imagine the dozens of the heated dockside conversations going on around New England at the time. Representative of many skeptics, Sargent (1818) would have none of Rich's theory: 'Tunny fish have been repeatedly seen last year and this year round our shores,' he writes in the concluding sentences in his pamphlet, 'and no doubt many people when seen the wake of this fish at a distance have mistaken it for the serpent but any discriminating person could readily perceive the difference between the agitation of the water and a real substance.' This 'difference' will be discussed in the next chapter.

The fact that the UMO returned to its old haunts the following year confirmed to believers in its existence and that Captain Rich's captured tuna had nothing to do with the Gloucester Sea Serpent. After the flurry of sightings in Boston Harbour in 1819, the *New York Daily Advertiser*, published the following editorial:

> We congratulate the public on the safe arrival of the 'Old Serpent' of the ocean at his usual station in Massachusetts Bay. This may now be fairly considered the monster's summer residence, and will probably be repeated from year to year, unless some hardy whalemen should succeed in fastening a harpoon to his scaly side...We have no hesitation in acknowledging our obligations to the 'Snake' for this visit; and hope at the same time that no evil-minded person will endeavor, by fair means or foul, to change this formable animal into an insignificant horse-mackerel – at least without first satisfactorily accounting for the twenty-three bunches on his back of the size of a ten gallon keg: it being physically impossible to place so many of them and of such size, upon the back of a fish not more than ten feet in length.

One cannot help but wonder what Captain Rich might have thought about being considered 'evil-minded.' Natural scientists continued to debate, in a more polite vein, the significance of Rich's creature. A contribution in the January 1819 issue of *Philosophical Magazine* by Professor Thomas Say from Philadelphia notes that he had received a letter from a physician friend which included the statement that the creature struck by Captain Rich's harpoon 'was acknowledged by all the crew to be the veritable Sea-serpent (and that several of them had previously seen and made oath to)' before continuing 'but when drawn from the water, and full within the sphere of their vision, it proved that this serpent, which fear had loomed to the gigantic length of 100 feet, was no other than a harmless Tunny (*Scombrus Thynnus*) nine or ten feet long!' This retelling is interesting in that all the individuals involved in the capture, including notably those whom had seen the UMO previously, shared a common belief that the creature they had pursued and harpooned was indeed the selfsame Sea Serpent.

The following year an article by Harvard Professor Jacob Bigelow was published in the influential *American Journal of Science and Arts*, edited by the noted (and unfortunately named) Yale University chemist, Benjamin Silliman. Bigelow expressed concern about how the events were being portrayed: 'I regret that many scientific journals in Europe have in good earnest treated of the absurd story of the Great Sea-Serpent, which is nothing but a result of defective observation connected with an extravagant degree of fear.' Coming from Bigelow, he of the baby sea serpent fandango, the phrase 'defective observation' in relation to Captain Rich and his crew is more than a little rich (pun intended). Bigelow continued to argue for the existence of the Gloucester Sea Serpent against its naysayers: 'This disappointment of public curiosity was attended at the time by a disbelief of the part of many, of the existence of a distinct marine animal of the serpent-kind, or of the dimensions and shape represented by the witnesses of Gloucester and elsewhere.' Ever the optimist, he concluded: 'It is hoped that the unsuccessful termination of Capt. Rich's cruise will not deter others from improving any future opportunities which may occur for solving what may now perhaps be considered the most interesting problem in the science of Natural History.'

In 1829, noted New York physician, naturalist, and sea serpent skeptic, Samuel L. Mitchell, in a comprehensive article in Silliman's *American Journal of Science and Arts*, offered some synthesizing comments on the Gloucester phenomenon, which he deemed to be characterized by a series of 'mistakes, deceptions and willful perversions on the subject.' After viewing the carcass in the New-England Museum in Boston, where it was displayed during September 1818 along with other curiosities that included a live sunfish, a live alligator, and several live turtles from the Ascension Islands, Mitchell confirmed that it was indeed a giant tuna, and offered the following comments of great significance on the fruits of Rich's efforts: 'After capture of the fish, the persons who, when they saw him in the water, declared positively that he was the sea-serpent, now changed their minds, and swore he was not.' So, it seems that the crew themselves recanted their initial suppositions about the alleged sea serpent. Next, Mitchell writes: 'At length the man of successful exertion [by which he means Captain Rich] arrived with his prize; and unexpectedly and unfortunately drew upon him the displeasure of his employers for attempting to impose upon them a *Horse-Mackerel* (as they call it) for a *Sea-Serpent!* He told me the story himself.' The reference to the 'story' being told raises a question in that it could be interpreted as implying, in a conspiratorial manner, that the Captain confided to Mitchell that he had 'impose[d]' upon his 'employers' (which must mean those putting up the reward money) a hoax. Is this what Mitchell meant, in the most circumlocutious way, by using the word 'story'; i.e. as in a work of fiction, or rather was it just another way to offhandedly refer to a factual 'account'?

Three decades after the events, Colonel T. H. Perkins, himself an eyewitness to and believer in the Gloucester Sea Serpent, reflected back upon the controversy in an 1848 article in the *Boston Daily Advertiser* that was reprinted in the international journal *Zoologist* in 1849 (the timing inspired by the famous *Daedalus* sea serpent sighting that was then gripping worldwide attention). He began by criticizing Captain Rich and his accomplices who brought such ridicule to Yankee sensibilities, going as far as to accuse them of lying: 'As it happened, a circumstance took place which did not do much credit to the actors in it, but which served to fortify the unbelief of our southern brethren...With a view, however, to keep the joke from themselves, they [Captain Rich and his crew of whalers] determined to throw or attempt to throw it upon others, though at the *expense of truth!* ' Perkins casts dispersion on how visitors were 'induced' by advertisements to view the sea serpent 'or what had been mistaken for one,' and how this led to the disappointment of thousands. For though a 'great natural curiosity,' a giant Bluefin was in the end only a big fish; and therefore nothing more than a big yawn. His most serious concern, however, was that, for many, this 'joke' had put to bed the idea in the existence of sea serpents (*in* Gould 1930):

> Those who had declared their disbelief of the existence of the Sea serpent amongst ourselves were delighted to find their opinions were confirmed, and gave themselves great credit for their judgment and discrimination. The report spread from Boston to New Orleans, that what had been thought by some persons to be a sea-serpent had proved to be a horse macquerel, and even those who had been believers now supposed that those who had reported that they had seen the serpent had either misrepresented or had been themselves deceived.

Perkins, from the perspective of hindsight, considered that he and his ilk of believers had had the last laugh: 'As no report of the snake having been seen after the capture of the macquerel was made, during that year, Captain Rich had the laugh with him, until circumstances, which have transpired since, have borne rather against him. Thus much for the transactions of the past years.' He is referring, of course, and not without due reason, to the continued sightings of the UMO that occurred in years subsequent to Rich's capture of the titular Sea Serpent. This enigma will be discussed in Chapter 5.

Chapter 5.

INTERPRETIVE HISTORY OF THE GLOUCESTER SEA SERPENT

5.1 NATURAL HISTORY AND ILLATION

Rafinesque Schmaltz, in his 1819 *Dissertation on Water-Snakes, Sea-Snakes and Sea-Serpents,* sidesteps the scandal over the misidentified deformed snake while focusing on the giant animal observed in Gloucester Harbour. In so doing, he becomes the first to provide the Great Sea Serpent with a scientific name: 'It is evidently a real sea-snake, belonging probably to the genus *Pelamis*, and I propose to call it [one wonders if he heard a building drumbeat here] *Pelamis megophias*, which means Great Sea Snake Pelamis.' 'It might, however,' he continues, 'be a peculiar genus; in that case the name of *Megophias monstrosus* might have been appropriated to it.'

Ducrotay de Blainville's opinion, voiced in an 1818 article in the *Journal de Physique, de Chimie et d'Histoire Naturelle* (*in* Oudemans 1892), is that, while he agreed that there was 'an animal of very great length, very slender, and swimming with rapidity, in the sea near Cape Ann', 'that it is a true snake, this is doubtful.' He particularly takes issue with the assumed, and unlikely, amphibious nature of laying eggs on the shore.

For Harvard professor, William Peck, 'The testimony is ample of the existence of such a serpent, in the portion of the Atlantic which washes our shores.' He praises the men of Gloucester for their 'fair and unblemished character' which 'renders undubitable' earlier evidence from the region (Peck 1819):

> In the testimonies above referred to, the imagination seems to have had no influence, and we certainly know from them, that the existence of the animal to which they relate is indisputable; we know that it moves by vertical undulations, at least while near the surface of the sea; that it is laterally as flexible as other serpents; and that its motion, at times, is very swift; but our knowledge is circumscribed by these limits. It is to be hoped, that if it again visits our shores, some successful means may be devised of taking it and presenting an opportunity of completing our knowledge of so interesting a link in the chain of animated beings.

Because the UMO was seen in Long Island Sound, Peck hypothesizes that it was progressing southward on migration, and 'may pass the winter season in Mexico or South America.'

Bostonian Dr Francis Boott, whose brother had seen the UMO off Nahant, unequivocally believed in the existence of the Sea Serpent. A letter of his to the naturalist Sir William Jackson Hooker was published in 1827 in the *Edinburgh Journal of Science* (*in* Oudemans 1892): 'The question as to the real appearance of a large serpent off the coast of Massachusetts, was put to rest by that publication [i.e. of the Linnaean Society]. There could be no doubt of the fact, and the testimony of thousands who saw the animal *for one or two years afterwards*, must have been sufficient to satisfy the most incredulous.' Hooker, in same edition of the journal, wrote the following:

> When we remember the numberless impositions concerning Natural History, which at various periods have been detected, it is not surprising that doubt should be a principal, nay, a necessary, qualification of the student of Nature. Yet we cannot but think that the scientific world in general has been too incredulous concerning the sea-serpent, seeing the mass of concurrent testimony which has been adduced to prove its existence. It is certainly true that vague reports had been spread abroad with regard to this enormous animal long ere any just foundation was afforded for them, and indeed before we has heard of any who professed to have seen it. This may have very far conduced to produce that skepticism which now is perfectly unwarrantable. We are so accustomed, whenever the subject is introduced in conversation, to couple it with the preposterous fables of the *Kraken*, that it would be extremely difficult to break down the barriers against belief which prejudice has so long assisted to support. The accounts of the most credible witnesses have thus been rejected, although...the generality of them have been taken under oath.
>
> So many wonderful discoveries, both in the arts and Sciences, have been made within the last century, that it is astonishing how the existence of the sea-serpent has been supposed either so marvellous or impossible. Time has satisfactorily proved the veracity of Bruce [I assume that this refers to James Bruce and his claims about his exploratory travels in Ethiopia], and we must leave it to time to do the same office with reward to the beholders of this 'wonder of the deep.' Is this monster more disproportionate to the extent of the sea than the elephant to that of the land? Or, it may be asked, has it a solid bulk...nearly approaching that of the whale?...While the recent discoveries of the *Plesiosaurus* and *Megalosaurus* have made demands upon our powers of credence far greater than the *serpent*, the descriptions of the latter animal have received very little trust, and even ridicule and contempt.

The problem, Hooker goes on to suggest (Oudemans 1892), is that people do not object to the size of the reported creature so much as they mistrust the validity of the evidence. Nevertheless, it is his opinion that the Linnaean Society *Report* provides the best evidence for the existence of the creature: 'The unadorned and unexaggerated style in which their statements were worded is of itself perfectly sufficient to win over to all unqualified trust.... – facts which have already completely satisfied some highly scientific gentlemen, who before were entirely skeptical.' As a result, the sea serpent 'can no longer be considered in association with hydras and mermaids, for there has been nothing said with regard to it inconsistent with reason.' 'We cannot suppose,' he concludes, 'that the most ultra-sceptical can now continue to doubt with regard to facts attested by such highly respectable witnesses.'

In a paper entitled 'The History of Sea-Serpentism,' read before the New York Lyceum in 1828 and published the following year in the *American Journal of Science and Arts*, noted physician, zoologist, and popularizer of science, Samuel L. Mitchell, is one of the first literati to cast doubt on the subject. He refers to 'the alleged monster of the deep' that haunted the coast around Gloucester, and notes that, in the absence of its capture (thereby dismissing Rich's efforts and claims), the evidence for the existence of 'the enormous reptile' consisted only of affidavits 'of what the several deponents 'believed they had seen.' On the surface such solemn declarations may seem to have presented a strong case. Because 'there was nothing better to show than those statements upon paper, which were, in no sense of the words, proofs of the fact, but merely expressions of the opinions formed by the deposing witnesses of what they had observed in the water,' Mitchell remained unconvinced. 'I who was a believer in the first instance, was gradually sworn into scepticism, which finally ended in incredulity,' he states.

Mitchell (1829) goes on to mention meeting up with General David Humphreys during the 'stage of panic' that enveloped Gloucester. Humphreys requested that Mitchell read a draft manuscript of his series of letters which were about to be sent to Sir Joseph Banks, President of the London Royal Society. Mitchell states that Humphreys' purpose in scripting the letters was to 'take the lead of the Linnaean Society, and to acquire the honor and glory of making the wonderful intelligence known first to the scavans [i.e. wise persons] of Europe.' Mentioning that the interview was 'very pleasant,' Mitchell states that he purposely remained silent in the face of Humphrey's belief in the monster. He agreed to submit the manuscript to a local bookseller 'to cause it to be put to press without delay,' as Humphreys had to journey to New Haven, where one presumes there were to be no printers. About this encounter, Mitchell writes, somewhat cattily, that 'I made a contract on his behalf, and directed the proofsheets to be sent to him there. I had a lucky escape from an association with the extraordinary creature.' Mitchell then follows this by a quick diss of the Linnaean Society's attempt to foist the deformed common snake as being the progeny of the giant UMO believed to have been observed in the harbour.

The bulk of Mitchell's paper is occupied by a review of seven tall 'tales' or hoaxes concerning what were believed to have been sea serpents or lake monsters, four of which relate to the Gloucester Sea Serpent. One that took place coincident with the opening of a steamboat service in Boston, was the invention by 'some wag' (i.e. person fond of making jokes) about a sea serpent who was consuming all the small fish in the area. This encounter was published in a local gazette. The narrative, writes Mitchell, was 'given with such grave diction and imposing seriousness, [that it] was received by many as an actual and literal occurrence, and credited accordingly.' Another encounter took place in Long Island Sound, about which 'on this fiction I am well satisfied of the particulars that follow,' Mitchell comments, before continuing:

> An active young fellow who had become weary of ploughing the land, bought a little sloop of about fifteen tons, which I remember to have seen [from which we can infer that Mitchell was familiar with the parties involved and received this tale first hand]; and resolved to try his luck in ploughing the waves. He named the vessel the *Sea-Serpent*. She was mostly employed in carrying country produce to the New-York market and in bringing manure back, with the advantage of passengers when any offered [one might presume more of these on the trips into than out of the City given the nature of the transported material!]. This boat was on her way from Mamaroneck harbor or thereabout toward the city, and was met by a sloop from that place, a short distance from City-island [Manhattan?]. The captain of the latter, on arriving at home, was eagerly interrogated by a quidnunc [i.e. an inquisitive and gossipy person] for news; and being a man of some humor and fancy, told his neighbor, the querist, he had just seen the sea-serpent. He then describe how (alluding to barrels on deck) he had seen the bunches on his back; how high the head (meaning the bowsprit) was out of the

> water; how the black and white colours (meaning the painted waist) were variegated; how he saw the lashing of the tail (meaning the motion of the boom in jibing as she was going along equaling at least from five to six knots an hour, which made all white before him (meaning the foam at the bows). The good man took the joke in real earnest, went away and told of it to a sensible acquaintance. This latter wrote a formal and solemn account of it; which, travelling an extensive round in the sheets of intelligence, was finally embodied in the aforesaid book, where it is registered as a part of the evidence.

As to which 'aforesaid book' Mitchell is referring to, this is unclear, since his description of the tall tale above does not seem to jive with the testimonials in either the Linnaean Society *Report*, or to the anecdotes relayed in Humphrey's letters.

Mitchell recounts being sent a piece of skin thought to have been from the body of a purported sea serpent:

> It was about this period of these transactions that I received from Boston an ichthyological production, enclosed in a letter, respectfully written, and with postage paid, submitting to me whether that article was not a piece of a sea-serpent's hide? It had been found on the shore of the region which the alarming visitor frequented; and was supposed to have been separated from his body by one of the musket balls which had been fired at him and washed ashore. To this serious communication I returned for answer that it was simply a portion of skin with closely adhering scales, belonging to the bony scaled pike (*Esox osseus*), and inhabitant of the Atlantic Ocean.

Mitchell (1829) states that 'so much curiosity and excitement were now raised about the sea-serpent, that he was a prominent topic of conversation. The feeling was more intense, inasmuch as it was confidently declared he had been frequently observed near boats and vessels. It was at length concluded to fit an expedition, expressly for the purpose of catching him.' He then goes on to review the saga of Captain Rich, as previously excerpted in Chapter 4, which, in the present context, can now be understood to be, in Mitchell's opinion, nothing less than an elaborate hoax perpetrated on a gullible public. In this light, we can better interpret Mitchell's use of the word 'story' in reference to Rich's account as being indicative of him believing it to all have been a created fiction.

The last three examples include a multi-humped UMO in Lake Ontario that turned out to be a mother duck and her fledglings bobbing up and down in the waves, another in Lake Erie that turned out to be a floating tree with sinuous roots and branches, and a carcass exhibited in New York City that turned out to be a basking shark.

Mitchells' shot across the bow of those claiming to have seen sea serpents is stated with bold assurance:

> The history of Sea Serpentism is a very memorable part of the sayings and doings in this enlightened age and country. For the benefit of the present generation, and of posterity, it ought to be written. In proceeding to pen a short sketch of it, I must premise, that I am one of the last persons in existence who would presume to put a limit to creative power. I admit that the allmighty being could make a watersnake as easily as a fish; and that such an animal might be as big as a *Kraken*, as easily as the diminutive size of the *Stickleback*. Yet, on reviewing these legends of the times, there is found such a propensity towards the strange and the marvelous, that the men of the present day show a credulity very

> much resembling that of the remote ages, when the terraqueous globe was peopled with gorgons, mermaids, hydras, dragons, and all the monsters of fabulous zoology.

And if that wasn't enough to produce a collective 'Ouch!' from later generations of cryptozoologists, Mitchell wraps up:

> Now, after all these mistakes, deceptions and willful perversions [triple ouch!] on the subject, every person of consideration may admit that the gambols of porpoises, the slow motions of basking sharks, and the yet different appearances of balaenopterous whales, all of which have fins on their backs, may have given rise to those parts of the narrations, not already herein commented upon.

The trouble is that none of Mitchell's earlier illations, nor these thrown out casually at the end of his article, can be invoked to describe the UMO that was observed by the thousands of eyewitnesses in New England and New York from 1815 to 1824. Mitchell's dismissive arguments, though not without interest and credibility, simply do not fit the corpus of 'data' about the Gloucester UMO. This was recognized immediately at the time, wherein the journal's editor, Professor Silliman, rebutted the generality of Mitchell's illations: 'We give place to the *skepticism* of the learned author, although not ourselves *skeptical* on this subject. We do not see how such evidence as that presented by Dr Bigelow Vol. II p. 147 of this Journal – particularly in the statements of Capt. Little of the Boston Frigate, and of Marshall Prince and family, and of Mr. Cabot, can be set aside.' Silliman concludes: 'Although we have no doubt that there have been on this subject both error and imposition; and we are far from believing that every thing that has been called a sea-serpent has really been such.'

Expressing a faddism prevalent in the protoscience of natural history of the time – as Heuvelmans (1968) states 'people were beginning to know about the plesiosaur and were later to be obsessed by it' – the increasing knowledge and acceptance about the fossilized remains of antediluvian 'monsters' by the middle of the century led many to suppose that sea serpents, including the one that visited Gloucester decades before, were plesiosaurs (e.g. Cogswell 1847; Gosse 1861; Newman 1848). This led some to re-examine and question the legitimacy of past narratives that did not seem to match current theories. One such is the article by W. Cooper of Worcester MA published in *The Zoologist* of 1848 (*in* Heuvelmans 1968), in which he comments on the account from Joseph Woodward of May 1818 which had been originally published in 1818 in the *Quarterly Journal of Science, Literature and the Arts of the Royal Institute at London* and reprinted at the time of the *Daedalus* fervor in 1848 in *The Zoologist*. 'I have waited anxiously to see whether any more competent person than myself would offer any observation upon the statement of Captain Woodward...relating to the Great 'Sea-serpent.' As no one has done so, I beg to offer you the following,' Cooper begins, taking Woodward to task for not specifying the year of the May sighting. He then tries to diminish the credibility of the witness as he, Cooper, had not heard of him (exactly why this is the fault of Woodward and not of Cooper is not explained). To the contrary, it is because of his ignorance of Woodward, that Cooper believes he is justified in stating the following: 'It is, therefore, upon the later ground that I venture to attack his statement, and I do so because in a disputed question it is necessary to throw aside all evidence that will not stand the strictest scrutiny.'

Cooper proceeds to admonish Woodward for not providing exact geographic details of where the sighting took place, nor of the precise time, before launching into his more substantive critique:

> The most extraordinary part of the statement will appear from this: Captain Woodward says, the beast moved with *extreme*, or, as he afterwards expressed himself, *astonishing*

> rapidity; that when he fired at the monster it was sixty feet at most from the bow of the ship,...but after he fired the best advanced towards his ship; that he had caused the cannon to be reloaded and pointed at its throat, – of course while it was advancing towards his vessel, – but before he could fire his crew were seized with terror; that he tacked and got out of the way. So here we have an animal sixty feet from the ship, moving with astonishing rapidity *towards the ship*, which it appears was also moving *towards the animal*, and yet allowing time to load a cannon, point it at its throat, and afterwards to tack to get out of its way. Truly a most accommodating serpent! But again, the animal remained five hours near the ship, allowing itself to be minutely examined, but yet no further attempt to kill the beast! And what is almost equally strange...[is that] yet no description is given of any scales, or anything else, to account for what is before stated, that Capt. Woodward and his crew 'distinctly heard the ball and bullets strike against his body, from which they rebounded as though they had struck against a rock.'
>
> It is much to be regretted that these inconsistencies did not strike you before you made the statement in question; it is also to be regretted that no one better able than myself to point them out has undertaken to do so. But it is highly desirable, in the present state of our ignorance upon this subject, that none but the most inexceptionable evidence should be received. Let us have 'the truth, the whole truth, and nothing but the truth.'

Cooper wraps up by saying he bears no unfriendly feelings towards Captain Woodward, and announces that he himself is a believer in the existence of such 'monsters of the deep,' which he hopes will one day be caught and 'lodged in one of our museums, to be at once the wonder and admiration of naturalists.'

But, as the century wore on, it seemed that for every individual clinging to a belief in the Great Sea Serpent, there were two people of prestige who expressed agnosticism. For example, in his widely popular *The Romance of Natural History*, Philip Gosse (1861) is very much of the 'show me the body!' school of skeptics, at least with regard to sightings made by untrustworthy Americans: 'The frequency with which the sea-serpent has been supposed to have appeared near the shores and harbours of the United States, has led to its being specified as the 'American sea-serpent;' yet out of the two hundred vertebrae of every individual that should have lived and died in the Atlantic since the creation of the species, not one has yet been picked up on the shores of America.' He then goes on to cast derision on the diminutive, co-called baby snake of the Linnaean Society. Henry Lee, director of the Brighton Aquarium and notable for producing one of the first skeptical summaries about sea serpents in his 1883 book *Sea Monster Unmasked*, in which he proposed many to have been misidentified giant squids, expresses his frustration in being baffled about the Gloucester Sea Serpent in particular. 'Of this I can offer no zoological explanation,' he concludes. Of interest is the fact that his words 'no zoological explanation,' as will be seen, strike close to what I believe to be the truth about the Gloucester UMO.

5.2 CRYPTOZOOLOGY AND CONFIRMATION BIAS

Oudemans (1892) devotes considerable space to expressing his interpretations of the Gloucester UMO. He concurs with eyewitnesses who believed the prong or spear to be 'nothing else but the animal's tongue.' In this, as I will argue in a latter section, he is wrong. But in his comments about the multiple humps, he is correct in emphasizing that they were not merely due to undulating motion: 'The animal may remain quite still on the surface of the water, keeping, however, its coils, or joints, or bunches,' he says at one point, and reiterates similarly later in relation to another account: 'We have here again the statement that the animal is able to keep its bunches, when it lies extended on the water.' This is important in that it means that the so-called humps of the UMO cannot be explained away, as Captain Rich attempted, as being an illusion caused by the wake of a rapidly swimming animal.

Mostly, however, the impression left by reading Oudemans' interpretations of the Gloucester UMO is of bemusement or frustration in consequence of his blinkered preoccupation with attempting to force-fit all the accounts into his preconceived belief that they are examples of a single type of paleo-mammal. In order to accomplish this task, he has to undertake many slights-of-hand to selectively exclude various data that are in obvious contrast to his thesis. For example, he cherry-picks and focuses on the few accounts in which the Gloucester UMO is described as having smooth skin and neither scales nor dorsal bunches.

Oudemans' second method is to resort to inventing what he believes the various eyewitnesses truly saw, irrespective of what they themselves had *actually* reported having seen! For example, in relation to the Gloucester UMO, which in many ways is a major problem for Oudemans, he writes: 'As to the number of bunches which Mr. Allen reports, viz. fifty, I believe that he has not *counted* them; he says: I should say fifty. Pontoppidan tells us that the greatest number ever seen was twenty-five, and I believe that this is indeed the case...Of course Mr. Allen is also mistaken as to its scaly surface. The roughness, however, may have been the result of the rippling of the water.' And here is another, even more blatant example: 'It is also probable that the animal's external characters, though Mr. Pearson says he believed it to be a serpent, made on him the impression of a mammal.' For Bennett's sighting in June 1819, he believes that in relation to the notable wake that was observed and ascribed to the tail, 'it is far more probable that it was caused by the motion of the hind-flippers, supposed the animal nearly touched the surface of the water with the hinder part of the body.' As for Mr. Story's account of the UMO lying motionless for half an hour:

> It's head and tail were not visible, says Mr. Story, and yet the animal remained about half an hour in this position, which I think may thus be accounted for: the animal's head, neck and back were in a straight line just above the surface of the water, so that its nose was also above it, which enabled the animal to breathe and to remain motionless, but at a distance of forty rods, though with a good spy glass, these particulars cannot have been distinctly seen by one who was not acquainted with the animal's external features, and so he believed its head was invisible. That its tail was under water, I will believe with him.

'This is the habit of Pinnipeds,' Oudemans adds to finish this particular bit of convoluted logic.

Oudemans third method is to use the type of egregious circular reasoning and overt confirmation bias that has become characteristic of much of cryptozoology:

> Remarkable in this testimony is again the considerable wake the animal leaves behind it when swimming rapidly. Easy it is to explain why the speed is greater under water behind it than when partly visible above the surface of the water. Those parts viz. which are above the surface must be borne by the body hidden under water, consequently this carries a burden, and the speed, it is evident, cannot be so rapid as when the animal is quite under water, in which position each part of the animal's body is carried by the water itself, and not by the individual. It has no burden to carry, it is specifically lighter,

> and the speed can reach its maximum. Remarkable too is the fact that the animal, when swimming under water, does so just below the surface, and causes the rippling of it.

He delves into further fantasy in trying to turn the Gloucester UMO into his relict paleo-mammal:

> Now, my readers will probably say that I have not yet explained why none of the eye-witnesses of the animal seen near Cape Ann saw a mane. I hope my readers will be satisfied when I tell them that I am convinced that the female Sea-Serpent has no mane, and that the mane is only a character of full grown males. So most of the eye-witnesses saw a female. It is only the individual witnessed by Messrs James Guion and Thos. Hertell, which was most probably a male and had a mane. Seen from a distance its back was uneven, and deeply indented.

In his early chapter, 'Cheats and Hoaxes,' Oudemans spares no mercy in excoriating what he imagines to be the sloppy scholarship of Mitchell (1829). He takes issue with Mitchell's statement that the sea-serpent 'first' haunted the coast of Massachusetts, thereby ignoring all the Nordic accounts. Next, he accuses Mitchell of being dishonest in attempting to claim that it was he who determined that the specimen purported by the Linnaean Society to be the progeny of the creature in the harbour was a deformed common snake. Both of these are moot red herrings it must be said. He does correctly note that the incident with the sloop *Sea-Serpent* is not found within the Linnaean Society *Report*. Ignoring the rest of the contradictory evidence – what will become the favourite *modus operandi* of later cryptozoologists – Oudemans pulls no punches in dismissing Mitchell's whole paper as being 'an unscientific, deceptive dissertation, unworthy of notice, and that the way in which he ridiculed the endeavours of the Committee was unfair.'

Also found in this chapter is a discussion of the 'hoax' perpetrated by Captain Woodward. Oudemans comments in respect to endorsing Cooper's (1848) dismissal of Woodward's account are very succinct: 'Here ends the history of this hoax, utterly smashed! Mr. Edward Newman [the editor of *The Zoologist* who was taken to task by Cooper for publishing what he regarded as being rubbish] has never answered to this attack!'

In this chapter on hoaxes, Oudemans also places the saga of Captain Rich, who at the start does receive his support: 'After perusal of this work [i.e. Rich's first encounter with the UMO] my readers will know why I am disposed to believe that the animal struck by Captain Rich was really a Sea-Serpent. As far as I can judge, after having read all that I have found about the sea-serpent, this is the only time that the animal was struck by a harpoon.' But, as for the so-called capture of the UMO, which turned out to be a giant Bluefin tuna, Oudemans will have none of it, accusing Rich repeatedly of perpetrating a 'Yankee-trick' to dupe the public. He questions how anyone could misconstrue a tuna for a sea serpent: 'How to make the animal's head (which is like that of a snake, a seal, a walrus, a sea-lion), its long neck, its four flappers, its enormous long pointed tail, agree with the general outlines of a tunny, even of nine or ten feet in length?' Putting aside Oudemans biased and unfounded assertion about the Gloucester UMO possessing a long neck and flour legs, I will posit an explanation to answer his question in the next section.

In his conclusions at the end of the book, based on compiling sightings he deemed to be worthy of consideration – a process which has incurred the criticism of many ever since – Oudemans (1892) continues to be very selective in what he considers to be valid observations. For example:

- All mentions of a tail being longer than about twenty-five metres are dismissed, thereby shrinking UMOs to make them more easily fit into his hypothesized single candidate.
- An absence of a 'mane' is explained away as resulting from either poor observation or the 'fact' that females are not so-whiskered, so as not to contradict his mammalian candidate animal theory.
- Undulations may result from muscular contractions of the dorsal skin and subcutaneous layer of fat.
- Failure to observe flappers is because they can be tucked into the body or are invisible beneath the surface of the water.

As can be seen, there is no end to inventive explanations that can be generated when one abandons the need to draw inference from an objective perusal of available data, and instead replaces it with dynamic explanations to force data into preconceived conclusions.

Oudemans proposes twenty-three possible explanations for sea serpents (Table 5.1). Of these, he states that 'it is obvious that of all animals, now living or extinct, the outlines of the *Plesiosaurus* fit best to the descriptions of the great sea-serpent.' For those finding difficulty in reconciling drawings of skeletons and reconstructions of said animal (Figure 5.1) with the physical attributes that have been described for sea serpents such as that from Gloucester, no worry, Oudemans has the answer. 'Plesiosauri with such an enormous tail as the sea-serpent has, are hitherto unknown to palaeontologists, but, as to me, this

Table 5.1. Various explanations advanced by Oudermans (1892) to explain sea serpent sightings.

A row of porpoises in a straight line.
Scoliophis atlanticus, a new species of snake with pronounced ridges on its back.
A large bluefin tuna
A true sea-snake of very large size
A gigantic eel
A fable from Norse mythology
A basking shark, or row of sharks
A baleen whale, or row of them
An *Ichthyosaurus*, or a similar saurian
A *Plesiosaurus*, or similar saurian
Not a saurian
A row of sperm whales.
A *Basilosaurus* or ancient whale
A *Saccopharynx* or *Ophiognathus*
A large boa snake
A sea elephant pinniped
A large piece of sea weed
A large ribbonfish
A large floating dead tree, piece of bamboo, or weed draped log of wood
A mass of flying waterfowl
A large sea turtle
A gigantic squid
A manatee

Figure 5.1. Oudemans' (1892) drawing of a plesiosaur which some have posited as a possible candidate for sea serpents.

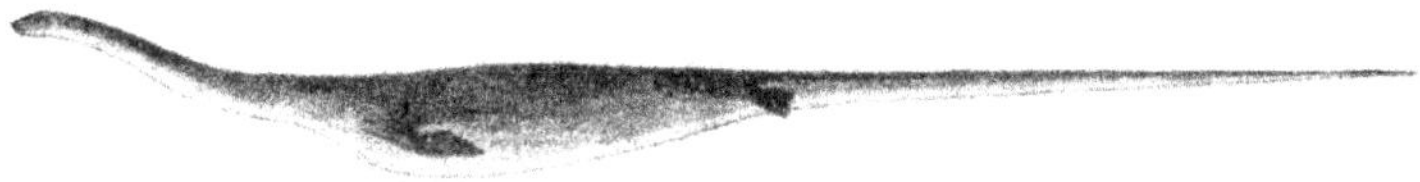

Figure 5.2. Oudemans' (1892) drawing of a long-necked seal which he believes to be the most likely explanation for all sea serpents including that observed in and around Gloucester.

cannot be of much importance; for there is no reason why in the course of ages this appendage should not have been developed to gigantic dimensions.' However, he shies away from selecting this prehistoric saurian as the most likely candidate because its neck and backbone could not possibly bend 'in such vertical undulations' as has been observed for sea serpents. The animal that is finally settled upon is a type of long-necked paleo-seal that resembles an extinct zeuglodon (Figure 5.2). 'I firmly believe that it belongs to the Order of *Pinnipedia*,' he concludes with great confidence.

For his own interpretations of the Gloucester UMO, Gould (1930) makes the valid point that all serpents, 'of course, progress by curving their bodies laterally – they have no power of humping them vertically, as a caterpillar does. Yet the creature seen at Gloucester seemed, to most observers, to progress by the latter method – vertical undulations – and sometimes to exhibit the same wavy outline of back even when motionless.' Furthermore, in regard to the UMO's swimming behaviour, 'if, in addition to this extreme lateral flexibility, the creature was also able to curve its back vertically into a series of humps, the structure of its backbone must have been truly astonishing – it must have looked like a series of universal-joints.' So far so good, in terms of commenting on the eyewitness accounts, but Gould continues by stepping into the woolly world of cryptozoology through venturing that as the UMO's back was only a few inches above the water and sometimes showed a smooth outline, devoid of 'bunches,' 'it seems much more probable that the latter were curvatures of the skin, produced at will by muscular contraction, and not actual undulations of the backbone.' As for Mathew Gaffney's shooting and striking the head of the UMO, this is easy to explain away: 'He missed his mark altogether. Or his bullet may either have glanced harmlessly off the wet skin, or perforated a non-vital portion.'

Gould ponders whether the many sightings of 1817 were made in a state of what he terms 'expectant attention': 'In such circumstances a whale, a shark, or a school of porpoises might well provide the raw material out of which the excited imaginations of expectant onlookers might conjure the creature which they so greatly desired to see.' But then just as quickly as the possibility is raised, it is rejected based on observations made from other locations and by people who were not part of a large crowd. He concludes: 'Unless sheer prejudice induces us to reject this evidence altogether, we must (I submit) concede that the witnesses actually observed, sometimes at close range and for considerable periods, a single living creature quite unlike any at present known to science.' Furthermore, he believes that the creature:

> was not a whale, or a shark, or anything with which naturalists, experienced seamen, or fishermen were familiar. This, indeed, is obvious from the attention which the Gloucester reports (of 1817) aroused. A town whose population chiefly consists of fishermen is not usually electrified by the visit of a whale, or a shark, or a school of porpoises, or any other common object of their daily life. There can be no doubt at all that the thing seen, whatever its nature, was a most unusual – a unique – sight.

Gould (1930) concludes his chapter with a tease, implying that should other evidence not have come to light elsewhere, he would warrant deducing 'that the existence of a marine animal of unknown species with serpentine head and neck, and about a hundred feet in length, was an established fact.' But, as is shown elsewhere in his book, because many of those other sightings were of what were obviously very dissimilar creatures, Gould is reticent in completely endorsing the existence of the Gloucester Sea Serpent.

Ivan Sanderson, co-conceiver, along with Heuvelmans, of cryptozoology (Sanderson 1947), suggests, in a letter to the latter, that the humps that were observed by so many for the Gloucester UMO:

> may actually be hydrostatic organs, air-filled sacs connected to the larynx and inflatable at will. They would lie between the skin and the muscles, like the similar complex of sacs around the neck, chest and armpits of orang-utans and siamangs [black gibbons]. The number of these sacs, the size of which would be limited for mechanical reasons, would increase with the size of the individual. This would explain the great difference of opinion of witnesses about their number and relative size. As to their function, they would serve primarily as reserves of air for prolonged dives, and also to lighten the enormous mass of these animals. Their longitudinal arrangement would provide great stability.

To all of this, Heuvelmans (1968) states simply: 'This ingenious theory would explain much and is most attractive.'

Heuvelmans (1968) accepts the quality of the testimony given in the Linnaean Society *Report* from the wide diversity of eyewitnesses interviewed: 'There is nothing to suggest the products of imagination.' He singles out the testimony of Solomon Allan as being the most distinct from the others (i.e. the noted presence of up to fifty humps), and considers, without irony given his own predilection for confirmation bias, that Allan 'may have had preconceived ideas about the nature of the beast.' The most characteristic feature of the Gloucester UMO is considered to have been its apparent undulations in the vertical plane which caused the creature to resemble a 'string of buoys' even when lying motionless. As such, 'this appearance is not necessarily the result of undulating motion. When the beads on the necklace are many and close together, as they are here, they must clearly be humps. They can only be undulations of the body if they are large and thus relatively few. This is one point established.' All very true. Nevertheless, he concludes by stating that a multi-humped animal could certainly propel itself by undulating up and down.

Heuvelmans writes that 'the way the animal turned round by first making a fairly large curve with the front part of the body and then swimming with the front parallel to the tail but going in the opposite direction shows that the animal's body was not all equally flexible and that it must have a rather more solid thorax joined to a long and more flexible tail.' From this, the sound judgment is made that 'this should have raised doubts whether it was really a snake.' He continues, mentioning that the ability of the beast to rapidly submerge straight down without twisting its body in the least, 'could only be explained,' he supposes 'if it had fins or flippers on the side of its body and hidden under the water. Nobody suspected that there were any such members, which would also have ruled out the snake theory.' The numerous humps rising behind the head, 'as if they were coils or folds in the body,' further suggests comparisons being made to a snake. However, Heuvelmans counters jumping to such a conclusion by stating that 'a reptile, because of the structure of its spinal column, is quite incapable of such tight bends. The beast's undulating movement increased the illusion of a snake's reptation [i.e. movement analogous to snakes slithering], but as it was vertical it could not have been done by a reptile.' From all this, Heuvelmans concludes emphatically (and correctly) that 'the Massachusetts sea-serpent could not possibly have been a snake,' before adding that 'had the Committee of the Linnaean Society but realized this they could have saved themselves from a dreadful blunder.'

Heuvelmans follows others in dismissing Captain Woodward account as being a hoax, his reasoning seeming to be due to the declaration being signed by two witnesses rather than a Justice of the Peace. And he also considers Captain Rich to have perpetrated a hoax:

> It seems clear from Norton's letter that Rich's men, having caught a tunny, and not wanting to be ridiculed for coming back with no sea-serpent, foolishly tried to have a joke at the expense of the Boston journalists. But the joke misfired, and the outcome was that the sailors, instead of being accused of a hoax, were blamed for a mistake of quite incredible stupidity. They had thought they had harpooned a sea-serpent, but when they had got it aboard it had proved to be a tunny. Rather than be accused of a cheat, even Captain Rich eventually admitted to this version. For the sceptics the matter was settled, the sea-serpent was just a tunny – though no tunny could ever carry its head above the waves and show a row of humps.

'Few hoaxes have discredited the sea-serpent more than this,' he notes, 'because here the truth was so strangely mixed with fiction.'

Cryptozoologists love to engage in the classification of what I call 'the nomenclature of nonsense' (France, in prep.). Throwing aside centuries of established taxonomic procedure that necessitates examination and cataloguing of reference specimens or parts thereof in accredited museums, these pseudoscientific, knights-errant of the unknown, go about ascribing impressive sounding Latinized names to imaginary animals in a mimic of Linnaean legitimacy, and even go as far as to sometimes postulate phylogenetic flights of fancy that describe inferred and non-existent evolutionary relationships.

Oudemans (1892) creates an elaborate phylogenetic system for his long-necked seal. However, due to the tradition of precedence given to the name initially proposed, the best he can do is to tweak Rafinesque Schmaltz's designation, such that 'In my opinion,' he states, 'the only name to be given to the sea-serpent is that of *Megophias megophias* (Raf.) Oud.' Note, how, by altering the species name, he gets to ascribe his own initials to the 'discovery' and formal description of the new species. Such things matter in cryptozoology just as they did and still do in natural history.

Heuvelmans (1968) brags that 'I cannot claim to have entirely solved the problem of the great sea-serpent, but I have cleared up a good deal of it.' He goes about this, rather than by force-fitting all sightings into a single typology as did Oudemans, instead by creating nine groupings which he imagines to be separate species, arranged in decreasing prevalence of sightings as: Long-necked; Merhorse; Many-humped; Many-finned; Super-otter; Super-eel; Marine saurian; Father-of-all-the-turtles; and Yellow-belly. He then goes on to offer explanations for a suite of physical traits for the Many-humped species, some of which pertain to the Gloucester UMO. In this regard, the appearance of a string of buoys may be due to:

Figure 5.3. Heuvelmans' (1968) drawing of the Many-humped typology of sea serpents of which he considers the Gloucester Sea Serpent to have been an example.

- Vertical bends in a very long animal.
- Vertical bends in a long animal happening to swim on its side.
- A row of solid humps or air-filled sacs along the back.
- Transverse folds in a very fat body.
- Turbulence waves in such a body.

And the appearance of rough or scaly skin may be due to:
- Skin that is really wrinkled or bumpy.
- Smooth skin covered with parasites.
- True scales.
- Bony dermic plaques beneath the skin.
- A reticulated pattern on the fur or skin.

Table 5.2. Selected anatomical, behavioural and biogeographical traits of the Many-humped sea serpent espoused by Heuvelmans (1968).

Elongated shape; very great size; many humps along spine
Ox-like shaped head
Medium-length neck; shoulders bear small fin on adult males; ridge of dorsal humps
Bilobate tail like that of cetaceans
Anterior pair of flippers
Skin generally smooth, rough if encrusted with parasites; scales or dermal plates sometimes present
Length generally between 60 and 100 feet; may reach 115 feet, the size of Blue whales
No visible breath
Vertical undulation like caterpillars leads to erroneous identification with serpents
Considerable surface speed of 25 to 46 mph
Seen primarily along the New England and Atlantic Canadian coasts, with 'a marked predilection for Massachusetts Bay'
Inhabits the Gulf Stream and is sometimes carried by it to Iceland and Scotland
North Atlantic sightings generally take place in late summer
Follows the Gulf Stream southward during the winter

Heuvelmans (1968) offers a drawing of what he imagines to be the archetypal Many-humped sea serpent (Figure 5.3), giving it a more prominent head and a distinctively bi-lobed tail, both of which were not widely reported features of the Gloucester UMO. But again, here, he is faced with the near impossible task – though not as much so in this regard as was Oudemans since Heuvelmans is dealing with a more limited (i.e. pre-screened) subset of all sightings – of attempting to jam in a group of disparate observations derived from many locations and times. He continues by generating a suite of anatomical, behavioural, and biogeographical attributes of this particular type of sea serpent (Table 5.2), formed, it must be noted, to a substantial, but not exclusive, degree from the 1817 Gloucester sightings. Heuvelmans concludes by stating that 'The scarcity of sightings since the beginning of the [Twentieth] century shows that the species is obviously becoming rarer.'

On the basis of displaying traits such as movement by vertical undulations, a bilobate tail taking the place of hind legs which have presumed to have disappeared through time, and the presence of smooth skin with dermic plates in places, Heuvelmans (1968) proposes the Many-humped sea serpents to be archaic cetaceans displaying serpentine form; in other words, resembling 'the fossil Archaeoceti with their thin necks.' And what should this newly invented beastie be called? Well, why not *Plurigibbosus novae-angliae*, or 'that-with-many-humps-of-New-England.' And so now we have a third name ascribed to the Gloucester Sea Serpent.

Heuvelmans justifies this bit of Linnaean classification as follows: 'I think the name *Megophias monstrous* must be abandoned once and for all. It was originally used by Rafinesque to describe the Massachusetts Bay sea-serpent, that is the Many-humped, but without sufficiently detailed description for it to be recognized. And when Oudemans took it over and modified it to *Megophias megophias* it was based on descriptions of several different animals.' By 'different animals,' Heuvelmans means what he imagines to be different species. And so, whereas Oudemans only got to add his initials as a footnote to Rafinesque by modifying the latter's species name, Heuvelmans goes one better by being able to put his name beside a newly named genus. Alas, if he hoped that his assumed 'sufficiently detailed description' of this living fossil of an archaic whale would be enough – in the absence of a registered specimen – to allow his own name to ride into the zoological history books alongside his foray into Latinized nomenclature, he must have been sorely disappointed. Ignored by the scientific community, his ideas have fuelled generations of nothing more than cryptozoological flights of fancy.

Heuvelmans (1968) adds, by way of caveat, one final supposition to his attempt to identify the Gloucester Sea Serpent, namely that it might have been a giant manatee:

> We cannot however, rule out the possibility of the Many-humped being a peculiar form of sirenian, the only sea-mammals which are herbivorous. It would then be like Steller's sea-cow, but without being restricted, as the sea-cow is [that should read 'was'] with its diet of seaweed and sea-kale, to coastal waters. It would be a great sirenian feeding on vegetable plankton. This would, of course, suppose considerable specialization.

It is sufficient to note that as physically impressive as the Steller sea cow was (up to ten metres long), and sad that it was brought to extinction in historic times, it is hard to reconcile drawings and skeletons of this magnificent marine mammal (Figure 5.4), with eyewitness descriptions and representations made of the Gloucester UMO. Discovering a 'lost' population of Steller sea cows living hidden in some remote corner of the Arctic Ocean remains a dream, realized only in fiction.

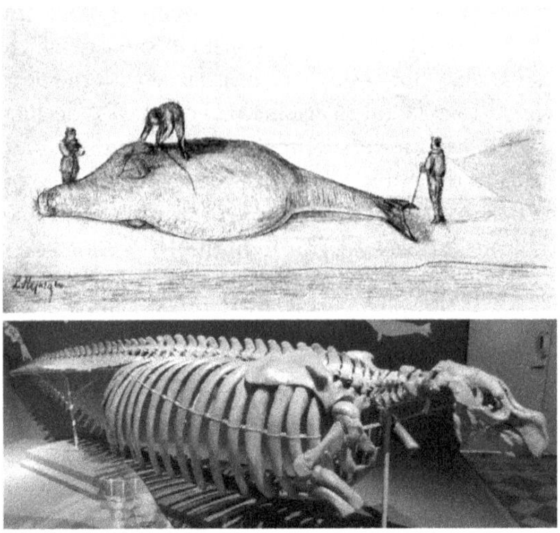

Figure 5.4. Heuvelmans also suggests that the Many-humped Gloucester Sea Serpent might be some form of sirenian such as the recently extinct Steller sea cow (Wikimedia Commons).

Others have grappled with developing classification systems for sea serpents. Champagne (2007), for example, builds upon Heuvelmans' model and adds a whole suite of what he imagines to be discriminating ecological and behavioural attributes. In contrast, Magin (1996) lambasts the shoddy scholarship and overt confirmation biases implicit in Heuvelmans' dabblings with his 'nine imaginary beings' and dismisses the entire process as a ludicrous enterprise to put a scientific veneer on what are really cultural entities like unicorns and mermaids. Here, for brevity, I will introduce two, more recent attempts at fitting the Gloucester Sea Serpent fits into cryptozoological classification and nomenclature schemes.

Loren Coleman and Patrick Huyghe (2003), in their imaginatively titled *The Field Guide to Lake Monsters, Sea Serpents, and Other Denizens of the Deep*, propose a new classification system for aquatic cryptids. They begin by lumping Heuvelmans' three groups of Many-humped, Super-otters, and Super-eels, into a new category they call the Classic Sea Serpent. 'We maintain,' the authors state, 'that in classifying the general serpentine form of unknown aquatic animals, a single type is best,' before going on to state that they 'remain unperturbed that this animal may turn out to be a mammal rather than a giant reptilian serpent.' An illustration is shown of the archetypal 'sea serpent' (Figure 5.5), which is accompanied by a discussion of various attributes of the imagined creature (Table 5.3). From this, the authors reject the early popular notion that these creatures were giant snakes, instead believing that 'there is much to recommend the zeuglodon hypothesis,' which their illustration certainly does resemble. Next, a handful of representative reports and source material are presented 'that highlight some of the most detailed, more credible, or best-evidence cases.' Although herein is where one would expect to find a brief synopsis about the Gloucester Sea Serpent, this is not to be. Perhaps Coleman and Huyghe were scared away from the Gloucester example, given their citing of a quote from Ellis (1994) concluding that those sightings 'remain one of the great unsolved mysteries of sea-serpent lore.' So instead, six sightings are examined in detail, only one of which, from 1958, occurred in New England; in this case, from Maine.

The book concludes with helpful tips for sea serpent watchers. In a section on the top eight places to look for a sea serpent, Gloucester Harbour and environs is disappointingly listed as only the seventh most promising location. This is because Coleman and Huyghe believe that 'it is doubtful that the New England Sea Serpent still haunts the shores near Cape Ann.' How different things were back in those halcyon days of 1817.

Keeping in mind Loxton and Prothero's (2015) criticism that cryptozoology suffers from an abundance of untrained specialists, Michael Woodley is, unlike either Coleman or Huyghe, a practicing biologist. But before one's hopes are raised for a sudden infusion of reality therapy to the nomenclature of nonsense

Table 5.3. Selected anatomical, behavioural and biogeographical traits of the Classic Sea Serpent espoused by Coleman and Huyghe (2003).

- Serpentine form, upwards of 100 feet; with distinct head and tapering tail
- Series of what appear to be dorsal humps or protuberances that resemble a strong of buoys, which 'may be more a manifestation of the motion of the animals than a physical characteristic'
- Slender neck of medium length
- Global range but with several specific concentrations such as one in the Gulf of Maine
- Historical sightings include those from Nordic countries in the eighteenth century, as well as New England, such as those from Gloucester in 1817

Figure 5.5. Coleman and Huyghe's (2003) archetypal or 'classic' sea serpent which they propose to be a zeuglodon which has escaped extinction (as well, I might add, a corresponding continual representation in the fossil record). Illustration by Harry Trumbore, copyright © 2003 by Harry Trumbore; from 'The Field Guide to Lake Monsters, Sea Serpents, and Other Mystery Denizens of the Deep.' Used by permission of Tarcher, an imprint of Penguin Publishing Group, a division of Penguin Random House LLC. All rights reserved.

regarding sea serpents, it is important to remember that Huevelmans, and Oudemans before him, were also both practicing biologists. The division of what constitutes science or speculation (Arment 2004), and who are crackpots or eggheads (Regal 2011), in reference to mystery animals and to the aspirations of those who seek them, is not clearly delineated.

Woodley's objective is to re-evaluate and modernize Heuvelmans' existing categories. Along the way he engages in one of the most blatant displays of cryptozoological fancy in terms of interpreting the ephemerality of sightings of several imagined typologies of sea serpents, not due to sequential faddism and expectation patterns among observers, but in consequence of ecological competition imagined to exist between the two types of 'sea serpents.' After this ridiculous bit of science fiction, Woodley reaches the conclusion that 'It is obvious that Heuvelmans' categories are incomplete as they stand, and are therefore not truly representative of the full potential diversity of marine cryptid forms.' Among the changes recommended is to retain the Latin names for both the Many-humped sea serpent and Super-otter but to switch their identity from archaeocetes (extinct whales or sirenians) to being members of the lutrinae (otters).

Super-otters, like regular sea otters, argues Woodley, possess highly flexible bodies, but also have long tails 'along which undulations propagate giving rise to the many 'humps' that witnesses describe.' He reasons that Super-otters are largely aquatic like their 'sea otter cousins,' rather than amphibious. 'After all,' he states, 'a member of the lutrinae of such vast proportions, and weighing many tons is going to have considerable difficulty manoeuvring on land or ice.' Differences between this hypothetical cryptid and the traits Heuvelmans believes to characterize Many-humped sea serpents are explained away: webbed-feet operate like pseudo-flippers and can look like cetacean fins; the bilobate tail could result from the fusion of hind limbs, as occurs in pinnepeds; the 'static' humps 'could simply be undulations propagating down the length of their highly flexible spines'; the dorsal fin, which Heuvelmans believed to be a dimorphic trait of large males, is a 'simple misidentification of something else, and can be safely dismissed on that basis'; and

> The supposed absence of fur, or the presence of smooth or 'scaly' skin in the many humped sea serpent, could perhaps be due to the secretion of large amounts of squalene (a sebaceous gland derived insulating chemical) coupled with the development of a more rubbery skin, and larger deposits of subcutaneous fat, as a way of overcoming the need to be hairy...A more plausible explanation is that these cryptids are in fact hairy...it is just that those who have been lucky enough to witness them, simply haven't been close enough to see the hair, which may also be slicked against the cryptid's skin, thus giving the erroneous impression of a smooth, hairless body.

A representative illustration is provided (Figure 5.6), and Woodley concludes by suggesting that Heuvelmans was correct in placing the two typologies into different genera, the reassessment being to consider them both to be otters: 'Whereas the super-otters possess a more 'traditional' body plan, the many-humped sea serpent would seem to be the more advanced of the two in terms of perceived adaptations to an exclusively aquatic lifestyle.' Some further ink is spilt on suggesting what the closest allies to these cryptids are among existing lutrinae species. The fact that most of these known species of otters are less than one metre in length is apparently irrelevant.

The Many-humped sea serpent sighting that Woodley (2008) selects as a representative example is the first by European in New England, that by John Josselyn in 1639. Brief mention is made of the 1817 Linnaean Society investigation, accompanied by the famous iconic illustration (Figure 3.11), before he concludes:

> It is interesting to note than many (although not all) of the east coast sea serpent sightings were classified by Heuvelmans as belonging to his many-humped category, yet in general these east coast sea serpents seem to have large mobile humps, and although they differ from the 'pawed' super-otters, they seem to bear little resemblance to the 'static humped' many-humped sea serpent of Heuvelmans. This lends added credence to the idea that the many-humped category simply needs to be reclassified, and a more detailed and discriminatory study needs to be made with a view to disentangling the various sightings that Heuvelmans places in his 'many-humped' category.

And so here, at the end of what might be called the apogee of cryptozoological theorizing about sea serpents, the Gloucester UMO still remains a puzzle. It simply defies biological explanation. In particular, the problem lies in that it seems impossible, even for believers in sea serpents, to reconcile the presence of 'large mobile humps' with any straightforward, strictly biological causality. In his use of the word 'disentangling,' Woodley unknowingly and ironically comes close to what I propose to be the most parsimonious and likely explanation for the Gloucester Sea Serpent, and its perplexing 'large mobile humps.'

Figure 5.6. Woodley's (2008) theory of the Many-humped sea serpent being related to a Super-otter believed to inhabit coastal New England.

5.3 CONSERVATION BIOLOGY AND INCONVENIENT TRUTHS

For cryptozoologists, the Gloucester Sea Serpent is often touted as being the gold standard for proof of existence of such imaginary creatures. For example, 'Of the very strongest cases...one of them, the Gloucester sightings by innumerable witnesses, is by far the best-documented evidence that sea-serpents exist' (Bauer 2013).

Richard Ellis is a popular writer about marine biology who has questioned cryptozoological claims. 'For the most part', he writes, 'I remain skeptical. There are rational explanations for many monster sightings,' including his favourite likely candidate, the giant squid (Ellis 1994). This is not a new hypothesis, given that Henry Lee proposed the same a century before (1883) in his influential *Sea Monsters Unmasked.*

In the beginning of the last chapter of his *Monsters of the Sea*, entitled 'The New Mythology of Monsters,' Ellis confronts the enigma of the Gloucester Sea Serpent head on:

> What are we to make of the sightings around Gloucester in 1817 and again in 1819? Shall we assume that hundreds of reputable citizens were deluded or victims of mass hysteria? Were they all participants in some enormous practical joke? If so, by whom? Or shall we accept the more parsimonious explanation that they saw *something* and their descriptions of a 100-foot-long snakelike creature are based on some sort of reality? Lacking anything but eyewitness accounts that are almost two centuries old, we obviously cannot 'prove' that the Gloucester animal actually existed (or that it didn't), but it will remain one of the great unsolved mysteries of sea-serpent lore.

To which, Ellis adds, as an afterthought, that 'the identification of a deformed three-foot blacksnake as its 'spawn' added nothing to the verisimilitude of the Massachusetts monster.' At the end of the chapter, after he has examined and proposed more parsimonious explanations for a number of the more famous sightings of sea serpents, and just before an appendix that presents the Linnaean Society's questionnaire of 1817, Ellis returns to the Gloucester enigma:

> The Gloucester sea serpent is considerably more difficult to explain away. None of the contemporaneous descriptions would appear to have any connection with giant squid – or any other known animal, for that matter – and the fact that it was seen regularly by so many people mitigates against hoax or hallucination. I have tried to analyse the sighting data with the idea that many of them can be explained by existing phenomena, but I can make no suggestion about the true identity of the Gloucester monster...The giant squid does not explain everything...Some monsters have been oarfish; others, basking sharks, whales, or their deformed carcasses. There are some, like the Gloucester monster, that simply cannot be encompassed by any rational explanation.

The present chapter sets out to propose one such 'rational explanation' for the Gloucester Sea Serpent that eluded Ellis (1994).

There are two things, and two things *only*, that we know with any degree of certainty about the Gloucester UMO: (1) residents and visitors from a variety of different backgrounds definitely saw *something* in the harbour, all of the sightings of which cannot be explained away as being a simple case of mass-hallucination al à the witches of Salem or the devils of Loudon; and (2) the nature of what was seen was *different* enough from common experience that it – whatever 'it' was – could not be easily dismissed as anything more unusual than a frequently observed, non-mysterious object unworthy of further discussion. That's it. All else is speculation. To postulate further relies to varying degrees on the level of faith that one places in the verisimilitude of the testimonies and narratives describing what eyewitnesses saw...or more accurately, what they *believed* they saw. This faith issue is not science; what it is, is very much a sliding scale of acceptance depending on one's individual need for a particular degree of accepted surety contrasted against a fog of uncertainty.

Up until this point I have worn the hat of an environmental historian, collating and commenting upon the known facts, or at least the *purported* facts in regard to the eyewitness accounts. Henceforth, we need to leave the intellectual security of that scholarly shore behind, and venture out into the uncertain waters of the harbour, churned up into a speculative froth by the actions of both the UMO and its many commentators. This particular section poses and examines six questions concerning the environmental nature of the UMO, about which there can be some degree of confidence in the illations that are reached. The next section, however, dealing with the biological nature behind the UMO, ventures further out into even murkier waters of speculation, straying perilously close to those dangerous whirlpools of cryptozoology into which many a zoologist has been drawn, swirling to a demise with respect to his or her professional credibility.

So, if you are one who is only comfortable with the surety of facts, stop reading at this point where history ends and speculation begins, allowing the mystery of the Gloucester UMO to persist. If you are of bolder nature, then continue warily ahead into the present ethnozoology section dealing with developing an environmental explanation for what was observed. However, for those brave enough to proceed onto the next, and final zoological section of this chapter, wherein I attempt a biological interpretation behind the Gloucester UMO, I feel the need, should I have been an artist or transportation engineer, to draw and post a big yellow, triangular warning sign stating: 'Danger: Sea Serpent Interpretation Ahead.'

Is believing seeing, or is seeing believing? That is the question that lies at the heart of all attempts to understand what lies behind sea serpents and lake monsters in the absence of direct physical proof. Cryptozoologists give great credence to anecdotal evidence in the form of first-person eyewitness testimonies, arguing that given this is enough to sentence people to life imprisonment or even execution, then it ought to be more than adequate to prove that monsters exist (note in that wording the absence of the standard scientific approach of statistical inference in confirmatory data analysis vis a vis disproving the null hypothesis). Furthermore, scientists, those self-imagined 'high priests of truth and wisdom' who are responsible for the entrenched protectionism of the status quo, should stop persecuting the genius of the great innovators and visionaries who believe that they have seen a cryptid. But how much weight can we really place in such 'evidence'? Sir Richard Owen, aka 'the sea serpent slayer' (Regal 2012), dismissed all such 'evidence'.

Visual perceptions of the natural world are by no means straightforward, the retina operating independently from our mind's eye (France 2001b; Sewell 1999). When challenged by confronting novelty, we can and do make mistakes. Tuan (1990), for example, tells the charming story of how African bushmen, when removed for the first time from their dense jungle home and told that the animals seen across the distant clearing were hunted, broke out in laughter. How could such teeny, tiny animals, they asked, be hunted? It was simply too funny to consider. As Kaschak (2015) explains:

> There is a mysterious alchemy involved in what we call normal sight, the well-functioning eye/brain duet. Every human eye has a blind spot near the center of the visual field. This is not about peripheral vision or a view from the margins. It is right at the center of experience. The eye does not know its own blind spot, mistakes it for vision. Nor does the mind's eye and so gives meaning to what is not really there....The eye is both guide and trickster; vision and trompe l'oeil intermarried simultaneously.

'Seeing is believing' lies at the core of the scientific method, but the reverse is how we live much of our day-to-day existence, wherein we unconsciously manipulate our beliefs to satisfy and to mess with our expectations of what it is we expect or want to see. As Walsh (2013) elaborates: ''Believing is seeing' is a belief in the knowing what is already there; a knowing derived through the inherent felt qualities and not through our thought form projections'; in other words, a belief based on the quality of the essence of that which is imagined to be real, irrespective to whether it truly is so.

Therefore, it is important to reiterate here that what follows, though at times incorporating the language of natural history and even science, is, at the core, something more akin to developing a legalistic argument, based as it is on an examination of the accounts of what it was the eyewitnesses believed it was they *thought* they saw. This can either be criticized as building a precarious house of cards or be countenanced as a possible pathway toward formulating an educated guess about the nature of the mysterious creature in question. If one believes in the fidelity of the eyewitness reportage in capturing what it was they observed, always keeping in mind the caveat about the mind playing tricks on the eyes, then the transcribed words in the accounts can be examined and reinterpreted through the lens of two centuries of accumulated scientific knowledge. But keep in mind, however, (pun intended) that a scientific mindset is itself also a reality filter.

Finally, it is worth noting that the surface areas of the two locations where the Gloucester UMO was most frequently sighted – especially Gloucester Harbour, but also Boston's harbour as well – are both considerably smaller in size than some of those freshwaters – Loch Ness, Lake Okanogan, and Lake Champlain – where aquatic cryptids have been reported to have been seen to the same degree by hundreds of eyewitnesses. This is yet another example wherein the sightings of the Gloucester UMO have more in common with repeated sightings of so-called lake monsters than they do with other sightings of reputed sea serpents observed on only a single occasion. As the former are now known to have been misidentified natural phenomena (e.g. Binns 2017), therein too may lie the parsimonious and rational explanation behind the Gloucester UMO. Certainly, it is a good place to start.

5.3.1 Inanimate illusion or real object?

Most sightings of what are reported to be aquatic cryptids are merely misidentifications of natural phenomena. Even cryptozoologists, if they are honest, acknowledge this; as for example, Table 5.1 from Oudemans (1892), and Coleman and Huyghe's (2003) textural listing of the usual suspects:

> And there is no doubt that what *most* people are seeing are *not* Sea Serpents and Lake Monsters but (take a breath, this list is a long one) such things as boats, waves, fish feeding close to the surface, tricks of light on the water, fish swimming in closely packed formations, waterspouts and water devils, currents uprooting debris from the bottom, Indian dugout bubbling to the surface, masses of floating seaweed, birds skimming the water, prototype navy submarines, and dead trees. From a distance, just about anything

> might look like a Lake Monster or Sea Serpent. And then there is psychological contagion – if people are told to look for monsters, they will see them.

Naturalists and biologists are even more skeptical. William Evans Hoyle, a marine zoologist on the *Challenger* Expedition of 1873-76, wrote the entry on sea serpents for the 1902 edition of *Encyclopedia Britannica.* He concluded that 'with very few exceptions, all the so-called 'sea serpents' can be explained by reference to some well-known animal or other natural object' (Table 5.4). He did, however, leave the window open by speculating that there still might be something out there. Most troubling to Hoyle were the perplexing accounts of the *Pauline* and *Osborne* sightings, which, incidentally, as seen in Appendix Two, I have proposed a new parsimonious explanation for, one that also pertains to the Gloucester UMO (France 2016b).

Campbell (1997) refers to certain natural phenomena as being 'sources of deception,' in that they tend to deceive the inexperienced into thinking that they are manifestations of aquatic cryptids. As Baker (1960) remarked, if observed at a distance and under certain conditions 'all inanimate objects can become most convincing 'monsters'' (Campbell 1997). Many scientists and cryptozoologists have been misled into thinking that floating logs or various parts of trees were lake monsters. Photographs of gnarly pieces of floating trunks and roots of mature trees, such as shown in Binns (1984) and Meurger (1988), can closely resemble every schoolchild's image of what a plesiosaur should look like. It has been proposed that logs lying on the bottom in the profundal zone of lakes can become buoyant through a build-up of anaerobic decomposition gasses and rise through the water column to emerge temporarily on the surface before de-gassing and sinking back down again, and in the process becoming

Table 5.4. Natural phenomena giving rise to 'sea serpent' sightings according to Hoyle (1902); excerpted from the entry for sea serpents in the Encyclopedia Britannica.

- A number of porpoises swimming one behind another may, by their characteristic mode of half emerging from and then re-entering the water during respiration, produce the appearance of a single animal showing a succession of snake-like undulations.
- A flight of sea-fowl...produced the appearance of a snake swimming at the surface of the water.
- A large mass of seaweed has on more than one occasion been cautiously approached and even harpooned under the impression that it was such a monster.
- A pair of basking sharks...have a habit of swimming in pairs, one following the other with the dorsal fin and the upper lobe of the tail just appearing above the water, and, as each animal is fully 30 feet long, the effect of a body of 60 or more feet long moving through the water is produced.
- Ribbon-fish, from their snake-like form and great length (sometimes as much as 20 feet), have been suggested as the origin of so-called 'sea serpents'.
- A gigantic squid...[with] the general form that of a serpent...and, furthermore, that their mode of progression is by means of a jet of water forcibly expelled from the siphon, which would impart that equable motion to which several observers allude as being evidently not produced by any serpentine bending of the body...[suggests that] numerous other accounts seem to be explicable by this hypothesis.
- A sea-lion...[has been] suggested...but afterwards was rejected.
- A plesiosaurus, or some other of the huge marine reptiles usually believed to be extinct, might certainly have produced the phenomena described, granting the possibility of one having survived to the present time.

misidentified as lake monsters (Gould 1934). Additionally, such surface water phenomena as floating logs or emerging boulders are even more likely to be misinterpreted as aquatic cryptids under certain atmospheric conditions. Air temperature inversions over large bodies of water in north-temperate latitudes can create optical illusions through refraction, whereby objects that would otherwise be easily recognizable as mundane, can be distorted, mirage-like, into all sorts of twisty shapes that resemble legends of yore (Lehn 1979; Lehn and Schroeder 1981). Binns (1984) shows photos of aquatic monsters that are created this way literally out of thin air.

But surely, you might argue, the observers of the Gloucester UMO saw the darn thing move. Logs mostly, and boulders certainly, don't shift about in the water in a way suggestive of locomotion. But, in this reasoning you would be wrong, for as with all things cryptozoological, believing is seeing. Oudemans (1892), to his credit, singles out several interesting examples of how the credulous can be duped into thinking they are seeing sea serpents when observing floating logs. In the 1860 pages of the *Zoologist*, A. Adams (*in* Oudemans 1892) recounts the following encounter that occurred in south-east Asia:

> An incident occurred on board the vessel of which I am surgeon, which, I think, deserves to be recorded as an illustration of optical delusion that might have become a source of error, and given rise to yet another sea-serpent....There was little wind, and the gentle ripples covered the surface of the sea...[when] my eye rested on a long dark object apparently masking its way steadily through the water. After observing it some time in silence I was sorely puzzled and could make nothing of it. It was neither a seal nor a diver [i.e. loon] nor a fishing cormorant, for with their forms I was familiar; so I went on deck and consulted other eyes than mine. Sundry glasses were brought to bear on the suspicious object, and the general scrutiny seemed to decide that it was a large snake, about ten feet long (or much longer according to some), working its way vigorously against the tide by lateral undulations of the body. So strong was this conviction that the course of the ship was altered, and a boat got ready for lowering. With a couple of loaded revolvers, some boathooks and a fathom or so lead-line, I made ready for the encounter, intending to range up alongside, shoot the reptile through the head, make him fast by a clove-hitch, and tow him on board in triumph!

Up to this point, the anecdote could easily be believed to have occurred in Gloucester Harbour in 1817. That is until, Adams continues with his revelation: 'By this time, however, a closer and more critical inspection had taken place, and the supposed sea-monster turned himself into a long dark root, gnarled and twisted, of a tree, secured to [i.e. entangled in] the moorings of a fishing net, with the strong tide passing it rapidly, and thus giving it an apparent life-like movement and serpentine aspect.'

Then there is the encounter that occurred near Sumatra, as told by E. Pringle in an 1878 letter published in *Nature* (*in* Oudemans 1892):

> One of our party suddenly pointed out an object on the port bow, perhaps half a mile off, and drew from us the simultaneous exclamation of 'The sea-serpent!' And there it was, to the naked eye, a genuine serpent, speeding through the sea, with its head raised on a slender neck, now almost buried in the water, and anon reared just above the surface. There was the mane, and there were the well-known undulating coils stretching out behind.

Once again, except for the 'mane', this reads like the shipboard sightings of the Gloucester UMO, until that is, Pringle continues:

> But for the opera-glass, probably all our party...would have been personal witnesses to the existence of a great sea-serpent, but, alas for romance! one glance through the lenses and the reptile was resolved into a bamboo, root upwards, anchored in some manner to the bottom – a 'snag' in fact. Swayed up and down by the rapid current, a series of waves undulated beyond it, bearing in their crests dark coloured weeds or grass that had been caught by the bamboo stem...Ignorance of the shallowness of the straits so far from land, and of the swiftness of the current, not doubt led us to our first hasty conclusion, but the story...shows how prone the human mind is to accept the marvellous, and how careful we should be in forming judgments even on the evidence of our senses.

Finally, Oudemans quotes physician Andrew Wilson, commenting on all this in his 1879 textbook *Leisure Time Studies*, which includes sections on the studious pursuit of natural history: '[These sightings are] an example of a case, the details of which, when related, tempt people to maintain without further parley, that sea-serpents always resolve themselves into inanimate objects of one kind or another.' Wilson continues with an example of one such deception: 'Floating trunks and roots of trees, serving as a nucleus around which sea-weed has collected, and to which barnacles and sea-acorns – producing a variegated effect by reason of their light colour – have attached themselves in great numbers, have also presented appearances closely resembling those of large marine animals, swimming slowly along at the surface of the water.'

Before we return to the particulars of the Gloucester Harbour and its mysterious UMO, let us examine one other example of natural phenomena that has been suggested to be responsible for certain sea serpent sightings: floating masses of seaweed. In an article from 1849 in the *Zoologist*, republished in Oudemans (1892), the commander of a ship makes the following statement, from an area in the southern Atlantic, not far from where the famous *Daedalus* sighting of a sea serpent had taken place the previous year:

> The captain...perceived something right abeam...stretched along the water to the length of about twenty-five or thirty feet, and perceptibly moving from the ship with a steady, sinuous motion. The head, which seemed be lifted several feet above the waters, had something resembling a mane, running down to the floating portion, and within about six feet of the tail it forked out into a sort of double fin...A boat was lowered down...[and] under the direction of the captain, they approached the monster. Captain Herriman standing on the bow of the boat armed with a harpoon, to commence the onslaught.

Change the name of our Captain Ahab-esque hero from Herriman to Rich, and we have a scene that could have played out at Cape Ann in 1818. The result was equally deflating for those believing in sea serpents:

> The combat, however, was not attended with the danger which those on board apprehended; for on coming close to the object it was found to be nothing more than an immense piece of sea-weed, evidently detached from a coral reef, and drifting with the current...which, together with the swell left by the subsidence of the gale, gave it the sinuous, snake-like motion...What appeared to be the head, crest, and mane of the

> *immesnsum volmen*, was but a large root which floated upwards, and to which several pieces of the coral reef still adhered.

The writer regrets discarding the seaweed-coral concretion once it began to rot, as it could have been exhibited in the Thames where 'the conflicting motion produced by the tide and steamers would in all probability give it a like appearance [to a sea serpent].'

A letter from Captain Smith was published in 1858 in the *Zoologist* (*in* Oudemans 1892), and recounts a similar story, in this case on a voyage from Burma:

> [We saw] a very extraordinary looking thing in the water, of considerable length. With a telescope we could plainly discern a huge head and neck, covered with a long shaggy-looking kind of mane, which it kept lifting at intervals out of the water. This was seen by all hands, and declared to be the great sea-serpent. I determined on knowing something about it, and accordingly lowered a boat...I watched them very anxiously, and the monster seemed not to regard their approach. At length they got close to the head. The seemed to hesitate, and then busy themselves with the line, the monster all the time ducking its head, and showing its great length.

Once more, we have a motif similar to that of the Gloucester saga, until, however, the following revelation:

> It was hoisted on board. It appeared somewhat supple when hanging, but so completely covered with snaky-looking barnacles, about eighteen inches long, that we had it some time on board before it was discovered to be a piece of gigantic sea-weed, twenty feet long, and four inches diameter; the root end appeared when in the water like the head of an animal, and the motion given by the sea caused it to seem alive...So like a huge living monster did it appear, that, had circumstances prevented my sending a boat to it, I should certainly have believed I had seen the great sea-snake.

When his observational abilities were challenged by supporters of the *Daedalus* sea serpent, Smith again took pen to paper, this time in a letter to the *Times* in 1858 (*in* Oudemans 1892):

> That in the water, before being divested of its extraordinary-looking appendages, the diameter of my marine creature was above three feet. Some buckets full of splendidly-coloured blue and crimson crabs, varying from the size of a shilling to that of a man's hand, were collected from it, and that this quantity of such animal life could be furnished with a refuge in the mats of snaky-looking creatures which constituted the moving monstrous-looking external will assist those who read my account in believing what I before stated that even when the object was laid on deck we had difficulty in making out what it was.

A final account in the same vein comes from a letter republished in *Nature* in 1881 from Captain Taylor (*in* Oudemans 1892), and concerns events that happened off Cape Town:

> While I was in my ship at anchor in Table Bay, an enormous monster, as it appeared, was seen drifting, or advancing itself...into the harbor. It was more than one hundred feet in length, and moved with an undulating snake-like motion. Its head was crowned with what appeared to be long hair, and the keen-sighted among the affrighted observers declared they could see its eyes and distinguish its features. The military were called out,

> and a brisk fire poured into it at a distance of about five hundred yards. It was hit several times, and portions of it knocked off. So serious were its evident injuries, that on its rounding the point it became quite still, and boats went off to examine it and complete its destruction.

Again, it is Gloucester 1817-1819 redux; until, however, the phycological revelation: 'It was found to be a specimen of the sea-weed...and its stillness after the injuries inflicted were due to its having left the ground swell and entered the quite waters of the Bay.'

The first lesson from these examples is that if one is looking for a sea serpent, then there seems to be no lack of opportunities to find one, curtesy of Mother Nature, vis a vis simple deceptions of mundane natural phenomena. The second lesson – and one that will be returned to later – is that it is remarkably easy to be deceived if one has, to use an ecological term, a 'search image' for one's pre-targeted species of prey (e.g. Ishi and Shimada 2010), in this case a sea serpent. What is most remarkable is that the eyewitnesses above, many of whom were experienced mariners, had to essentially get close enough to stick their noses right up against the UMO in order to discern that it was not a living animal. What the implications of this are for interpreting the Gloucester UMO will be elaborated on subsequently.

So, the question is: can such natural phenomena of floating wood or seaweed be invoked to explain the particular case of the Gloucester UMO? Gosse (1861), an earnest believer in the existence of sea serpents, is dismissive of the seaweed hypothesis: 'But surely it must have been a very dim and indistinct view of the floating and ducking object, which could have mistaken this [piece of seaweed] for a living animal; and it would be absurd in the highest degree to presume that of such a nature could be the creatures, going rapidly through the water at ten or twelve miles an hour, with the head and neck elevated, so distinctly seen.' While Gosse's criticisms may be apt for sightings in the open ocean, they ignore that fact that coastal waters have notoriously complex patterns of water circulation. From the above narratives, it seems that the likelihood of floating objects being misconstrued as sea serpents may depend on a perception of locomotion as created by flowing water.

Several locations in Massachusetts are known for the strength of their tidal currents. For a person who grew up in the interior of the continent, and learning to canoe on raging rivers in the Boreal forest, it was the oddest experience to go white water kayaking in the ocean, which in my landlubber naivety I had imagined to be an unexciting place in which to paddle unless there was a storm churning up the shore-pounding waves. But upon moving to Massachusetts, I soon found out otherwise. There are in fact many places wherein the once open bays have become tidally-chocked due to construction of environmentally insensitive roads – a subject I was to write about in my section in a book on road ecology (Forman *et al.* 2002), the first to be published on the subject. In such cases, the tidal surge is squeezed between two sides of a bridge, producing a strong river of incoming flood or outgoing ebb current, enough to make kayaking a challenge. And had there been any long strands of entangled wood or seaweed present, such extensions would have indeed resembled vigorously swimming sea serpents. And there are also natural places, such as Woods Hole on southern Cape Cod, which are notorious for their extremely strong alongshore currents, in some cases up to four knots. My first time driving there to launch a kayak was intimidating in consequence of seeing all the boat salvage operators along the waterfront who obviously made a healthy living rescuing water craft wrecked due to the current. Later, after capsizing in the current, I was being pushed out to sea, and only stopped myself – so I thought in panic at the time – from being swept all the way to the Azores, by colliding with and clinging desperately onto a large buoy until I could be rescued. Coastal currents such as internal tides therefore depend on a combination of topography and bathymetry. Internal tides or waves are subsurface oscillating surges produced by gravity and dependent upon water column stratification patterns due to density differences produced by temperature or salinity (Sutherland 2010). *Seiches* are waves that occur in enclosed and partially-enclosed bodies of water (lakes, harbours, bays, estuaries, and lagoons such as Venice's, which I have investigated (France 2008, 2011b)). They are produced when strong winds push water from one end of the basin to the other, such that, when the wind stops, the water oscillates back to the other side of the enclosed area; i.e. like water sloshing about in a bathtub (Csanady 1973; Hutchinson 1957). So, what about Gloucester Harbour in this regard?

Gloucester Harbour is open to the sea, with a mouth extending more than a kilometre wide. In consequence, the tidal height is modest, between one and two metres. Therefore, unlike other locations in coastal Massachusetts, no kayaker seeks out Gloucester as a location in which to go for playing in tidal currents. Although internal waves occur elsewhere in Massachusetts (Haury *et al.* 1979; Scotti and Pineda 2004), they are not noted for Gloucester Harbour. Low rhythmic seiches in harbours are small, generally resulting in standing waves of only several centimetres in height and a duration of only a few seconds. Only in large lakes can seiches produce irregularities in surface waters of a scale to perhaps create impressions of giant monsters (Teresi 1998), but even there the effect is more likely to be water merely piling up on the windward shoreline rather than as a series of isolated, hump-like standing waves, something we've known about for a good while (Jackson 1833).

Dynamic surface water conditions in Gloucester Harbour only really develop during onshore storm surges. It is therefore very unlikely that a strand of seaweed with entangled wood, floating on its own in the harbour, being shifted here and there by the slowly moving waters on the calm days in which the UMO was often sighted, could have been misinterpreted as being a sea serpent. Most of the harbour shoreline had long ago been denuded of trees, and the predominant form of SAV or submerged attached vegetation along the New England coast is short-stalked eelgrass (France 2013), not long-fronded kelp. As such, an ambient source of wood and seaweed to form a sea serpent surrogate would have been rare. Also the likelihood of atmospheric-induced illusions seems remote. Climate records indicate that summer conditions from 1817 to 1819 were not atypically hot for the region (you will remember that 1816 was the year when summer did not arrive), suggesting that temperature inversions causing light refraction – rarer anyway for such a mid-latitude location compared to what the Norse experienced on their voyages of Viking exploration in the arctic – would not have been an issue (interestingly, this is in contrast to the situation of the extremely hot year of 1933 in northern Scotland when the bulk of Nessie sightings occurred – Binns 1984).

And then, of course, there is the observed behaviour of the Gloucester UMO itself. Unlike the aforementioned narratives of undulating fronds of seaweed or entangled wood, which bobbed up and down and gave the impression of movement in a linear direction in response to tidal flows or ocean currents, the Gloucester UMO zipped about the harbour every which way, twisting upon itself as if tracing the pattern of a giant pretzel on the surface of the water. Short of being caught up in a set of river or tidal bore rapids and whirlpools, water just does not move about this way. It is therefore impossible to reconcile the movement patterns of the Gloucester UMO as described by eyewitnesses with any conceivable water circulation patterns known to exist for the harbour. But that still leaves one further, aqueous-based, inanimate possibility to explain the UMO: surface water disturbances such as waves created by wind and/or wakes produced by boats.

Could the Gloucester UMO have simply been due to misidentified surface water turbulence, produced either naturally by wind or anthropogenically by boats, as has been suggested for other locations? Cheezum (2007), for example, provides a detailed environmental history of 'Chessie', the sea monster thought to have inhabited Chesapeake Bay from the 1960s to 2000. Longtime residents there inevitably posited mundane explanations for the 'monster' sightings made by freshly-arrived newcomers to the Bay:

> Holechek implied that the Frew sighting was based on ignorance of the water: 'Like the Frews,' he condescended, 'I live on the bay and look out over its surface in all kinds of weather. Just when you think you've seen all of its characteristics, something new occurs.' He recounted seeing a wave 'rise mysteriously 2 feet over a flat calm surface, then speed toward the shore...Old-timers acknowledge the strange waves, but none can explain them. Ship and boat wakes do look like the backbone of a spiny snake, especially at certain times of the day.

And Cheezum (2007) provides another report from a local newspaper: 'Many local watermen say that on some days a boat can go by and as long as an hour can pass and all of a sudden you may see three or four perfectly aligned bluish-black humps in the water. At first you don't know what to make of it, but after a while it gets closer and then eventually breaks on your boat, just wakes! Many agree that they sometimes appear to be a spiny sea serpent.'

Waves and wakes occur in a myriad of different forms. Through decades of conducting limnological research, I have spent thousands of hours along the shorelines, in boats on the surfaces, and immersed during diving, on/in lakes of all shapes and sizes around the world. From that considerable experience, I well know that the perceived relationship between an Aeolian (wind) event and a water effect is often not a simple and obvious one. Wind can have profound effects on the hydrodynamics of lakes – something which I have studied (France 1997) – and can produce all number of complex, and sometimes surprising, wave patterns (Hutchinson 1957). The same phenomena pertain to semi-enclosed estuaries and marine bays and harbours as well. Add to this, the intersecting and shoreline rebounding actions of multiple boat wakes – again something which I have observed in the Venice lagoon where the structural erosion of saltmarshes and buildings caused by such is a serious problem (France 2008, 2011b) – results in a patchy mosaic of surface water irregularities that could easily be construed as being related to an inferred sea or lake creature of monstrous proportions. Then, consider the *soliton* or 'wave of translation,' for example, a self-reinforcing solitary wave that is able to maintain its shape while it propagates at a constant velocity, sometimes traveling over considerable distances without either flattening out or steepening and toppling over (Drazin and Johnson 1989). The soliton phenomenon develops from a cancellation of nonlinear and dispersive effects and was first described in 1834 in Scotland by naval architect and shipbuilding engineer, John Scott Russell, who explains it thus:

> I was observing the motion of a boat which was rapidly drawn along a narrow channel by a pair of horses, when the boat suddenly stopped – not so the mass of water in the channel which it had put in motion; it accumulated round the prow of the vessel in a state of violent agitation, then suddenly leaving it behind, rolled forward with great velocity, assuming the form of a large solitary elevation, a rounded, smooth and well-defined heap of water, which continued its course along the channel apparently without change of form or diminution of speed. I followed it on horseback, and overtook it still rolling on at a rate of some eight or nine miles an hour, preserving its original figure some thirty feet long and a foot to a foot and a half in height. Its height gradually diminished, and after a chase of one or two miles I lost it in the windings of the channel.

Waves and wakes have often been commented upon with respect to Loch Ness and its putative cryptid. Campbell (1997), for example, found that whereas in rough weather boat wakes are soon broken up and dispersed, they may linger and grow in prominence during calm periods:

> A wake consists of a series of individual wave-fronts which move at an angle to the line of the wake. Consequently, from a low angle, the wake can appear as a series of humps travelling across the surface...In calm conditions a wake may travel a great distance. In fact there is a type of wave called a soliton which travels great distances without losing much energy...Complications set in when two very obvious wakes intersect; even though the individual wakes may not have been very obvious, when they cross constructive interference can cause an obvious hump of water that will move in a direction different from that of either wake, appearing to make its own wake. Baker and Westwood [two researchers who all the way back in 1962 effectively debunked Nessie] noticed that such a hump, after appearing to remain stationary for some time, would suddenly leap forward across the water, giving the impression of a bow wave and following wake. They noticed that very impressive 'monsters' with one or more humps were formed, and that the effect occurred up to half an hour after a boat had passed. In spite of regular traffic through the lake, it was rare enough to take even the most experienced observers by surprise. They concluded that it was most likely that all the multi-humped N [Campbell's designation for 'Nessie'] were really such wake effects.

Furthermore, observers have found that when several wakes intersect in the loch, 'an eruption of humps,' 'a trail of obedient humps,' or 'an animal-like brown and glistening hump with foam at one end like a lashing tail' can develop, sometimes not appearing until the boat has passed out of sight (Campbell 1997). The 1962 investigation determined that wave-like deceptions predominate in Loch Ness. The famous Dinsdale film of the monster is really 'nothing more than waves plus a little imagination,' and the famous MacNab photograph shows that 'a wake effect appears to be visible *ahead* of the 'monster', suggesting that it was indeed caused by a boat or boats' (Binns 1984). This led Binns to conclude that most, if not all, of the multi-humped sightings of Nessie were seemingly 'washes from nowhere' caused by boats which may be twenty minutes away up the loch by the time the 'monster' is created and sighted. Even an ardent 'Nessie' believer such as Mackal (1976) was forced to dismiss fully 90% of all sightings in Loch Ness as having nothing to do with what he believed to be the 'real' monster. And Drinnon (2010) reanalysed Heuvelmans' tabulated sightings of sea serpents, concluding that many could be explained by wave actions. So what about Gloucester Harbour and its UMO?

As I have personally observed on several occasions, when a nor'easter blows, storm surges can enter southwest-facing Gloucester Harbour and produce waves. But the effect is much less dramatic, and nowhere near as impressive as occurs down the coast in Marblehead, for example, whose much narrower harbour in orientated toward the north-east and in consequence directly points into the prevailing wind and therefore develops whitecaps that crash about the buildings there. In short, Gloucester Harbour is relatively well protected from winds of gale force intensity or otherwise – a factor in its selection by settlers as the repository for their fishing fleet – and its waters are less likely to generate wave patterns that might be misconstrued as being a sea serpent compared with some other locations.

For Loch Ness, a statistical analysis of hundreds of sightings throughout the 1960s (Olaf-Willums *in* Binns 1984) revealed that the vast majority were made in calm surface conditions, and of these only a fraction referred to any movement of the object being observed. A wind-induced set of standing waves therefore seems to be a believable supposition for producing an impression of multi-humped Nessie. With respect to the Gloucester UMO, not enough background detail is provided to know the ambient weather conditions at the time of the bulk of the sightings. What is known, is that many eyewitnesses reported the UMO to display, for extended periods, rapid and multi-directional movement, some purposely mentioning the presence of calm surface waters at the time of the sightings. Wave-induced apparitions of the UMO resulting from ephemeral Aeolian provenance therefore seems unlikely.

During 1817 to 1819, Gloucester Harbour would have experienced considerable daily traffic from schooners heading out or returning from fishing trips. Added to this, the harbour would have been chock-a-block full of dories, skiffs, and other small coastal vessels powered by oar or sail. This would have been particularly the case during the times when many were competing to bring in the prize of the Sea Serpent for the offered reward. In consequence, even on calm, windless days, it is likely that parts of the water surface would have been repeatedly disturbed by intersecting wakes generated from the many water craft plying the waters and rebounding from the complex, variegated shorelines. Is it possible that sightings of the UMO may have resulted from such misidentified wakes? Except for but a few such occurrences, I would suggest that the likelihood is low. The frequently observed notable speed and sudden reversals in direction of the UMO while disported itself on the one hand, and the motionlessness with which it floated upon the water on the other, makes it difficult to imagine any Venn diagram combination of conflating wakes and waves that might explain *all* instances of such behavioural contrasts. Even more dissuading for accepting the wake-wave theory is that the Gloucester UMO, unlike sightings of the phenomenon in Loch Ness, was seen for extended periods of time, sometimes for the duration of an entire day. That, and the fact that the sightings were made simultaneously by dozens of people, significantly from many different vantage points, makes it very hard to imagine that the happenstance collision of waves and wakes were responsible for the body shape that was repeatedly observed on all occasions. Finally, the aforementioned wave and wake theories for Loch Ness were generated in the modern era of motorized boat traffic. The speed in which water craft moved about Gloucester Harbour in the early nineteenth century, powered by wind and oars, makes it less likely that water disturbances were of the same magnitude as observed today (see Appendix One for my own sighting of a wave and wake-induced Gloucester UMO).

However, where surface water disturbances *do* make a contribution to the Gloucester UMO is in their possibly being responsible for exaggerating estimates made of its size by some of the eyewitnesses. As Binns (1984) believed in relation to the Loch Ness Monster, any wake can give an illusory impression of being one or more solid humps behind the structural one caused by an animal's body. A significant proportion of the eyewitnesses mention seeing a wake associated with the Gloucester UMO, something not surprising given the reported speed of the creature. It is possible, therefore, that what some eyewitnesses believed to have been the terminal-most 'humps' of the UMO, especially in those accounts where many such body segments were observed, may in fact have been disturbed water in the trailing wake. But it might not be as simple as that, as eyewitness Colonel Perkins (Account 35) suggests the opposite. After mentioning that the UMO caused an agitation in the water, such as the wake produced by a small vessel, he goes on to state his belief that the multi-humped, 13 metre-long body he observed must be of a greater length than what was shown due to the extent of the 'considerable wake in his rear.' So, here we have an example of an eyewitness who seems able to differentiate between an UMO body and an UMO-generated wake.

And let us not forget the puzzling circumstances of Captain Rich's stalking and capture of the purported sea serpent. As for the likelihood of whether the tuna that was eventually caught was the actual UMO, this will be discussed in a later section. At present, we will restrict ourselves to the creature's wake. To recap, Rich and his crew spent at least an entire day stalking what he, and they, believed to be the UMO, before finally hurling a harpoon that stuck in the creature, resulting in the boat being pulled about for a brief period before the harpoon pulled free from to whatever it had become temporarily affixed. Later, after the capture and display of the giant Bluefin, and in response to questions as to how anyone could be so daft so as to suppose that, big though it was, such a fish could bear any relationship to the UMO, Rich responded to his discreditors. In his letter to the editor, he stressed that of his eleven-man crew, eight had personally seen the UMO on previous occasions, and as such could vouch for its authenticity. The day of the capture was stated as being perfectly calm, from which we can assume that there were with no interfering, Aeolian-produced waves. Again, the experienced eyewitnesses verified that the 30 metre-long creature which they pursued and eventually captured was the selfsame UMO that had been seen hitherto by themselves and many others. Rich's account goes on to address the perplexing issue of how a 30-metre sea serpent could suddenly shrink to the proportions of a Bluefin tuna. And here we have to read between the lines as Rich is either being deliberately vague or simply sloppy in presenting his evidence. To reiterate, he states:

> If I am asked – how is it possible for a Fish like this to produce such a wonderful appearance, I can only answer: His peculiar movement and his velocity produced a greater deception than I had ever saw before, and that describing his body as being like kegs fastened together, struck me so forcibly, that had I not followed it up and discovered the deception, I should have given my Account to the long list already given, of the existence of a Sea Serpent on our coast.

'I used no deception,' he emphatically concludes. From Rich's statement we may infer that what he, his crew, and all the other eyewitnesses, had described as being a series of multiple humps – resembling ever so much a string of floats or kegs tied together – were in fact not so. All apparently had been duped. The many-humps of the so-called sea serpent were not structural at all but simply an artifact of the peculiar movement and velocity of said giant Bluefin. As to why a swimming fish's movements might be so described as being 'peculiar' is not elaborated upon. And as for the 'velocity' aspect, from this we may infer that Rich was referencing the large wake created by the rapid transit of the creature through the calm surface waters. His reticence in clearly stating this is at best puzzling, raising more questions than answers.

The Bluefin tuna-cum-sea serpent link will be explored further below, following a discussion as to why I believe, in direct contrast to Rich, that what so many eyewitnesses described as being the body of the UMO was in fact a physical entity and not merely an aqueous illusion. That said, from the above arguments what we can be certain of is the following illation:

Conclusion: The Gloucester UMO was a real object, and a living creature.

5.3.2 Multiple animals or single creature?

In the Linnaean Society *Report* (Davis *et al.* 1817), sandwiched between the eyewitness testimonies and the description of the putative baby sea serpent, are several accounts of other sightings presented to demonstrate the precedence for such creatures. One is a letter from Reverend William Jenks concerning a New England sighting of a twenty metre-long sea serpent that occurred in in 1809: 'A gentleman of intelligence (Rev. Alden Bradford of Wiscasset, now Secretary of the Commonwealth) inquired of Mr. Cummings whether the appearance might not be produced by a number of porpoises following each other in a train; but Mr. Cummings asserts, that the animal held its head out of the water.' Oudemans (1892) presents a series of letters exchanged between the Reverends Cummings and Bradford at the time, which were later printed in several issues of the *American Journal of Science and Arts* in 1820, at the height of speculation about the Gloucester phenomenon. In one, Rev. Cummings retorts that: 'Who ever saw fifty or sixty porpoises moving after each other in a right line, and in such a manner that those who formed the rear were no larger than haddock or mackerel, and none but the foremost shewed his head?... Now who ever saw a porpoise swim so great a distance without emerging at all?'

The idea that a group of animals observed swimming in tow might be mistaken for a sea serpent (Figure 5.7) has been frequently discussed in the literature (reviewed by Oudemans 1892), and is a favourite parsimonious explanation advanced by skeptics, such as Lee (1883), who believed it applied to every case cited by Pontoppidan. For example, an entry about an 1805 Danish account in the 1837 natural history book, *Essai sur la physionomie des Serpens*, by German herpetologist Hermann Schlegel, mentions 'In summer porpoises approach the coasts and fjords. They often meet in the open sea in troops of several scores, and when the weather calm and fine, they range in a line after each other to play and to tumble; they can have the appearance of a chain of little eminences floating on the surface of the water; some fishermen of the North, seeing them at a distance, took this resemblance to an immense animal and gave it the name of sea-serpent.' Schlegel goes on to state (*in* Heuvelmans 1968) that he too might have been a believer in such fantastic creatures had not he had the following encounter:

> I too should be silent, when the doubt by which I always felt had not been turned into certainty by a little observation I made in the spring of 1826. Once when hunting on a stormy day along the coasts of the sea, I suddenly saw a sea-animal of great size swimming before the mouth of the Rhine-river. I was about to fire at this animal which I took for a shark, when I distinguished through the fog several others closely following each other. For the greater part hidden by the water, the upper part of this creature could be distinctly seen only for a short moment, when it was carried on the top of a wave, and plunged down into the precipice formed before it. The illusion caused by the continuous agitation of the waves indeed contributed to make doubtful the appearance of a great number of black objects, appearing together out of the water, disappearing the moment afterwards, and the whole of which deceitfully represented the simultaneous movements of the undulations of one single body. Convinced that the animals were unable to swim in vertical undulations, I kept looking at this spectacle, till I knew this monstrous creature to be composed of a little troop of porpoises.

Newman (1848) recounts a similar encounter, in this case of a group of harp seals playing 'follow-my-leader':

> While viewing this singular proceeding (and I had many opportunities of doing so), I could not but be struck with the plausibility of one of the suggested explanations of the appearance which has obtained so widespread a notoriety under the name of the 'great sea serpent'...I could quite understand any person,

Figure 5.7. How animals swimming in a row can be construed as a multi-humped sea serpent (Oudemans 1892).

> not an unromantic naturalist, on witnessing for the first time such a sight as I have tried to describe, honestly believing that the mythical monster was actually before his eyes.

Sea serpent sightings have persisted into the present century (Coleman and Huyghe 2003; Harrison 2001; Hebda 2015; Leblond *et al.* 2014), and it is still possible to be fooled, as the following account from Daniel Loxton reveals (*in* Loxton and Prothero 2015): 'Sitting in a window seat of the tiny ferry that carries foot passengers between Seattle, Washington, and Victoria, British Columbia, I saw Cadborosaurus [sic; the local name for presumed sea serpent] in broad daylight. When I noticed its rolling coils, my heart leaped into my throat. There I was, a 'professional skeptic,' and I wanted to shout out loud. It was only with sustained observation that I was able to break down the illusion – and even then, my brain kept *trying* to interpret those porpoises as a sea serpent' (I, too, while taking a ferry between Vancouver and Victoria, British Columbia, have been temporarily fooled into believing I may have also caught a glimpse of the elusive 'Caddy' sea serpent).

Cryptozoologists have always dismissed the parsimonious multiple-animal hypothesis as being improbable given the remote likelihood of a train of like creatures swimming conveniently one directly behind the other, in a well-behaved straight line. Loxton and Prothero (2015), however, demonstrate through photos of small lumps of modeling clay placed randomly on a table, that when viewed from a low angle perspective, such as might occur from a small boat or the shoreline, both at a height near water, or in this case, from the level of the table surface, an illusion is created – 'presto, a sea serpent' – of lumps being in a straight line when in fact they are laterally displaced from one another.

And so, is the multiple-animal hypothesis tenable for explaining the Gloucester UMO? The answer is no, for two simple reasons. Firstly, as annunciated by Loxton and Prothero (2015), an illusion so produced is, by its very nature, extremely fleeting. The aforementioned accounts above, as well as others so identified in the literature, were all over in but an instant. Not so, many sightings of the Gloucester UMO, which lasted for minutes and even hours, and which were simultaneously observed by people from many different vantage points and elevations in relation to the surface of the water. And secondly, Oudemans (1892) is correct in dismissing this hypothesis for the Gloucester sightings for the very good reason that according to eyewitnesses 'the sea-serpent has the property of keeping his bunches [on his back] when lying quite still.' Therefore, the humps could not have been produced by a group of actively moving animals, appearing, disappearing, and reappearing sequentially above the water, wherein 'the connecting undulations beneath the surface hav[ing] been supplied by the imagination,' as Lee (1883) phrases it.

Conclusion: The Gloucester UMO was a single, serpentine animal.

5.3.3 Giant serpent or something else?

Body size is one of the most important and fundamentally defining of all biological characteristics (Peters 1986), through its influence on, and in its response to, physiology, ecology, and evolution. Abnormally large organisms, those outside the size ranges considered to be normal for any particular species, most certainly do exist. Near where I live in Nova Scotia, for example, a small town museum features details about the life of native Anna Swan (1846-88) who, at four and a half metres (7.5 feet) tall, was one of the tallest women to have ever existed. Even the most cursory of online searches can reveal remarkable photographs of dogs of such enormous size that they stagger belief ('Giant George' for instance, recognized by the Guinness World Records as the tallest dog ever, stood over four metres tall). The study of such abnormal development is referred to as *teratology*, the first part of the word, interestingly, being derived from the Greek word for 'marvel' or 'monster'. Then there are entire species, related to those familiar to us of more Lilliputian proportions, all of whose individuals are typically of a size that seems 'monstrous' but is really just normal for that particular species. Consider, just these few, for example: salamanders (*Andrias japonicas* and *A. davidianus*) in mountain streams of Japan and China that reach the size (two metres) of crocodiles, freshwater crayfish (*Astacopsis gouldi*) inhabiting rivers in Tasmania that are the size (one metre) of lobsters, earthworms (*Megascolides ssp.*) in Australia and South Africa that can approach a length (up to two metres) that makes them seem as if they are invaders from the planet Arrakis, and, my favorite, if only due to the online photos which are guaranteed to instill nightmares, land crabs (*Birgus latro*), endemic to some Pacific islands, that at more than a metre long, are the size of garbage cans or domestic pets. And those are just a few examples of species existing today. When one considers the paleontological biodiversity of species that once existed, knowledge about which we have through the fossil record, as for example, beavers and turtles as large as automobiles, clearly we need to reconsider what we refer to as being 'monstrous' and 'normal'. The world is indeed a marvellous place where, zoologically, the safest motto might be to expect the unexpected. Species of large marine animals are still being discovered (e.g. Reyes *et al.* 1991; Taylor *et al.* 1983; Wada *et al.* 2003), with the possibility that there exist others yet to be found in the future (Paxton 1998, 2009; Raynal 2001; Woodley *et al.* 2008).

So back to Gloucester's putative Sea Serpent. Despite Theodore Roosevelt's 1910 offer of a financial reward for a snake of 30 feet in length, no one has yet claimed the prize (although there is an unverified report that 32-foot (10 metre) reticulated python was captured in Sulawesi in 1912). The largest freshwater snakes are anacondas, pythons, and constrictors, all at less than ten metres in length. And the largest known sea snake is the diminutive three metre-long coral reef snake *Hydrophis spiralis*. Until recently, the largest snake ever known to have existed was the 11 metre-long *Gigantophis garstini*. In 2009, this record was broken with the discovery, in Columbia, of fossils of *Titanoboa cerrejonensis,* which indicate it to have been 14 metres (48 feet) in length (Head *et al.* 2009). A model (Figure 5.8) displayed in several locations including Grand Central Station, a TV documentary, and much fanfare and press reportage around the world, demonstrates the public appetite for the topic of monstrously proportioned serpents. Most interesting, was a lively debate in the scientific press (Denny *et al.* 2009; Head *et al.* 2009) about how the size of the snake could, or could not, allow inferences to be made about the climate during the Paleocene epoch 60 to 58 million years ago when it lived. One theory has it that the giant size of the animal reflects a warmer paleo-climate that enabled ectotherms such as *Titanoboa* to achieve much larger sizes than today's snakes (i.e. presently, the largest snakes are found in tropical wetlands, with smaller sizes existing on the land and in the waters of temperate regions such as Massachusetts). The alternative theory is that unless Palaeocene temperatures were lower than they are today, the metabolic processes of an animal the size of *Titanoboa* would have produced lethal overheating. To counter this threat, the snake would have had to spend much of its life immersed in the shallow, cooling water of its wetland habitat

Figure 5.8. Model of Titanoboa*, the largest serpent known to have existed (national Geographic for the Smithsonian Institute).*

for thermoregulation, as well, of course, as a simple necessity in order to provide the buoyancy needed to support its estimated 1,100-kilogram weight. Notably, there are no examples from the fossil record of any marine sea snakes approaching the proportions obtained for these, their freshwater wetland contemporaries. It is important to note that no extant or extinct species of snake (Feldman *et al.* 2016; Maritz *et al.* 2016; O'Gorman and Hone 2013) has ever reached the sizes that eyewitnesses ascribed to the Gloucester Sea Serpent. Sightings of enormous aquatic snakes belong to folklore (Miller 2018), not to science.

The dissimilarity in the lengths between existing and extant species indicates that what was observed in Gloucester Harbour and nearby environs in the nineteenth century was not a serpent. Further evidence in support of this conclusion comes from reports of the observed locomotory behaviour of the Gloucester UMO.

Whether explicitly inquired about during the Linnaean Society investigation, or offered independently by other eyewitnesses, the majority of all those who observed the type of swimming motion of the Gloucester Sea Serpent stated that its undulations were in the vertical, not horizontal, plane. Swimming snakes, like swimming eels, lampreys, and other elongated fish, all move through the water through the process of side-to-side locomotion. Such *anguilliform* movement depends on horizontal body waves initiated by the lateral displacement of the head followed by propagation of this displacement along the body axis to the tail (Tytell and Lauder 2004). The hydrodynamic result is a backward push of the laterally thrown coil of the body against the water, thereby leading to forward propulsion. Pelagic sea snakes (Hydrophiinae) have more flattened sides than other snakes, in order to provide larger propulsion surfaces (Graham *et al.* 1987). In contrast, the swimming behaviour of leeches (Hirudinae) depends on the vertical movement of a wave traveling from anterior to posterior along the length of the animal's body. Such a single, *metachronal* wave represents a balance between forward thrust and vertical stability, and results in a progression of peaks and troughs (Beckman 2016). More thrust could be obtained through half-waves, but this would produce too much pitching of the head. During swimming, the leech flattens its body and progresses forward through a series of quasi-sinusoidal dorso-ventral contractions that progress from the head to the tail (Chen *et al.* 2011; Harley *et al.* 2013).

The Gloucester UMO was not observed to move via horizontal anguilliform progression, and as a result, cannot have been any type of sea snake. This is notwithstanding, of course, the tiny inconvenient truth that no sea snakes actually exist *anywhere* in the Atlantic Ocean due to cold temperatures preventing them spreading beyond their Indo-Pacific distribution (Lillywhite *et al.* 2018). The UMO did display locomotion resembling that of leeches, only one species of which, however, the giant Amazon leech (*Haementeria ghilianii*), grows to more than a half-metre in size. Although perhaps a blood-sucking creature of such a size might properly be regarded as being 'monstrous,' it is unlikely, to put it mildly, how any such could have been mistaken for the Sea Serpent.

Finally, the great velocities noted by observers of the Gloucester UMO completely negate the possibility of the creature being a sea snake of any type. Aqueous propulsion is achieved through the application of the body against the density of water in order to create forward thrust. That's why fast-moving cetaceans and fishes have developed tails that are proportionally large for their respective body sizes. (Incidentally, Michael Phelps is a 'biomechanical freak of nature' (Siebert 2014), given his well-above average sized fin-like feet and paddle-like hands to provide propulsion. Add to this, his abnormally long arms and double-jointed elbows and ankles, makes this 'monster' almost unbeatable). The noted absence by all of any lateral fins, and by most, of a caudal appendage, for the Gloucester UMO, makes it physically impossible, by the simple laws of biomechanics, for any creature described as having caterpillar-like motion, to be capable of achieving the high speeds that were reported. Obviously, undulating a narrow tail up and down on the surface of the water, as displayed by the Gloucester UMO, is completely insufficient to explain the rapidity of its movement. Clearly, the force of locomotion must have come from elsewhere. And the only other place for this to be generated, is from below the surface of the water, hidden from view. As a result, we can confidently advance the following illation:

Conclusion: The serpentine Gloucester UMO was not a giant serpent.

5.3.4 Serpentine body biological or inanimate?

Today, it is widely acknowledged that many of the sightings of sea serpents and lake monsters in the past have been the consequence of the unrecognized observation of explainable, often non-living, phenomena. As recounted above, floating or moored objects in moving water, influenced by particular meteorological conditions, especially at high latitudes, can resemble an undulating multi-humped sea serpent. Waves in particular, caused by wind, intersecting boat wakes, internal waves, and seiches, are all known to produce similar illusions. But none of these external phenomena, however, are adequate to explain what was observed and reported by the thousands of eyewitnesses of the Gloucester UMO. These individuals, in boats or on the shore, at all times of the day, and over a period of weeks and even years, most certainly did see something that cannot be dismissed away as being a mere illusion; rather they saw something that was very much a real entity. So what could it have been?

From the eyewitness accounts, it can be determined that what was in fact observed on the surface of the water was most certainly *not* a living organism (Table 5.5). The UMO was mute and paid no attention to what was going on around it, being completely oblivious to the, at times, flotilla of water craft of enthused serpent watchers that would have filled the harbour during the last days of August 1817. On a handful of occasions the UMO came within a few metres, actually touched, and at times even passed underneath, various vessels, doing so on one occasion, it was noted, without providing any consequent lifting of the hull. Clearly this was a non-sentient, floating object that offered no resistance to being depressed beneath the surface. The further, and most convincing proof of the assertion of the inanimate nature of the above-surface portion of the UMO comes from its imperviousness to direct assault. On several occasions, muskets were fired at the object, the balls seen to actually strike, and even rebound without penetrating. Hurled harpoons bounced off and even a cannon ball rebounded from hitting the object. Obviously, the elongated body of the Gloucester UMO that was observed by hundreds on the surface of the water was in no way a living, breathing, sentient creature (unless, of course, one believes that cryptids are invulnerable to everything but kryptonite!). And so from this the following illation can safely be made:

Conclusion: The visible, serpentine 'body' of the Gloucester Sea Serpent was actually an inanimate object.

5.3.5 'Body' a single serpentine object or formed by component parts?

Superficially, the Gloucester UMO seems to fit the description of the classic, multi-humped Sea Serpent, seen over the centuries and in many locations around the world (Coleman and Huyghe 2003; Heuvelmans 1968; Oudemans 1892). Several of the less fantastical drawings made by eyewitnesses do in fact resemble renderings of such presumed creatures. But on closer examination of the accounts (Table 5.6), we find that the 'humps' characterizing the body of the Gloucester UMO are certainly not the coils shown in rendered drawings of imagined sea serpents, either for the one seen in New England or for others observed elsewhere. Rather, what was imagined to be the body of the Gloucester UMO was composed of a series of discrete 'protuberances' that were discontinuously separated by water, thereby giving the overall impression of a multi-jointed object, not unlike those articulated plastic toy snakes played with by children (Figure 5.9). Indeed, it will be remembered that the sequence of rigid dorsal protuberances was such a prominent characteristic of the UMO, that when an unfortunate black snake was found with a markedly deformed backbone, this led to the major blunder by the Linnaean Society in linking together the two beings.

In fact, there are dozens of accounts in the literature of UMOs that closely match the descriptions given for the one (for the time being we'll assume it was a 'one') that frequented Massachusetts and New York between 1815 and 1824. For example, Figure 5.10 shows the outline of the back of an UMO observed by a Reverend Brown in 1879, off the coast of Australia. Not only does the compartmentalized body resemble that of the Gloucester UMO, but so does it's the behaviour (from a letter to *Nature, in* Oudemans 1892):

> [There was] not a breath of wind, and the sea was as smooth as a glass...I noticed ahead of me what looked like a black log of wood a stone's throw from the shore, nearly end-on to me, and apparently more buoyant at that end; getting nearer, I noticed that it was *drifting*...and soon discovered that it was moving, leaving behind it a very long, narrow ridge on the smooth water...At the sound of my coo-ee [to call his horse] the fish started off seawards out of sight (under water), and doubled again in shore, but so rapidly as to leave both outward and inward 'ridge' on the water distinctly visible at once, like a wide V with quite a sharp corner. It gave me the idea of two fishes, the one darting outward, the other crossing its track inward at the same moment...The fish again came to the surface, showing gradually more and more of its length, till, when he was almost at rest, and all apparently in view, I estimated the length to be 60 feet, straight and taper, like a long spar, with the butt end, his head and shoulders, showing well above the surface. I can only describe the head as like the end of a log, bluff, about two feet diameter; on the back we noticed, showing very distinctly above the water, several square-topped fins...What its propelling power was I cannot say from observation; I saw no lateral fins and no fish-tail. When it started away at the sound of my voice, it was with the rapid movement of a pike or sword-fish, and yet the thick bluff head had but little resemblance to a snake.

Even Oudemans, in his commentary upon this sighting, places quotation marks around the author's use of the term 'fins' to describe the square box-like body components, implying in so doing that he does not consider them to be true dorsal extensions from a real body. So, if not fins or coils per se, or what I have been referring to, for convenience, as 'humps,' do these objects represent something else entirely? In this respect, the Gloucester UMO does have a bit of an identity crisis in that it displays traits characteristic of not one, but two, typologies that have been suggested by cryptozoologists for sea serpents.

Often the deeply-indented and irregularly-shaped components of the Gloucester UMO were described as being rough and variable in size, with one-third to one-half a metre exposed above the surface of the water. The scaly nature of the observed dorsal ridges might suggest the 'Many-finned' cryptid thought by Heuvelmans (1968) to be a living fossil of an ancient whale species. He believed that the appearance

Table 5.5. Descriptions of the Unidentified Marine Object (UMO), imagined as a sea serpent, observed between 1815 and 1824 in Gloucester Harbour and elsewhere in New England and New York, and indicative of an inanimate object that is oblivious to the surroundings and impervious to assault.

- 'The animal did not utter any sound, and did not appear to notice any thing.'
- 'I fired at him...he did not [appear more shy, afterwards]; but continued to play as before.'
- 'We immediately rowed from him, and at first concluded to pass by his tail; but feared we might strike it with the boat, concluded to pass around his head, which we did, by altering our course. He remained in the same position, till we lost sight of him. We approached so near to him that I could have reached him with my oar.'
- 'The serpent passed out under the stern of our boat...then he stood in towards us again, and crossed our bow.'
- 'The musket balls appear to have no effect on it.'
- '...two [musket] balls were thought to hit his head, but without effect.'
- '...he did not appear to alter his course in consequence of being so near the vessel.'
- '...he did not appear to be at all disturbed by the vessel.'
- '...he has never showed any indications of a disposition to chase or attack any moving object or living creature.'
- '...neither...did he appear frightened, or in any degree disturbed by their sudden and near approach.'
- 'I fired it at the head of the monster; my crew and myself distinctly heard the [cannon] ball and bullets strike against the body, from which they rebounded, as if they had struck against a rock...he almost touched the vessel...he dived, but in a moment we saw him appear again, with his head on one side of the vessel, and his tail on the other, as if he was going to lift us up and upset us. However, we did not feel any shock.'
- '...several whalers got within a few yards, but their harpoon bounced off the animal.'
- '[The serpent] passed under his boat.'
- '[The monster] received, with apparent unconcern, seven or eight salvos [of musket fire] and went on playing on the surface.'
- '...reported firing upon it, but the balls did not seem to penetrate.'

of rough skin might be due to: 'skin that is really wrinkled or bumpy; [or] smooth skin covered with parasites'; and that scaley skin might be due to: 'true scales; bony dermic plaques beneath the skin; [or] a reticulated pattern on the fur or skin.' The saw-toothed crest observed along the spine in this type of sea serpent, is often found to point forwards, about which Heuvelmans, in good cryptozoological fashion, has no problem explaining: 'This quite abnormal angle shows that they cannot be dorsal fins, for this would not agree with the laws of hydrodynamics [he is correct here]. They are more likely to be lateral excrescences of the body, as certain Archaeoceti [i.e. paleo-whales] are believed to have had [now known to be incorrect based on more recent paleontological evidence]. This crest along the sides is almost essential for the stability of an animal without a large tail or much of a ridge along the back.' By shifting the ridge from being dorsal to lateral, does this mean that the Many-finned UMOs viewed by many eyewitnesses were all swimming on their sides? If so, what does that say about the presumed 'stability' of such extensions? Untroubled, Heuvelmans (1968) plods on:

> The skin is smooth and like tanned leather, but covered, or rather supported, in places by very large scales, probably huge dermic plaques, forming, on the dorsal surface at least of the animal, segmented and jointed armour like that of the armadillos, and is believed to

Table 5.6. Descriptions of the UMO, imagined as a sea serpent, observed between 1815 and 1824 in Gloucester Harbour and elsewhere, and indicative of having a body composed of multiple segments, likened by some, to resembling barrels.

- 'I saw perhaps thirty or forty of those protuberances or bunches, which were about the size of a barrel.'
- '...about the size of a halfbarrel, apparently having joints from his head to his tail...fifty distinct portions... appeared rough and scaly.'
- '...he appeared to me to have joints, about the size of a two-gallon keg...the bunches...appeared about six inches above the surface of the water.'
- '...raising himself as he turned, six or eight inches out of the water.'
- '...his body as large as a barrel.'
- 'I judged him (in the largest part) about the size of a half barrel, gradually tapering towards the two extremes.'
- '...he saw 60 or 80 black substances on the water.'
- '...his humps plainly to be distinguished.'
- '..the humps plainly to be seen.'
- '...could plainly distinguish his humps as before described.'
- '...appearance of humps.'
- '...saw his humps plain.'
- '...a continuity of the parts from one end to the other.'
- 'I saw bunches on his back about a foot in height, when he lay extended on the surface.'
- '...appeared...in distinct bunches.'
- '...occasionally exhibiting parts of his body, which appeared like rings or bunches...he appeared jointed.'
- 'I distinctly saw bunches on his back...his size I judged to be about the size of a half barrel...ten or twelve distinct portions.'
- 'It was not, to be sure, a continuity of body, as the form from head to tail (except as the apparent bunches appeared as he moved through the water) was seen only at three or four asunder [i.e. apart from each other].'
- '...of the size of a barrel about the body, which he formed into parallel rings, which – when he is on top of the water – are so prominent.'
- 'I should judge his body about the size of a half barrel in circumference.'
- 'From what appeared out of the water I should suppose he was two and a half feet in circumference.'
- 'He describes the irregularity and unevenness of his back.'
- 'His back...appeared to be irregular, uneven, and deeply indented...ridges or humps on his back.'
- '...which at first view resembled a ship's buoy.'
- '...the body was equal to a 70-gallon cask.'
- 'We counted twenty bunches, and we supposed them on average about five feet apart...His body was of the size of a sixty or eight-gallon cask.'
- '...his back appeared to be composed of bunches or humps, apparently about as large as, or a little larger than a half barrel.'
- '...thirteen bunches on his back...'
- 'I perceived at a short distance eight or ten regular bunches or protuberances, and at a short interval three or four more...his bunches appeared to be not altogether uniform in size.'
- 'The apparent bunches on his back they consider as arising from the construction of his body.'
- 'It was difficult to count the humps, or undulations on his back, as they did not all appear at once.'
- 'I could plainly see what appeared a succession of humps upon his back.'

Table 5.6. Continued.

- 'I counted fourteen bunches on his back, the first one, say ten or twelve feet from his head, and the others about seven feet apart. They decreased in size towards the tail...The protuberances were not from his motion, as they were the same whether in sloe or rapid movement.'
- '...three yellow collars [were] observed on his neck...'
- 'There were several protuberances on his back, the highest point of which appeared to be seven or eight inches above the level of the water.'
- 'The circumference of thee animal was about the size of a common barrel...the protuberances on his back were about six feet apart.'
- '...five distinct humps...'
- '...followed by seven or eight bunches, the first about six feet from the head, all were about six feet apart, and raised about six inches above the water.'

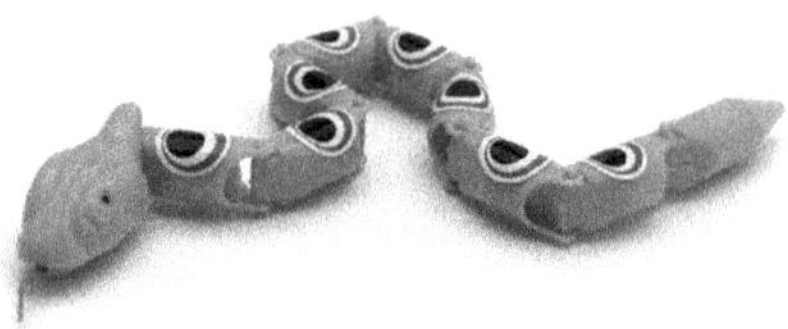

Figure 5.9. Plastic toy snake whose articulated body segments resemble the discrete protuberances reported for the UMO observed in and around Gloucester (Wikimedia Commons).

Figure 5.10. Redrawn sketch from Reverend Brown of the UMO he observed in 1879 and whose description is similar to that of the UMO observed in New England and New York between 1815 and 1824 (from an original Nature *article reprinted in Oudemans 1892).*

> have existed on some Archaeoceti. This sectional armour may make the animal look as if it has projecting ribs, and no doubt it also forms the saw-toothed crest along the spine... The Many-finned moves in vertical undulations of its body, in which it is well served by its lateral fins. But this arrangement prevents it from using its flippers to turn, and it is obliged to roll right on its side to do so, thus showing its many fins above water.

And so now it all becomes clear (I write that facetiously). Heuvelmans goes on to state that the presence of armour implies that the animal frequents rocky or coralline waters, a conclusion 'confirmed [yes, he uses that word!] by its speckled colour, which agrees with this kind of habitat.' This is the kind of teleological claptrap that plagues much of evolutionary biology theory (Peters 1991).

Numerous sightings have been made of UMOs displaying compartmentalized and seemingly armoured bodies. Several such, presented in Heuvelmans (1968), will suffice to recount. In 1921, a carcass, 20 metres long and one metre wide, was found washed up on a shore in an advanced state of putrefaction. Missing a head, the UMO was 'formed of successive segments almost all alike one another. Each segment was 2 feet long and 3 feet wide and had a pair of appendages 2 feet 4 inches long. The teguments [i.e. covering] were of a remarkable consistency and rang like sheet-metal when hit with a stick.' Another sighting was of an UMO with 'an appearance of annulations, or ring-like divisions of the body...[which were] distinctly perceptible.' In 1878, a sighting was made of a ship-sized and flipperless UMO that floated motionless on the water, and which was characterized by having an eyeless and mouthless head that resembled a tree trunk, and a series of forward-pointing dorsal protuberances.

The fact that the Gloucester UMO was not described as having numerous lateral fins does not necessarily preclude it fitting into this cryptozoological category should one be so inclined. Coleman and Huyghe (2003), in good cryptozoological fashion, suggest that such appendages might be held near the body and be used for heat dispersal. They propose calling such creatures 'Great Sea Centipedes,' whose most notable feature is the presence of 'segmented, jointed armour of bony dermal plaques.' One representative sighting described the lateral fins which lined the body as 'turning over and over' in association with considerable spray during the UMO's movement. To me this tumbling about does not sound like the actions of a *bone fide* flipper or fin involved in generating forward thrust, but rather that of an inanimate object passively responding to moving water, the force of the UMO's propulsion originating from elsewhere. Woodley (2008), however, interprets this as being indicative of the fins being 'independently movable.' In contrast, I choose to interpret such 'fins' as being mere lateral extensions of the inanimate components that comprise the elongated UMO. And although Woodley believes that such cryptids could conceivably be giant invertebrates, even going so far as to propose a Latin genus and species name for such creatures, there is here, like for so many cryptozoological theories, no evidence to support such a contention. He is correct only in dismissing the idea, held by some, that a giant polychaete worm might be a possible explanation for this particular physical typology. Further digression down such dark corridors of phylogenetic fantasy are without merit in terms of elucidating the mystery behind the Gloucester UMO.

Many of the descriptions about the morphology of the Gloucester UMO liken its component parts to resembling some type of cask or barrel. It would be wrong, however, to immediately interpret such statements as being literal confirmation that the 'body' parts were actually liquid containers *per se*. As was often the case, the components were likened to barrels in terms of their estimated size (Table 5.6). In other words, barrels of various sizes (Figure 5.11) were not only used as physical containers for holding different volumes of liquid; they also formed specific units of measurement throughout the nineteenth century. Components of the imagined body of the Gloucester UMO were estimated to range in size from a 2-gallon keg or small barrel to a large 70-gallon cask or puncheon. Most estimates referred to body segments being like a half-barrel (about 16 gallons) or a barrel (32 gallons), with corresponding diameters of 17 and of 24 inches, respectively (*Wikipedia* searches, 2017). It is interesting to note that these particular container dimensions agree with estimates made of the height projected by the 'body' protuberances above the water. As such, the size of containers equates with the observed size of the UMO's 'humps.' Given that these humps are almost certainly to have been inanimate objects (see above), there is a possibility that, in the case of the Gloucester UMO, when certain eyewitnesses likened the segmented components to barrels, they may indeed have meant more than just causally referring to only the physical size; i.e. perhaps, to some, the so-called humps really did look like actual barrels. This remains conjecture, but is supported by the subsequent revelation and conclusion made below. At best, at this stage in the sequence of building my case, the illation we can have confidence in is the following:

Conclusion: The inanimate serpentine shape of the Gloucester UMO was composed of a sequence of discrete units.

Figure 5.11. Barrels used in nineteenth-century maritime activities. Photos taken at the Mystic Seaport Museum, the Maritime Museum of the Atlantic, and the Fisheries Museum of the Atlantic.

5.3.6 Natural phenomena or anthropogenic artifact?

In February 1857, Dr Biccard and seven others observed an UMO in Cape Town Harbour, the description and illustration of which would be published in the *Illustrated London News* later that year and again in 1895 in the *Strand Magazine* (France 2018). The 70 metre-long UMO had a white-splotched body that could be submerged for extended periods, and displayed a tail that was very thin and extremely flexible, capable of twisting upon itself in tight curves. The junction of the tail to the body in the accompanying illustration (Figure 5.12) is marked by a sharp discontinuity of form, the former being smooth whereas the latter being annelated and resembling a piece of twisted rope. Oudemans (1892) remarked that the behaviour and 'astonishing lateral flexibility' of the Cape Town creature closely resembled that of the Gloucester Sea Serpent, and that 'nobody can help but laughing when he sees this figure, representing something very much like a black buoy, with white streaks and spots, and glittering in the sun, having a long rope attached to it!' Demonstrating a degree of perceptive reasoning markedly absent from much of the rest of his tome, Heuvelmans (1968) states:

> Dr Biccard's picture shows something more like an immensely elongated tadpole or a gigantic spermatozoon than a serpent. One wonders why he called the only substantial part of it a 'head', when it looks more like a 'body', and the long twisted tail, that looks nothing so much as a floating length of coir rope, the 'body'. This so-called body is so unlike any part of an animal that one cannot help thinking that it may have been a net or rope towed by a shark of porpoise which had got caught in it and whose wounded body appeared to be what the doctor called the head.

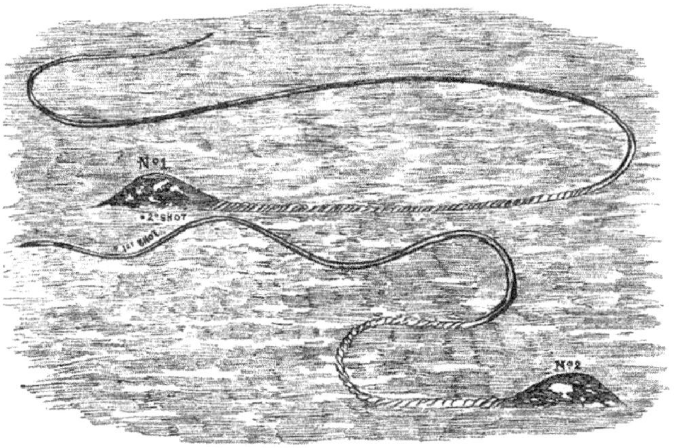

*Figure 5.12. A purported sea serpent, or simply an unidentified, surface-swimming and presumed air-breathing, marine mammal entangled and trailing a long rope of maritime debris? (*Illustrated London News *1857 and the* Strand Magazine *1895, reprinted in Oudemans 1892, Heuvelmans 1968, Sweeney 1972 and France 2018).*

My conclusion (France 2018) is that Figure 5.12 represents one of the earliest depictions of an animal entangled in maritime debris. Furthermore, that it was a 'fringe' cryptozoologist and not a 'real' marine biologist who first posited, in a Sixties' publication that marine animals could be susceptible to becoming entangled in fishing gear and be 'wounded' in the process, will no doubt be a little unsettling for some scientific purists.

In fact I believe that a not insubstantial proportion of *all* sea serpent sightings, particularly those UMOs described as being 'multi-humped' or characterized as the 'string-of-buoys' typology, such as the Gloucester Sea Serpent, may be animals entangled in fishing gear or other maritime debris that went unrecognized at the time. Before examples of these cases of misconstrued sea monsters are presented, it is necessary to first briefly consider the magnitude of what is, erroneously I believe, perceived to be the exclusively modern problem of plastic entanglement, and second to counter this misconception by reviewing the longevity of nineteenth-century, natural fibre hemp.

It is now widely acknowledged that we are residing in an era of environmental manipulation on a truly global scale: the 'Anthropocene' (Ruddiman 2013; Zalasiewicz *et al.* 2013). Massive alterations, for example, have taken place in the abundance and diversity of marine fauna in consequence of millennia of exploitative harvest and hunting (Holm *et al.* 2001; Lotze *et al.* 2006; Starkey *et al.* 2007). Chemical pollution now encircles the planet (France 2010), and though some of the most remote oceanic areas may be technically classified as 'wilderness' through absence of human inhabitation, nowhere is really free of the scattering and flotsam of human waste (Benton 1991; France 1992). Today, over six million tons of waste are jettisoned into the oceans and redistributed around the globe by currents and convergences each year (Barnes *et al.* 2009; Eagle *et al.* 2016; Wilcox *et al.* 2014). Plastic has become the predominant element of this litter flotsam (Figure 5.13; Derraik 2002; Sheavly and Register 2007; Vegter *et al.* 2014), leading some to declare the problem to be 'an ocean emergency' (Wabnitz and Nichols 2010).

Figure 5.13. Flotsam, a significant portion of it plastic, displayed in the foyer of the maritime museum in Barcelona (France 2016a).

Entanglement in maritime debris and fishing gear is recognized to be a serious conservation problem on a global scale that affects more than two hundred species of marine animals, and is thought to impact the abundance and biodiversity of marine-life everywhere (Laist 1997; NOAA 2014). Referred to by the ominous sobriquet of 'ghost nets,' lost or purposely abandoned fishing gear can continue to entrap animals for extended periods, and is an issue of particular concern (Brown and Macfayden 2012; Hall *et al.* 2000; Matsuoka *et al.* 2005). For example, one of Atlantic Canada's largest seafood companies, Clearwater, was taken to court and fined in 2018 in consequence of storing up to eight thousand lobster traps and accompanying lines in offshore waters, where they sat untended for prolonged periods as their by-catch increased.

Due to their life cycles and swimming, feeding, and migratory behaviours, some animals, such as sea turtles (chelonians) and whales (cetaceans) are particularly susceptible to entanglement (e.g. Carr 1987; Johnson *et al.* 2005; Nelms *et al.* 2005; Philo *et al.* 1992). Entanglement can result in immediate death from drowning or long-term disability due to impaired swimming and the consequent reduced ability to feed or avoid predation. The number of animals succumbing each and every year to lethal entanglement is truly staggering, amounting to many hundreds of thousands. It is one thing to read the unemotional statistics and unemotional text about entanglement in scientific papers. But to go online to see photos of the corpses of such animals, sometimes massed together in groupings of a dozen or more, or even worse, to read the harrowing first-person accounts of attempting but ultimately failing to successfully free animals (e.g. Johnson 2005), is to induce extreme heartbreak and outright anger. No murders, human or otherwise, are more gruesome, awful, disgusting, and evil, than slow death by strangulation.

Significantly, for purposes of my thesis about the explanation behind the Gloucester Sea Serpent, is the fact that thankfully by no means every animal unfortunate to become entangled, actually dies. Some can remain non-lethally entangled in fishing gear or maritime debris for extended periods of months and even years (Johnson 2005). Fama (2012), for example, mentions a North Atlantic right whale named 'Necklace,' who was observed in the Gulf of Maine for a decade with a fishing net wrapped around her tail stock. Another entangled humpback whale is known to have travelled more than four thousand kilometres (Lyman *in* NOAA 2014). Furthermore, the opportunity for an entangled animal to escape was much greater in days before the use of synthetic lines, which are much stronger and more elastic than hemp. In the past, a strong swimming whale or large fish could often muscle its way through gill nets made of cotton twine, 'leaving behind a big hole and a frustrated fisherman' (Johnson 2005), and sometimes in the process, making off with portions of the net and its accompanying floats in tow. Life for such escapees would not have been ideal, however. Notwithstanding the impaired swimming ability produced by pulling around a net and floats (Feldkamp *et al.* 1989), such animals, at least those today whom have struggled with synthetic lines, could experience ongoing health problems. Many entangled animals display wounds from the abrasion and cutting action of the debris (Derraik 2002), thereby increasing their long-term risk of infection. That fully eighty percent of all humpback whales surveyed in New England carry scars from entanglement with fishing gear (Johnson 2005), makes this an issue of deep concern.

There is widespread belief that entanglement is a modern problem with little to no occurrence prior to the advent of non-degradable plastic in the middle of the twentieth century (Derraik 2002; Gregory 2009; NOAA 2014). In fact, documented incidences of entangled whales only date back to the late 1980s, and to the early 1990s for many other species (NOAA 2014). In a survey of environmental management reports and fishing records, Balazs (*in* NOAA 2014) determined that 95% of all documented cases of entangled chelonians occurred after 1970, with none existing before 1950. From this, NOAA inferred, in their 2014 summary report, that 'the absence of entanglement records prior to 1950 could be from the low use of synthetic materials in fishing practices and land-based products.' Other conservation biologists echo such beliefs, as for example, Gregory (2009): 'Prior to the 1950s, rope and cordage used in all marine activities, including fisheries, was made of natural fibres – typically Indian or Manila hemp and cotton, and it was often strengthened with a coating of tar or strips of worn canvas. These materials lose their resilience in usage and if lost or discarded at sea tend to disintegrate quickly.' It is only in the mid-1980s that scientists began to recognize the severity of the problem of marine litter and its implications for wildlife conservation. Literature reviews by Vegter *et al.* (2014) were able to find only a single paper published in the 1950s about plastic pollution, and the earliest paper found by Nelms *et al.* (2014) pertaining thus to entangled chelonians was published in 1985.

Cordage production extends back to the Paleolithic, and for maritime uses, to the fourth millennium BCE (McCaskill 2009). During the great Age of Sail, hemp (*Cannabis sativa*) and then later abaca or manila hemp (*Musa textilis*) were the mainstay in all maritime operations, with ships each carrying many kilometres of such material (Goode *et al.* 1884; Wilcocks 1884). So insatiable was the navy's need for hemp, that in sixteenth-century England, farmers could be fined for not setting aside a portion of their fields to grow the plant (McCaskill 2009). When Europeans first arrived to New England they met Natives who used strong nets made of hemp to trap the giant sturgeon that still populated the coastal rivers at that time (Bolster 2012). The use of hemp in fishing nets grew substantially throughout the nineteenth century, and it was not until the 1950s to 1970s (depending on the region) when synthetic materials became dominant in the construction of ropes and fishing nets (Bekker-Nielsen and Casola 2010; Kristjonsson 1971).

Natural fibre ropes (Figure 5.14) and nets (Figure 5.15) always needed to be regularly repaired (Figure 5.16) due to the constant cycle of soaking and drying, swelling from salt intrusion, physical stress and fatigue, weakening from sunlight, and microbial activity (McCaskill 2009). But it would be wrong to believe that 'these products broke down quickly in the marine environment' and thus were incapable of entangling animals, as NOAA (2014) stated. Hemp actually becomes 'stronger when wet and does not rot easily in water,' owing to the hydroscopic nature of the fibre (McCaskill 2009). Also, hemp ropes were often impregnated with dye, as a mordant, or tar to slow down their rate of deterioration (Figure 5.17; Bekker-Nielsen and Casola 2010; Cardamone 2001). Under certain conditions, such as in the hulls of wrecks buried in sediment, hemp ropes can persist for centuries (Bekker-Nielsen and Casola 2010; McCaskill 2009).

Figure 5.14. Natural fibre ropes and the elongated rope-walks where the material was braided together. Photos taken at the Mystic Seaport Museum, the Maritime Museum of the Atlantic, the Charlestown Navy Yards, and the Fisheries Museum of the Atlantic (George Naas Collection).

Figure 5.15. Fishing nets constructed of hemp. Photos taken at the Mystic Seaport Museum and the Maritime Museum of the Atlantic.

Figure 5.16. The laborious reparation of hemp nets and ropes was an ongoing and necessary chore in the nineteenth-century fishing industry (Goode et al. *1884, reprinted in France 2016b). Photos taken at the Fisheries Museum of the Atlantic (George Naas Collection).*

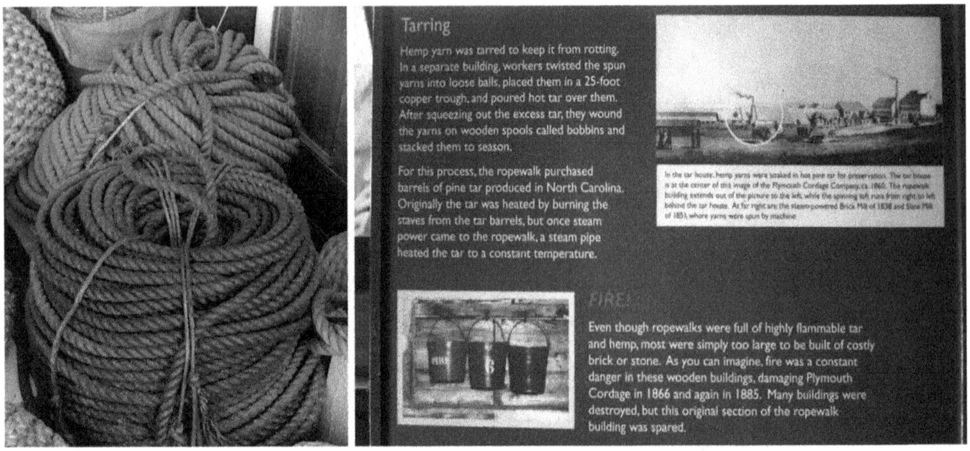

Figure 5.17. Natural fibre ropes were treated with tar (note darkening colour of front coil) as a mordant to increase longevity (as explained in the sign). Photos from the Maritime Museum of the Atlantic and from the Mystic Seaport Museum.

The percentage tensile strength and longevity of wetted rope can increase up to 20% and extend well beyond a year given proper treatment with preservatives (Aiken and Purser 1936). Synthetic ropes that absorb water actually lose about the same amount of tensile strength, and in certain circumstances, ropes made of manila hemp can have a durability equal to or exceeding those made of nylon or polyethylene (Krisjonsson 1971). Indeed, the rise in popularity of manila hemp during the nineteenth century was particularly due to its natural oils providing resistance to deterioration, thereby eliminating need for tarring to prevent rot (McCaskill 2009). In some cases, hemp ropes were even reinforced with braided wire (Goode *et al.* 1884; Wilcocks 1884).

Although there is no denying that water- and air-exposed ropes made from treated natural fibres in the nineteenth century will have generally deteriorated more rapidly than modern ropes constructed of synthetic material, the former obviously had enough longevity to justify their continued maritime use. It is my contention that any rope and net material of a durability as to be useful to the needs of fishermen or mariners will have been available for entanglement by animals. Today, it is known that the non-lethal entanglement of specifically recognized cetaceans can last for years (Neilson *et al.* 2009). It is possible that entanglement in the nineteenth century may have lasted, if not for half a decade or longer as is the case today due to plastic, at least for a duration of many months, and possibly even up to several years. This would certainly have been a sufficient length of time for the unfortunate animal to have been witnessed and misidentified as a sea serpent.

During the eighteenth and nineteenth centuries, nets were also set to specifically entrap marine mammals along the Northwest Atlantic coast (Mowat 1997). In particular, seals and porpoises, considered then as now to be competitors to fishermen with respect to cod and mackerel, were caught in the Gulf of Maine, one individual collecting a generous bounty for producing five hundred tails of porpoises over a two-year period (Bolster 2012). A large-scale commercial net fishery existed in the Gulf of St. Lawrence and southern Labrador for seal oil. Operating on the same principle as gill nets for fish, but made with stronger hemp twine, seal nets up to 83 metres long would be moored to the bottom two metres below with primitive anchors, and kept perpendicular by a string of corks along a surface headrope. As described by one fishermen, 'as the seals dive along near the bottom to fish they strike into the net and are entangled' (Bolster 2012). Another technique employed across narrow choke-points in shallow bays where seals were known to congregate, was to construct a semi-permanent frame of large nets to herd the pinnipeds into a pound which would then be hauled ashore via a capstan. In short, already by the early nineteenth century, an obstacle course of deployed fishing nets existed along the Northwest Atlantic seaboard in which to entrap any large marine animal unfortunate to become entangled.

Furthermore, with respect to fisheries that involved hunting animals such as whales, sharks, tuna, and other large fishes, with thrown harpoons (Figure 5.18), it is possible that some of these struck creatures could have survived for extended periods (Gardner 2007). Indeed, in pre-ballistic times, as many as a quarter of all struck whales actually managed to evade capture (Mowat 1997), as witness to the following account from the noted polar explorer Fridtjof Nansen (*in* Mowat 1997 and France 2016b) who had lost a small dory to an escaping whale:

> I felt pretty sore at losing such a good boat. Well I wasn't going to risk another boat, but I thought I'd be even with him all the same. Next year I took some petroleum casks with me. I rigged up three of these casks, fixing them on to three new whale-lines, and laid them all ready at the bottom of the boat. Then we started out again. Well, I made fast to a fish [i.e. a whale], and down he went in the same way. The first line ran right out and we chucked the first cask overboard. But he pulled it on down with him full pelt as before, without stopping a moment...Then the second line ran out, and we chucked the second cask into the water, but it just went after the first and disappeared in the same non-stop fashion; while the third line went running out as fast as though we hadn't had any casks at all. At length we chucked the last cask overboard; but I'll be handed if it didn't go under every bit as quick as the others. So we'd lost the whale and the lines and the casks, and we never saw them again.

One wonders if the next person who spotted this particular whale, pulling along its three intertwined ropes with large casks behind, imagined that s/he had caught a glimpse of the elusive sea serpent. There must have been plenty of opportunities to invent a sea serpent in this way. For it is worth noting that the customary technique used over the centuries in harpooning, be it for whales or for tuna, is 'kegging,' wherein the harpoon line is tied to a single (Figure 5.19) or a series of wooden kegs which the struck animal tows about until it becomes tired and slows down enough to be usually caught up with and subsequently slain (Johnson 2005). 'Usually,' but by no means always, as for example when the harpoon head is not buried too deeply and works its way out, the entanglement persisting with the rope line wrapped around the caudal flukes or fins. Furthermore, could such a harpoon, wiggling about every which way while entangled in a trail of hemp

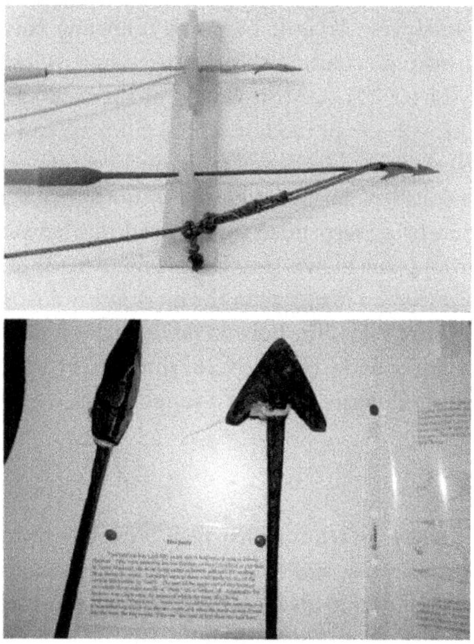

Figure 5.18. Fishing harpoons from the mid-twentieth-century at the Wedgeport Tuna Museum and Interpretive Centre and Trinity Museum.

rope and barrel floats, be mistaken for a sea serpent's 'tongue' such as ascribed by eyewitnesses to the Gloucester Sea Serpent and many other UMOs in the nineteenth century?

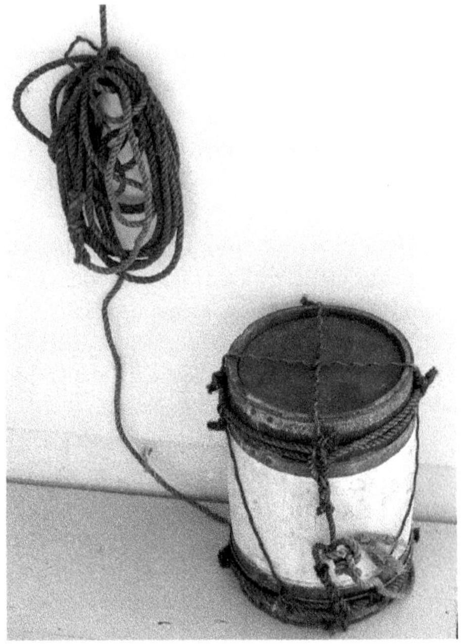

Figure 5.19. Kegging fishing gear (photo taken at the Wedgeport Tuna Museum and Interpretive Centre).

Examples of whales entangled in hunting gear abound, none described more dramatically so than by Melville (1851) in *Moby-Dick*: 'The White Whale so crossed and recrossed, and in a thousand ways entangled the slack of the three lines now fast to him, that they foreshortened, and, of themselves, warped the devoted boats towards the planted irons in him... Caught and twisted – corkscrewed in the mazes of the line, loose harpoons and lances...'

It is likely that there has been a by-catch of non-targeted species ever since fishing began (Alverson *et al.* 1994; Saila 1983), and that this may have played a role in the decline of certain species in medieval Europe (Bolster 2012). This, in turn, would have led to the attraction, exploitation, and eventual colonization of the North America seaboard (Pope 2004), given that herring were the staff of life in Christian Europe. Fishing by Europeans in the Northwest Atlantic began by following Native techniques in the use of weirs across rivers to trap anadromous fishes, before progressing to inshore ground fishes, and then to offshore pelagic species (Bolster 2012). Most fishing at the time of the Gloucester Sea Serpent sightings were undertaken through use of modest seines, gill nets, and hooks deployed from small boats (Figure 5.20). Baited hooks would be towed from outrigger poles and a 'rat's nest of lines' (Bolster 2012). Filter-feeding menhaden, unsusceptible to capture by mackerel jigs, had to be caught in beach seines (Figure 5.21), the nets being suspended from a headrope buoyed by floats, which during the nineteenth century shifted from being blown-glass balls (Figure 5.22) to larger cork floats (Figure 5.23). Casks and barrels of all sizes were a mainstay in the maritime industry (Figure 5.24) and would also be employed as net floats (Figure 5.25). The bottom line is that it is not difficult to imagine how the sight of any animal pulling a string of buoys (Figure 5.26) through the water could become misidentified as the undulating motions of the elongated body of a supposed sea serpent.

It is my contention that many nineteenth century accounts of sea serpents can be reinterpreted as evidence that marine animals have been susceptible to becoming entangled in fishing gear and maritime debris for periods far longer than what is commonly believed. For example, there is a long history of conflating or misidentifying cetaceans as sea serpents. In France (2016b) I present numerous accounts from the literature, some being initially identified as large cetaceans, others simply too strange to countenance anything other than a sea serpent (e.g. Figure 5.12, France 2018), all of which I believe to be consistent with my thesis of entanglement in fishing gear or maritime debris. A subsample of these cases are reiterated in Appendix Two. Cetaceans are by no means the only marine animals capable of becoming entangled in anthropogenic material and therefore being misconstrued as sea serpents. Entangled seals and sea lions, porpoises and dolphins, large fishes, aquatic birds, and sea turtles have also likely to have been misidentified as sea serpents. In particular, I was able to find more than a dozen accounts in which the observed UMO, whose described features are either specifically mentioned or can be inferred to resemble those of a sea turtle, also suggest early entanglement in fishing gear or maritime debris (France 2016a, 2017). A subsample of these are reiterated in Appendix Two.

It would be possible to go on to present many more examples from around the world of what I believe to have been entangled marine animals that were imagined (both at the time by the eyewitnesses, as well as thereafter by cryptozoologists) to be sea serpents. Many of these accounts follow the same train of reasoning displayed by the eyewitnesses as shown in Appendix Two: first, some creature is observed that is first thought to be a whale, or a turtle, or a seal, or a porpoise, or a fish, or a bird; next, with more careful scrutiny, a long and narrow body or tail is detected which is incongruous with any of those aforementioned known animals; this then, leads to the conclusion that the creature is unlike any recognizable type of animal; and finally, a large jump in logic is then made wherein the mysterious creature is assumed, *ipso facto*, be something completely different....in other words, a sea serpent or sea monster.

Figure 5.20. Small nineteenth-, and early twentieth-centuries shallops used for coastal fishing and as a ship's tender (photos taken at the Cape Ann Museum and the Fisheries Museum of the Atlantic).

The examples of entangled animals masquerading as sea monsters shown in Appendix Two are important for being able to construct a suite of observed physical and behavioural attributes (Table 5.7) with which to compare to those of the Gloucester UMO (Table 5.8). A striking concordance was found in the overall presence of like attributes, at the same time as differences existing in their relative prevalence amongst the three groupings of presumed whales, presumed sea turtles, and the Gloucester UMO. In terms of these latter differences, the finding that Spearman's correlation coefficients were not statistically significant among the rank-ordered attributes between those for the Gloucester UMO and those for either cetaceans or for chelonians, suggests that the entangled creature behind the Gloucester UMO may not have been either a whale or a sea turtle. Hypotheses about the identity of the mysterious

Figure 5.21. Nineteenth-century beach seine fishing (Goode et al. *1884).*

Figure 5.22. Blown-glass balls used as floats from the nineteenth-century (groups of floats, photos Fisheries Museum of the Atlantic and curtesy of the Mystic Seaport Museum Archive and Collections), and early twentieth-century (solitary float, photo taken at the Wedgeport Tuna Museum and Interpretive Centre).

Figure 5.23. Nineteenth-century cork floats (photos curtesy of the Mystic Seaport Museum Archives and Collections, and also taken at the Fisheries Museum of the Atlantic).

entangled animal observed in Gloucester Harbour will be discussed in the next section. On the other hand, the fact that there are shared attributes for all three groups of sightings, irrespective of their particular rank order of prevalence, supports my contention that what was observed in New England and New York in the years 1815 to 1824, is by no means a unique phenomenon.

Loxton and Prothero (2015) touch upon (but do not elaborate further) on what I believe to be the truth behind certain sightings:

> It is important to recognize that eyewitnesses do not usually report having seen animals shaped like serpents. Instead, they describe a series of discrete coils, humps, or dark rounded objects ('like a string of buoys' is typical) and infer that they are connected beneath the water's surface. The problem, of course, is that such sightings are by their nature ambiguous: a humungous serpentine animal might resemble a string of buoys, but a group of smaller objects (say, an actual string of buoys) also might resemble a string of buoys.

Nevertheless, this apt statement is important in that it is one of the very few mentions in the entire literature about sea serpents that countenances entanglement as a parsimonious explanation. This subsection began with Heuvelmans' (1968) similar interpretation behind the 1857 sighting in South Africa. I have been able to find another such explanation, in this case made by L. Sprague de Camp concerning the famous *Daedalus* sighting of 1848 (de Camp and Crook de Camp 1985):

> The senior author has a theory (or at least a guess) as to what this serpent might have been. If one looks at the *Illustrated London News* picture [possibly the most famous illustration of a sea serpent in history, and therefore easy to find on-line] squint-eyed, one sees the 'serpent' looks like a dugout canoe, of the kind used in primitive lands for fishing. The dark color of the 'serpent's' back would be the shadowed interior of the canoe, and the 'head' the expanded bow platform on which a fisherman places his foot in shooting or spearing a fish.
>
> Our surmise, then, is that some fishermen, fishing from their dugout off the coast of West Africa or eastern South America, rashly harpooned a whale shark. The harmless whale

Figure 5.24. Casks and barrels were a dominant feature of maritime activities in the nineteenth century (Goode et al. *1884). Photo taken at Mystic Seaport Museum.*

Table 5.7. Observed physical and behavioural attributes of UMOs posited to have been entangled cetaceans or chelonians. Numbers shown indicate the incidence of occurrence for 28 different sightings of putatively entangled whales (France 2016b, 2018) and 16 different sightings of putatively entangled sea turtles (France 2016a, 2017).

Attribute	**Cetaceans**	**Chelonians**
Narrow, tapering, sinuous, snake/eel-like shape, often with absence of a caudal fin or lateral appendages	14	10
Notable length	12	7
Extended body pulled down into water, thrown up into the air, or thrashed about on the surface	11	1
Body composed of a series of irregular, jointed component parts (multiple humps, coils, or a ridge)	6	8
Rapid speed of movement	6	-
Presence of a horn, spike, spine, mane, or other protuberance (sometimes identified as a head)	5	6
Notable flexibility of body	4	4
Vertical undulating movement of body segments	3	4
Body components likened to kegs or barrels, and sometimes scales or saucers	3	4
Obvious trailing wake or water disturbance	2	3
Overall body likened to a string of kegs or buoys	2	0
Oblivious of surroundings or impervious to disturbance	2	3
Floating motionless, gently swaying in waves, or moving very slowly	2	5

Figure 5.25. Casks used to mark deployed fishing nets in the nineteenth century (Goode et al. *1884), which from a distance and being pulled by an entangled animal, might look like a long tail of a presumed sea serpent.*

Figure 5.26. Strings of fishing net floats (Goode et al. *1884) which, from a distance and pulled behind an entangled animal, could easily give the impression of being the long body or tail of a sea serpent. Photo from the Cape Ann Museum.*

> shark is the largest living fish, reaching 45 feet, and is fairly common in tropical waters. The fish, naturally, took off. The harpoon line was belayed to the canoe. Failing to untie or cut the line, the fishermen dove overboard and swam for shore, leaving the shark to tow the boat about the South Atlantic for weeks or months until the line broke or the harpoon head tore out. This is just a guess, which can never be proved one way or another. But who has a more plausible one?

It is worth adding, that the *Daedalus* UMO was described as having a mane that 'washed about its back,' which to me bespeaks of evidence for the presence of entangled fishing nets or a harpoon line. Indeed, the supposition of de Camp is given some credence in that it reads remarkably similar to the actual encounter, noted previously, wherein the polar explorer Nansen became frustrated by the escape of a harpooned cetacean, mentioning having 'lost' a dory in such a way the previous year. In this regard, Sweeney (1972) notes that 'Fishermen who have harpooned whale sharks report that their staying power is prodigious. On being ironed or wounded, they will dive to a great depth or make a high-speed run, towing whatever vessel is on the end of the line.' The eyewitnesses of the *Daedalus* UMO may be forgiven for being confused since the whale shark (*Rhincodon typus*) was first described only in 1828, with a second specimen being found in 1848, the year of the famous encounter with the UMO.

Table 5.8. Observed physical and behavioural attributes of the purported Gloucester Sea Serpent (see 'Assembling the Composite Cryptid' in Chapter 4), which was almost certainly a misidentified entangled marine animal, ordered from top to bottom in relation to diminishing incidence of occurrence from all eyewitness testimonies and accounts.

- Considerable or 'monstrous' length (often estimated to be from 13 to 27 metres)
- Body composed of a series of irregular loosely articulated 'jointed' component parts that resemble humps of 'bunches'
- Rapid speed of movement
- Remarkable flexibility of body (i.e. exposed 'tail) which can twist and coil upon itself
- Vertical undulating movement of body segments, as if floating on the surface
- Body (again, the exposed 'tail') components likened to kegs, casks, barrels, or firkins
- Oblivious of surroundings (such as boats) and/or impervious to disturbance (such as musket shot, harpoons, and cannon fire)
- Overall body likened to a string or chain of buoys, corks or kegs roped together
- Large trailing wake
- Floating motionless on water, occasionally for extended periods
- Narrow, tapering, sinuous (i.e.' serpentine') shape with often absence of a tail or lateral appendages
- Extended body (i.e. 'tail') pulled down into water, thrown up into air, or thrashed about on
- the surface
- Presence of a 'tongue' that resembled a marlinspike or prong
- Observed in shallow water, and occasionally stretched out on the shore

Table 5.8 presents the behavioural and physical traits identified by eyewitnesses for the Gloucester Sea Serpent as summarized from Chapter 4. Table 5.9 presents quotations extracted from Oudemans' (1892) Conclusions section regarding traits he considers to be characteristic of the Great Sea Serpent. I have selected these based on my belief that they are demonstrative of animals entangled in pre-plastic fishing gear or other maritime debris. These summarizing statements by Oudemans are important for two reasons. First, they suggest that entanglement is ubiquitous. It is important to stress, however, that in no way do I propose that the entirety, or even the majority, of sea serpent sightings derive from unrecognized entangled animals. Rather it is only a certain subset of the entire corpus of all such sightings that I believe can be so explained. And second, the summarizing statements (compare Table 5.8 and 5.9) indicate the relative importance played by the Gloucester sightings in contributing to the overall conceptualization of how the Great Sea Serpent became regarded during the nineteenth century.

From Oudemans' summarization (Table 5.9), I propose that entangled animals – including that (or those) from Gloucester – which are misconstrued as sea serpents, can be described thusly: The so-called tail of such an UMO is composed of numerous discrete parts [A], of fixed position and shape [H], that resemble various containers [E] linked together in a long string [F]. These components are of various sizes [M] which emerge or float above the water [G] and taper to a point at the end of the 'tail' [N]. They are sometimes festooned with a 'mane' of flowing strands [O] and can resemble inanimate, floating objects [I]. The exposed 'body' or 'tail' of the UMO gives the impression of vermicular movement [D] though means of vertical undulations [C]. As the UMO swims about in circles [L], its 'tail' is either positioned just beneath the surface of the water [J], or on the surface where it is shown to be extremely flexible [B] and oblivious of its surroundings [K].

It is significant to note that, with the exception of [O] – presence of a 'mane', all these traits pertain to the sightings of the Gloucester UMO. This, as well as the many examples in Appendix Two, bespeaks of evidence that the Gloucester UMO was indeed an entangled animal. Note: I am very much aware of the circular reasoning of this argument, based, as it is, on the use of my own selected attributes throughout; i.e. using the presence and concordance of those attributes to develop a case that these are all examples indicative of entanglement, in the absence of independent and irrefutable evidence (but this is always the problem with endeavouring to interpret pre-photographic historical happenings that rely upon written or verbal 'proof'; see below for the actual words used by eyewitnesses to describe the entanglement of the Gloucester UMO, which comes as close as possible to being that 'smoking gun').

In summary, I believe that the evidence contained within the anecdotes and drawings noted above and in Appendix Two suggests that it is incorrect to regard entanglement in fishing gear or maritime debris as being only a recent phenomenon restricted to the age of plastic, rather than a stress with a much more lengthy environmental history. The decline, and in some cases, extirpation and even extinction of

Table 5.9. Characteristics of the Great Sea Serpent as directly quoted in the Conclusions *from Oudemans' (1892) analysis of 162 accounts, which I believe are demonstrative of entanglement. Numbers in brackets indicate the particular account (sometimes as a direct quotation) as presented in Oudemans' temporal sequence of tabulation, with those from # 34 to 71 being predominantly, but not exclusively, associated with the Gloucester UMO.*

- [A] It has the power to bend its neck, trunk and tail dorso-ventrally into numerous 'bunches'...[and] is able to bend in such a manner, for its observed lying perfectly still, showing, however, numerous bunches (34, 42, 61, 64, 67, 69, 106A, 154). In this condition it may even swim (60, 63). These bunches according to its body-length, may be of the size of a barrel (34), or from six to seven feet from each other, and from three to four feet high (154).
- [B] Its lateral flexibility is also astonishing. In turning it bends its body quite in the form of a horse-shoe, the head nearly touching its tail end (39, 41, 44); in turning twice immediately after each other or in playing, its body is bent in the form of the letter S (63). Also it may play in circles (39). Once, seized by a sperm-whale, evidently in its trunk, it wound itself laterally round the head and upper-jaw of the attacker (144). Its tail is said to lash the water (151, 158), and to wind itself up, and to rest for a moment on a part of the trunk (106A). In short it is as limber and active as an ell (44).
- [C] Generally it swims with vertical undulations (1, 2, 9, 10, 12, 18, 23, 28, 29, 30, 31, 39, 41, 42, 43, 44, 48, 51, 60, 63, 65, 66, 67, 70, 81, 82, 83, 85, 91, 92, 94, 95, 101, 102, 103, 113, 114, 115, 117, 119, 126, 128, 137, 138, 139, 150, 155, 157, see also the following figures 24, 26, 27, 32, 33, 34, 35, 39, 47, 51). The undulations may be large or small, so that their number differs, but also the animal's higher or lower position in the water is cause, that their number may greatly vary. Of course it is not always easy to account for a small number of coils. This number is mentioned to be two or three (102), three (113), three to four (138), three to five (117), four or more (139), seven (137, 157), not more than seven (137), seven or eight (9), not more than eight (41), at least ten (85), ten or twelve (44, 60), thirteen to fifteen (63), fifteen to twenty three (63), fourteen (69), twenty five (2). In our illustrations we find four ([Oudemans'] figs. 40, 47, 51), six ([his] figure 26), seven ([his] figs. 24, 39), eight (fig. 35), eleven ([his] figs. 27, 34) and twenty ([his] fig. 33).
- [D] The motion of the animal is said to be *wrongly* serpentine (29, 91, 103, 119, 157), like that of a snake (101, 115, 155), like that of an eel (117), and *rightly* vermicular (82), like that of a caterpillar (41), like that of a leech (94).

Table 5.9. Continued.

- [E] The coils are said to resemble or to be as large as ten-gallon kegs (41), half-barrels (60), flour barrels (60), large kegs (117), those of a dromedary (83), about three feet long (117).
- [F] The whole animal swimming with vertical undulations, and seen from afar, resembles a string of tuns or hogsheads, a large shoal of fish (read porpoises) with a seal at one end of it (29), a string of empty barrels tied together (60), a string of casks tied together (60), a string of large casks, gently bobbing up and down (114), a long chain of rocks (106A), a long chain of enormous rings (106A), a number of barrels linked together (106A).
- [G] The height of the coils above the water was, according to the animal's lower or higher position in the water, or according to its bulk, about six inches (41), eight or ten inches (39), at least three feet (114), only a few feet (106A); we also find the notices: 'apparently about one third of the upperpart of its body was above the water' (93), 'it partly raised itself above the surface of the water' (94).
- [H] 'The bunches appeared to be fixed' (60); 'his bunches appeared to be not altogether uniform in size, and as he moved along some appeared to be depressed and others brought above the surface, though I could not perceive any motion in them' (63); 'the protuberances were not from his motion, as they were the same whether in slow of in rapid movement' (69). See also # 137.
- [I] The instances in which the animal swims with its body in a straight line, propelling itself only with its flappers, are few in comparison with its swimming with vertical undulations (3, 18, 34, 38, 56, 59, 83, 93, 104, 115, 118, 120, 129, 130, 132, 138, 143, 145, 146, 147, 150, 160, see also figs. 28, 29, 31, 36, 45). The animal in this position resembled some drift of sea-weed (143), a mast of a vessel floating (83), an enormous log of timber floating (83), a trunk of a large tree floating (104), an unwrought spar (18), a long spar (150), a log of wood (150), and immense tree floating (157).
- [J] 'When moving under water you could often trace him by the motion of the on the surface, and from this circumstance I conclude he did not swim deep' (41), 'we could trace his course under water' (69), 'swimming below the surface so that merely a stripe indicated the rapid course' (117), 'in swimming under the surface the animal swims not deeply, for on the surface one can trace its course' (126), 'and moved away just under the surface of the water, for we could trace its course by the waves it raised on the still sea' (137).
- [K] There are instances that the animal is reported as taking no notice at all of men, vessels or other objects (29, 34, 48, 82, 83).
- [L] Like the seals and sea-lions in our Zoological Gardens, sea-serpents have often been seen amusing themselves for hours in a harbor, gracefully gliding in circles...Twice an individual just as in seals, showed its head and tail quite above the surface, the body slightly under (135, 162), stretching itself comfortably; at other times crooking up its back to sun itself (114, 137).
- [M] The animal generally swims in such a way that there is a little part of its back rises above the surface of the water, completely hiding its thickest part and its flappers, so that it makes the impression to be a serpentine animal without any appendages, and of a uniform size. So the animal is said to be ten or twelve inches thick (147), about twelve inches (113), about fourteen inches (102), fifteen inches (19), as thick as a halfbarrel (39, 41, 44, 48, 63), as thick as a common firkin (63), about twenty two inches (17), as thick as a barrel (34, 41, 80), as thick as a man's body (46), as thick as a wine barrel (2, 85), as thick as a stout man (94), as thick as a barrel of two hogsheads (12), three feet (17), as thick as a sloop's boom (24), three to four feet in circumference (25), as thick as a full-grown ox (79), about two feet in diameter (92), inconsiderable (95), as thick as a large horse (109), he is thickest just behind the head (103), several ells (115), as thick as our main mast (135).

Table 5.9. Continued.

- [N] The tail ends in a point (figs. 19, 20), and consequently is mostly said to resemble that of a serpent ort snake. It is also said to be as pointed as a boat-hook (8), or very pointed (12)...The tail itself is cylindrical (146), like that of a snake (101), and tapering to its end (8, 12, 146, 150, fig. 19).
- [O] The mane seems to begin near the occiput, and to extend over the whole length of neck and trunk, being thickest near the head, and diminishing gradually to the tail...The mane is said to have been visible on its head (135); at the back of the head (figs. 17, 24, 107)...on the neck (9, 11, 12, 101, 103, fig. 31), from the back of the head a mane commenced (91), just behind the head the mane was thickest and got thinner further backwards (91), close behind the head a mane commences along the neck (92), the mane stretched rather far hindwards (92), the head was provided with a mane hanging down (152A); evidently the mane extends from the head over the whole length of the neck and the trunk (18, figs. 28, 29). The mane near the head is long (9, 152A), tolerably long (92), two feet long, and all along the neck and back: not very long (91), that it is of some length, we must suppose, for it is said to wash about to and fro in the water (91, 118, 120), and to spread to left and to right floating on the water (92), when the animal swims. ...The mane resembles that of a horse (91, 92, 103, 118) or rather seaweed (118, 130).

marine animals over the last four centuries has primarily been through a consequence of non-sustainable, exploitative hunting (Mowat 1997). Added to this, though no doubt of a much less significant nature in terms of ultimate impact, it may very well be that ever since humans have been placing fishing nets and associated maritime equipment into the ocean, animals have been susceptible to becoming caught up in the hemp ropes that have fastened together such paraphernalia. Entanglement experts seem to agree, as reflected by several comments in Deedy's (2017) magazine article about my 'sea serpent' theory:

> 'No fisherman or mariner is going to use something which is going to break down overnight,' says France.
>
> When pressed with France's ideas, NOAA spokesman Jonathan Shannon says experts agree anything left in the ocean could present an entanglement hazard for marine life, but the problem has escalated with the use of plastic gear, which is stronger than earlier lines, and consequently brings a higher likelihood of injury or death for marine mammals.
>
> Scott Landry, director of the entanglement response team with the Center for Coastal Studies, says he has 'no doubt there has been an entanglement problem for as long as humans have been setting rope-based gear – plastic or not.'

And Johnson (2005) mentions that old-timers she interviewed acknowledged that animals would certainly become caught in traps and gill nets set along the New England coast in the early twentieth-century, long before the advent of plastic gear. Even today, it is still possible for animals to become entangled in natural fibre rope (Wegner and Cartamil 2012).

Finally, coming to the recent realization that entanglement can lead to known animals being labelled as cryptids (France 2016a,b, 2017, 2018), is perhaps not too surprising. After all, a teaser was there in the very title of a report published more than two decades ago: 'Marine debris entanglement and ghost fishing: A *cryptic* and significant type of bycatch?' (Laist *in* NOAA 2014; my use of italics). In this respect, the answer to that question, is 'yes'.

Gloucester Harbour was jammed with a flotilla of boats, the town filled with reporters, all drawn to observe the plight of the unfortunate entangled creature. August 1817, you think? No; fast-forward to October 1984 and you'd be correct. Meet 'Ibis,' a 10-metre humpback whale pulling a train of fishing nets who had wandered into the shallow waters near the mouth of the harbour. Although non-lethally entangled – she was known to have carried the nets for months while feeding – imminent disaster struck when the fishing gear around her tail and mouth became snagged on debris on the bottom (a location where trash such as broken engines had been illegally dumped for years), thereby anchoring the animal and threatening to drown her (Johnson 2005). Freed from her imprisoned tethering (Figure 5.27), and eventually from the entangling net, Ibis became a poster-child for a public just waking up to the consequences of entanglement (Figure 5.28; Himmelman 1990), something that fishermen had secretly been aware of forever.

Figure 5.27. First attempt at rescuing the entangled humpback whale Ibis who became snagged off the mouth of Gloucester Harbour in 1984. Photo curtesy of the Center for Coastal Studies, printed in Johnson (2005).

Fast-forward another sixteen years to the autumn at the start of the new millennium, and Gloucester was once again all a-buzz about the unusual sighting of a large creature within its harbour. This time the story is sadder. Meet 'Inland,' a two-year-old humpback whale who was so-named for her predilection to move in and out of the harbours of Boston, Salem, Beverley, and Gloucester over a four-month period (Capone 2011). Remarkably, the 9-metre youngster was often observed within sixteen to thirty metres, of the shoreline (Figure 5.29), in areas of sharp drop-off such as near Abbott Rock at the outer edge of Salem Harbour, where the depth is about seven metres, but also off Niles Beach, halfway down the length of Eastern Point in Gloucester Harbour, with water of a similar depth. The young whale was repeatedly observed surface lunge feeding, a trait characteristic of fast swimming whales, wherein the cetacean rushes into a fish school with mouth wide open, the prey often attempting to flee by moving to the surface (adult humpbacks, in contrast, are relatively slow-moving and known to feed by blowing bubbles to confuse and focus prey). Often Inland would lie on the surface, resting near the shore, enticing the hundreds, including reporters up from Boston, who had come out and lined the shoreline to see her, much as their forefathers had done back in 1817 in relation to the UMO at that time. The whale was last seen in

Figure 5.28. Children's book from 1990 about the plight and eventual rescue of Ibis.

Figure 5.29. The juvenile humpback whale Inland along the Massachusetts coast in 2000. Photos taken at and curtesy of the Whale Center of New England, some printed in Capone (2011).

the region near the end of December, before being found dead the following April off a Virginia beach, a gill-net wrapped around her head and tail. Today, the skeleton of this last notable marine visitor to Gloucester Harbour is displayed on the waterfront in the Whale Center of New England (Figure 5.30), a sad testament to the lethal repercussions of entanglement.

Since the 1980s when I was conducting limnological research at the famous Experimental Lakes Area (ELA) of northwestern Ontario (France 1983), in 1990 while I was a participant on the Trans-Ellesmere Island Ski Expedition (France 2010), and later that decade when I was frequently taking the ferry to Vancouver Island to visit family, I have been personally aware of the deception posed by animals entangled in debris resembling lake monsters or sea serpents of lore (ELA sighting described in Deedy 2017; France, unpubl. polar and west coast sightings). *In mari multa latent*, goes the old adage: 'in the ocean many things are hidden.' In a similar vein, the novelist Elizabeth Fama (2012), who was the first to suppose, in a blog, that the great New England Sea Serpent observed in Gloucester Harbour had been an entangled animal, supports her case with six abbreviated eyewitness accounts in which the answer to identifying the hitherto UMO was 'hidden in the plain language of their own descriptions.' Table 5.10 expands these descriptions to the entire body of eyewitness evidence. This is as close as it comes to a 'smoking gun' in terms of identifying the true nature of the Gloucester Sea Serpent. Apply the well-known 'duck test' for abductive reasoning (i.e. a form of logical inference which starts with an observation then seeks to find the simplest and most likely explanation) – in other words, if it looks like a duck, swims like a duck, and quacks like a duck, then it probably is a duck – and the conclusion is obvious and the mystery behind the Gloucester phenomenon is finally solved.

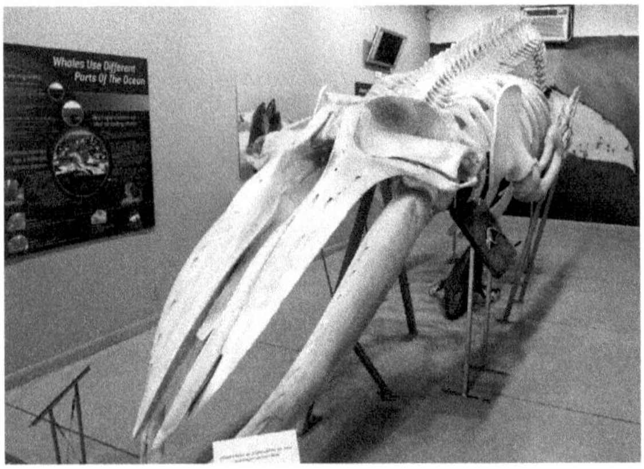

Figure 5.30. Although under natural circumstances Inland could reasonably have been expected to live to about 2040, today her skeleton is mounted in Gloucester's Whale Center of New England, a victim of entanglement.

The elongated body or tail of the Gloucester creature was thus likened to being jointed and to resembling a string or row of wooden buoys, empty barrels, water casks, or gallon kegs, tied together on a net rope 'as set in the water to catch herring.' As floatation devices, wooden barrels and buoys, the latter sometimes made of cork, were a standard feature of nineteenth-century fishing practices (Goode *et al.* 1884). It is not difficult to imagine how a view of the floats of such a net that had become non-lethally entangled on a swimming animal during the process of active setting or following abandonment, might be construed as being an enormous, undulating sea serpent if it was dragged along, bobbing up and down, over the surface of the water (Figure 5.26).

It may seem remarkable that so many nineteenth-century eyewitnesses, like those in Gloucester in 1817, failed to specifically identify trailing gear or debris in their sightings. However, one has only to go online today to find dozens of photographs showing unfortunate animals pulling long lines of entangled gear, such as widely separated buoys or floats, to see how these unnaturally elongated UMOs could easily be considered by the uninformed to be very different from any known animal. Furthermore, it is possible to observe in online photographs of entangled marine animals, examples of lengths of entwined rope and matted netting, some festooned with seaweed like kelp, that certainly do give the overt impression of solidity, making it easy, therefore, to imagine how such a structure could give the illusion of being a tail. Remember those cases described previously, which demonstrated how easily observers, even experienced mariners, can be fooled into thinking that long mats of floating algae were sea serpents. In some cases, it was not until they were within a few metres, and in one remarkable case, not until the 'serpent' had actually been captured, dragged behind the dory, hoisted on the deck of the ship, and then finally examined, that the true nature of the deception was finally recognized. Therefore, remove the use of spyglasses from many eyewitness accounts in the past, or telephoto lenses from today's photographic efforts, and misidentifications would be, and indeed still are, commonplace. In a zeitgeist wherein there is widespread belief in the existence of such antediluvian creatures, sea serpents abound; i.e. believing is seeing. If one wants to observe a sea serpent, this is always easy to accomplish, as nature, and in this case human action, provide many opportunities to provide believable doppelgangers. One lesson from *Twenty Thousand Leagues Under the Sea* (Verne 1869) concerning imagined sea serpents, is that it is easy to misconstrue the mundane for the marvellous: 'Given the ready tendency of the human brain to detect marvels everywhere it will not be difficult to understand the excitement produced all over the globe by the supernatural apparition' of the appearance of a denizen of the deep. The same pertains to the Gloucester Sea Serpent. 'Supernatural' here, in the sense of that which is extraordinary, not paranormal. Throughout history monsters have been regarded as metaphoric agents of portent (France, in prep.). And so, at such times if there was something going on in the world that was extraordinary or untoward, then that is precisely when the public would be most open to seeing monsters as interpretable signs from God. This was certainly the situation for New England in the *annuli horribilis* of 1816 and 1817.

The marine littoral is a social-ecological system (SES) characterized by complex interrelations among humans and natural resources (Folke 2006; France 2016c; Walker *et al.* 2004). The 1815 eruption of the Indonesian volcano Tambora, the largest or second-largest in recorded history, produced repercussions felt around the world (Klingamen and Klingamen 2013; Wood 2014). In particular, as Alexander *et al.* (2017) explain in a fascinating paper, 'Tambora and the mackerel year: Phenology and fisheries during an extreme climate event,' the volcano triggered a precise sequence of changes evident in the historical record that had a marked impact on the ecosystem and human population in the Gulf of Maine. Global temperatures dropped by 1 to 1.5 °C, with extreme weather events in northern latitudes, leading to 1816 being referred to as 'the year without a summer' (Stommel and Stommel 1983), as exemplified by abnormally low daily temperatures in Massachusetts (Van der Schrier and Jones 2008). An unprecedented environmental and social crisis ensued. Resulting crop failures were so severe that livestock were slaughtered for want of fodder to keep them alive through the subsequent winter, leading to a situation in 1817 wherein soup kitchens became established in urban centres and large numbers of rural people were forced to emigrate from the region to avoid famine (Alexander *et al.* 2017). By 1817, therefore, little protein was left in New England other than the seasonal arrival of spawning fish followed by their predators. In consequence, New Englanders, in the years 1816 and 1817, turned to the sea as never before for a reliable source of food. The problem, however, was that the spawning runs of anadromous fishes such as alewife (*Alosa pseudoharengus*) and shad (*Alosa sapidissima*) faltered in 1816 due to the fish being met with inhospitable temperatures due to many ponds and streams still being ice-covered late into the spring. Added to this impairment in reproduction, a regional drought may have acerbated the problem through impeding inland fish passage in developed rivers due to low base flows (Alexander *et al.* 2017). In contrast, marine spawning species such as mackerel (*Scomber scombrus*), herring (*Clupea harengus*), and menhaden (*Brevoortia tyrannus*) were not as adversely affected. In desperation, people turned to mackerel *in extremis* as they were the most abundant species near commercial distribution centres. From 1816, the 'mackerel year,' onwards, the fishery expanded. Henceforth, mackerel functioned as both a lucrative commercial product as well as a utilitarian fish for sustenance as had once alewife. The climatic emergency caused by Tambora therefore produced a rapid resource switch from riverine to coastal fisheries as a sociological response to a phenological mismatch (Alexander *et al.* 2017). And once the switch was made, it stuck; there was no

Table 5.10. Descriptions of the UMO, imagined as a sea serpent, observed between 1815 and 1824 in Gloucester Harbour and elsewhere, and clearly indicative of a marine animal entangled in fishing gear or other maritime debris.

- '...his appearance in this situation was like a string of buoys. I saw perhaps thirty or forty of those protuberances and bunches, which were about the size of a barrel.'
- '...looked like the buoys of a seine'
- '...with a good glass [I saw what] seemed like gallon kegs tied together'
- 'His body when out of the water looks like the buoys of a net, or a row of kegs, or a row of large casks'
- '...of the size of a barrel about the body, which...are so prominent, that they resembled buoys attached to each other'
- '[The body] appears in joints like wooden buoys on a net rope almost as large as a barrel, that the musket balls appear to have no effect on it, that it appears like a string of gallon kegs.'
- '...as he moved he looked like a row of casks following in a right line'
- 'He appears to be full of joints and resembles a string of buoys on a net rope, as is set in the water to catch herring. Others describe him as like a string of water casks...Two [musket] balls were thought to hit his head, but without effect.'
- '...resembled the link of a chain.'
- 'The first view I had of him appeared like a string of empty barrels tied together, rising over what little swell of the sea there was.'
- 'The back was composed of bunches about the size of a flour barrel, which were apparently about three feet apart...and looked like a string of casks or barrels tied together.'
- 'The body, which is formed into parallel rings, which – when he is on the top of the water – are so prominent, that they resembled buoys attached to each other.'
- '...and to seem jointed, or like a number of buoys or casks following each other in a line.'
- '...the curvature and bunches on his back. To some he appeared jointed, or like a string of kegs or buoys connected on a rope.'
- '...giving the appearance of a long moving string of corks.'

going back to small nets and traps straddling streams and rivers, when larger nets could be strung throughout coastal bays and offshore waters to collect more biomass.

So what does this have to do with the Gloucester Sea Serpent? Johnson (2005) describes the present-day situation in terms of the omnipresent threat posed to animals by fishing gear:

> By some estimates, there are currently over a million crab and lobster pots dotting the surface of the sea off New England alone. Early European settlers used to joke that one could walk across the water of the Northwest Atlantic on the backs of the cod. Now they make the same joke about the lobster pot buoys, and the Northwest Atlantic, including the offshore banks, has become an obstacle course for feeding whales.

Aided by the mackerel jig, which had been invented in Cape Ann in 1815 by simply adding shiny material to hooks to increase their attractiveness (Bolster 2012), the entire New England coast in 1817 became covered by a network of herring nets as never before. The result was a seascape primed to create a sea serpent through bycatch entanglement in fishing gear (France in prep.). Migrating animals unaccustomed to and therefore unwary of fishing nets suddenly faced a labyrinthine gauntlet that had to be navigated. On the other hand, here for the first time, conveniently collected in one place before them, was the novelty of a huge mass of their prey. And so, why not take advantage of it by swimming into the unrecognized net for an easy bite. A like situation developed in the summer of 2017 as I began writing this book. North Atlantic right whales (*Eubalaena glacialis*), of which only about five hundred are known to exist, switched their summer feeding grounds from the quiet waters of the Bay of Fundy, which is relatively unfished, to the Gulf of St. Lawrence, which is riddled with fishing traps and nets and filled with commercial boat traffic. The result was disastrous, with more than a dozen whales killed by entrapment and boat collisions, fully three percent of the entire species. So perhaps it is no coincidence that it was post-Tambora when the frequency of sea serpent sightings increased in New England. Alexander *et al.* (2017) describe the contingent response of fish and fishermen to this climatic event. Is this also a *geomythic* example (*sensu* Masse *et al.* 2007; Vitaliano 2007) of history's most significant volcanic eruption giving birth to history's most sighted sea serpent? I believe this to be the case.

The likening of the multi-humped nature of the Gloucester creature to being a string of entangled floats or barrels had been hinted at before in early imaginative portrayals of the creature: Craft's 1819 play, for example, referred to the 'thirty castles on his back he bore,' and later 'his back was scatt'd o'er with hills,' and came close to what I believe to be the truth of the situation:

> Sometimes at rest upon the wave he lay,
> Then cut his cables and ran swift away
> Apparently for exercise and prey.
> The fishes fled him, but they fled in vain,
> Heaven only knows, what shoals he does contain!

Rescuing marine animals by cutting them free of 'cables' of entangled debris to allow them to swim away is sadly how many of today's conservation groups spend an inordinate amount of their time.

Other imaginative efforts mentioned the string-of-floats motif. The 1817 poem 'The Sea Monster,' described the Gloucester Sea Serpent stretching out on the water's surface 'looking like rum casks swimming in a row' and with a 'ball-proof shell' (Burns 2014). In the 1823 futuristic story 'Boston. September 11, 1855* Sea Serpent Caught!,' the captured beast is examined and it is found to have 'swallowed a whole freight of casks,' which, due to its digestive system 'place the smaller ones at each of his ends, the largest in the center of his 200-foot length' (Burns 2014). And finally, tourist souvenirs were available in 1817 in the form of a 'Sea Serpent in miniature, which will move in any direction by the least touch, having joints throughout its body' (Burns 2014), which sounds like an early version of modern articulated wiggling snake toys (Figure 5.9). And more recently, Anderson's (2005) children's picture book stated that the UMO 'looked like a line of slippery floats.'

Conclusion: The serpentine shape of the Gloucester UMO was produced by a string of entangled fishing gear or maritime debris.

5.4 CANDIDATE SCREENING AND ATTEMPTS AT A TENTATIVE IDENTIFICATION OF THE ENTANGLED MARINE ANIMAL

5.4.1 What the observed characteristics of the 'head' can tell us about the identity of entangled animal behind the Gloucester UMO

Eyewitness descriptions of the head of an UMO can sometimes prove invaluable for identifying the real animal behind the legend. For example, descriptions of the 'Soay Beast' seen off the coast of Scotland in 1959 (see Appendix Two): 'The head was definitely reptilian 2 ft. 6 ins. high, with large protruding eyes. There were no visible nasal organs, but a large red gash of a mouth which seemed to cut the head in half and which appeared to have distinct lips,' from one observer; and 'The neck appeared to be cylindrical and...about 8 inches in diameter. It arose from the water about 12 inches forward of the body...The head was rather like that of a tortoise with a snake-like cranium running forward to a rounded face... When the mouth was opened I got the impression of large blubbery lips and could see a number of tendril-like growths hanging from the plate,' from another, have led to suggestion it was a sea turtle (Burton 1960), probably a leatherback (*Dermochelys coriacea*; Brongersma 1968), and one possibly entangled in maritime debris (France 2017).

What was believed to be the head of the Gloucester UMO was likened, in terms of size, to that of a horse, man, or hat crown, and in terms of appearance, to that of a dog, snake, serpent, sea turtle, horse, seal, fish, or egg (Table 5.11). The head was also said to be the size of a four-gallon keg, ten-gallon keg, or a barrel, with a shape described as being flat or flat topped, with a crest or bunch or slight hollow upon it, and a hollow underneath the chin. Numerically, the height of the presumed head varied according to different accounts, including and ranging from 6 or 6-8 inches, to 1, 2, 2, 2, 3, 3-5, 4-7, and 6-8 feet above the water, the head being from 2 to 12-14 feet long, and 3 feet in circumference. One account says an ox-sized eye was seen on the bottom of a bunch (presumably like those bunches of the 'body' or 'tail'), another stating the eyes to be horse-sized and ringed in white; one eyewitness stated there to have been no eyes, gills, mane, or breathing hole, whereas another mentioned the presence of open jaws. Clearly, the diversity of these descriptions precludes an easy identification of the animal behind the Gloucester UMO, presuming it was indeed, a single animal and not different animals observed over the years.

However, I do not believe that what was described as being the UMO's head, was actually the real cranium of the entangled animal. Rather, I concur with Fama (2012) in that 'it's likely that just by chance, the first keg or buoy in the line was different than the others, or was made of multiple objects lashed together, to create the illusion of a head raised above the body.' There are several lines of evidence or reasoning to support this contention:

1. Depending on the type of materials that comprised the anterior-most piece of debris, and the speed of movement through the water, the appearance of the 'head' might be expected to change. There are descriptions in the literature of reputed sea serpents which seem to be dragging all manner of bric-a-brac through the water (France 2016a,b, 2017), and it is possible to go on-line and find many examples of unfortunate animals entangled in a wide diversity of flotsam, including sails, boat spars, buckets, masses of knotted rope, and fishing floats, some of the latter sporting upright poles to which flags are, or at onetime were, attached (France 2017).

2. Several accounts specifically stated that no head was visible, and it must be remembered that most – actually two-thirds – of all the eyewitnesses failed to mention the presence of a head. Also, remember that many accounts mention the UMO as being completely unaware of its surroundings, taking absolutely no notice of people, boats, etc.

3. Then there is the odd behavior of the 'head' noted by several eyewitnesses. One account mentions it as moving moderately sideways; another, that it vibrated from side to side; another, that it shook its head, just like its tail, in an extraordinary manner; and a further account that it turned its head repeatedly slowly back and forth towards and away from the boat. To me, this all reads like a buoy floating on the water, rocking to and fro in the waves, the frequency dependent upon the turbulence created by wind or locomotion. I believe a similar false identification of a head to be the case for the famous Valhalla sighting of 1905 (France 2017; Appendix Two).

4. Finally, there is the presence of the mysterious 'horn' or 'tongue' of the UMO that was noted by several eyewitnesses. One describes a prong or spear arising from the head 'about twelve inches in height, and six inches in circumference at the bottom, running to a small point.' Another account describes the single horn being 'about nine inches to the foot in length, and in the form of a marlinspike.' And a third account mentions what the eyewitness thought to be a two-foot long tongue, resembling a harpoon, that was thrown out 'backwards several times over his head' to where it fell down into the water before being lifted up to repeat the cycle. I believe there are two possible explanations for this phenomena. First, again the duck test: if it looks like a marlinspike, it may have indeed been one. A marlinspike is a piece of equipment, a third to two-thirds of a metre long, made from hardened wood or metal, tapered to a rounded or flattened point, that is ubiquitously used by sailors to separate strands of rope during splicing or untying knots that have become tightened under tension (Figure 5.31). Most significantly, they are also used to create a handle in the middle of a length of rope through a marlinspike hitch (a temporary twisted knot), thereby allowing a grip for sailors that is more secure than simply holding onto the line by hand alone. A fid (Figure 5.32) is a similar, conically-shaped tool that is two-thirds of a metre long, traditionally made of wood or bone, and used in maritime rope-work (an activity that is often referred to as 'marlinspike seamanship') to separate rope strands or hold apart canvas openings during reparations. Either of these commonplace tools could easily be likened to a horn or tongue by one believing he or she was viewing a sea serpent. That it was observed in the anterior-most link in the entangled chain of debris, supports the idea that what was regarded as the 'head' was simply a slightly different-looking piece of flotsam. Second, it is possible that the pointy extension on the 'head' or first link in the entangled chain, which was noted by some to be larger than posterior 'bunches', was an upright pole as is commonly attached to a float at the end of a net (Figure 5.33). In France (2017), I give an online reference to a photograph of a group of entangled turtles in a jumble of ghost-net material that includes just such an upright pole. Could such a pole-float, or a marlinspike or fid, around which a cord of twisted hemp rope is wrapped, swaying side-to-side in the water, possibly festooned with algae, be mistaken for either the long neck noted by some, or the horn or tongue referred to by others here? I believe so. After all, whoever heard of a rigid two-thirds of a metre-long tongue being thrown out backwards, *behind* an animal's head?

Whereas morphology is disappointing as a diagnostic trait for identifying the true identity of the animal behind the Gloucester UMO (although it does provide additional support for the entanglement hypothesis), examination of the creature's exhibited behaviour proves very helpful in this regard.

Table 5.11. Descriptions of the head of the UMO, imagined as a sea serpent, observed between 1815 and 1824 in Gloucester Harbour and elsewhere in New England and New York.

- '...the head appeared to be about six or eight feet long, and where it was connected with the body was a little larger than the body. I could not discern any mouth...I could not discover any eyes, mane, gills, or breathing holes.'
- '...his head very plain of the shape of a dog.'
- '...with his head out of the water...about as large as a horse's head.'
- '...and frequently put his head out of water as if to swallow his game; his head was as large as a horse's.'
- 'His head appeared shaped much like the head of a sea turtle, and he carried his head from ten to twelve inches above the surface of the water. His head at that distance appeared larger than the head of any dog that I ever saw.'
- 'His head formed something like the head of the rattle snake, but nearly as large as the head of a horse...I perceived [no eyes or horns]...At times [he carried his head above the water] about two feet, then again he would carry the top of his head just on the surface of the water.'
- 'His head appeared full as large as a four-gallon keg...The top of his head was of a black colour, and the under part of his head appeared nearly white.'
- '...his head...was not visible.'
- 'Saw his head out of [the] water very plain [and it] seemed as large as a horse's head.'
- '...saw his head and neck very plain out of the water; and they had every appearance of a Serpent.'
- '...with his head above the water about a foot...His head appeared to be about the size of the crown of a hat...I saw no ears [and] horns...'
- 'As he drew near, and when opposite to me, there rose from his head or the most forward part of him, a prong or spear, about twelve inches in height, and six inches in circumference at the bottom, and running to a small point...his head appeared as large as a man's head.'
- 'The top of his head appeared flat, and was raised seven or eight inches above the surface of the water.'
- 'The head was flat in the water...I was struck with an appearance in the front part of the head like a single horn, about nine inches to the foot in length, and in the form of a marlinspike.'
- 'It had a head like that of a horse.'
- 'Neither his head nor tail were visible.'
- 'The head appeared to be the size of a ten-gallon keg, and six inches above the surface of the water. It was of a dark colour...The motion of the head was sideways and quite moderate.'
- 'His head was like a serpent's tapering off to a point. He threw out his tongue a number of times, extended about two feet from his jaws – the end of it resembled a harpoon – he threw out his tongue backwards several times over his head – and let it fall again – I saw one of his eyes as he passed; it appeared very bright, and about the size of an ox...There appeared a bunch above the eye...as he advanced, the head vibrating from side to side. The colour of his tongue was a light brown.'
- '...his head was more elevated above the water then the ridge or humps on his back.'
- '...a flat head with a bunch or crest upon it, his eyes about the size of the eyes of a horse, with a white ring around each of them, about three inches in diameter.'
- 'The serpent shook his head and tail in an extraordinary manner, and advanced toward the ship with open jaws...his head is full 12 or 14 [feet]...his ear-holes are about 12 feet from the extremity of his head.'
- '...with a head as big as a barrel.'
- 'His head was elevated from three to five feet...His head was of a dark brown color, formed like a seal's, and shined in glossy appearance...his head as large as a barrel...There was nothing that appeared like gills.'

Table 5.11. Continued.

- '...the head, which perfectly resembled a snake's was elevated from four to seven feet above the surface of the water.'
- '...his head was about as long as a horse's and was a proper snake's head – there was a degree of flatness, with a slight hollow on the top of his head...He turned his head two or three times slowly round towards and from the vessel, as if taking a view of some object on board.'
- 'The head [was] serpent shaped; it was elevated about two feet from the water, and he depressed it gradually, to within six or eight inches as he moved along. I could see under his chin, which appeared hollow underneath or to curve downward.'
- 'At one time his head was about two or three feet above the water, but soon disappeared to the level of the sea.'
- 'His head was out of the water...and then all his body would sink except his head.'
- 'His head is about three feet in circumference, flat and smaller than his body'
- '...and three yellow collars observed on his neck, which led some to believe it might be another individual and species.'
- 'He had a head about two feet long, and shaped somewhat like an egg, which he carried out of the water when he was moving.'
- '...with his head raised from the water about three feet...his head shaped like that of a horse...'
- 'I heard a noise in the water, and saw...something resembling the head of a fish or serpent, elevated about two feet above the surface.'

5.4.2 What the displayed locomotory behaviour can tell us about the identity of the entangled animal behind the Gloucester UMO

Eyewitnesses observing the putative Sea Serpent in Gloucester Harbour and elsewhere along the New England and New York coasts noted that it displayed three overt types of behaviour: lying motionless upon the surface of the water, twisting and turning about in tight loops, thereby giving the impression to some that it was 'playing', and suddenly and rapidly submerging.

Frequent observation of what we now recognize to have been a train of fishing gear or maritime debris lying 'perfectly still' on the surface, sometimes for durations of up to half an hour (Table 5.12), suggests several possibilities. One is that the entangled animal might have been resting or possibly sleeping. All cetaceans (whales, dolphins, porpoises) are voluntary breathers and need to consciously think about every inhalation they take. They accomplish this while resting through the act of *unihemispheric sleep* wherein they rest one half of their brain at a time while keeping the other awake to ensure breathing and to be alert to danger (Lyamin *et al.* 2008). Some sleep for short durations of not more than several minutes while lying on the sea floor; others can rest at the surface as, for example, humpback whales (*Megaptera novaeangliae*), who can doze for thirty minutes at a time. Dolphins rest at the surface in either a vertical or horizontal position, and while swimming slowly next to a companion animal. Deeper sleep usually occurs at night and is referred to as 'logging', as the dolphin resembles a floating log. Many pinnipeds (seals) prefer to sleep on land but can do so, again via unihemispheric sleep, while in the water. Some do so at the surface while assuming a posture called 'bottling' wherein their bodies are completely submerged except for their heads, and sometimes, just their snouts, being exposed (Lapierre *et al.* 2013).

Figure 5.31. Marlinspikes used in maritime rope-work which might be mistaken for a sea serpent's tongue if inserted in a piece of entangled line, such as described by several eyewitnesses of the Gloucester UMO. Photos taken at Trinity Museum, Cape Ann Museum, and curtesy of the Mystic Seaport Museum Archive and Collections.

Figure 5.32. Wooden fids used in maritime rope-work which might also be mistaken for sea serpent tongues or horns, such as described by several eyewitnesses of the Gloucester UMO. Photos taken at Maritime Museum of the Atlantic.

Other species rest at the surface, floating on their side, with one flipper raised up to feel the wind or water motion. As for fishes, based on analysis of brain-wave patterns, they technically do not sleep. Certain species, however, can reduce their metabolic rate, slow their movement, and lower their response to stimuli (Robertson and Sheldon 1979). But pelagic species, such as tuna, never stop swimming. One theory is that the open ocean is such a monotonous place that such species do not need 'down time' to rest from processing memorable features seen. Furthermore, because Bluefin tuna, unlike other fishes, are warm-blooded, they must be in constant motion in order to maintain their elevated body temperatures higher than that of the ambient water or else they will expire (Ellis 2008). For sharks, it had long been thought that they too needed to be in continuous motion so as to ensure that a constant supply of oxygen was kept moving across their gills. It is now known that certain bottom-dwelling, (sub)tropical species of sharks that rely upon ambush predation, can lie motionless on the seafloor for extended periods due to the presence of spiracles that force water across the gills (Carlson and Parsons 2001).

Figure 5.33. Poles attached to a barrel floats used to mark deployed fishing nets. Photos taken at the Cape Ann Museum. Illustrations from Goode et al. *(1884). Could such an extension attached to the barrel-like head identified by several eyewitnesses have been what was described as the tongue or horns of the Gloucester Sea Serpent?*

Table 5.12. Descriptions of the UMO, imagined as a sea serpent, observed between 1815 and 1824 in Gloucester Harbour and elsewhere, and indicative of an object displaying the ability to either lie motionless on the surface or to rapidly submerge, such as might occur through a result of a string of fishing gear or maritime debris either floating loose or being pulled underwater by an entangled animal actively engaged in pursing underwater prey.

- '...he stopped, and lay entirely still on the surface of the water...It remained motionless for five minutes or more...He often disappeared, and was gone for five or ten minutes under water. I thought he was diving or fishing for food.'
- 'When he disappeared, he sunk apparently directly down, and would next appear at two hundred yards from where he disappeared, in two minutes.'
- '...did not turn down like a fish, but appeared to settle directly down, like a rock.'
- 'When immersed in the water, his speed was greater...when moving under water, you would often trace him by the motion of the water, on the surface, and from this circumstance I conclude he did not swim deep... when he disappeared, he apparently sunk directly down like a rock.'
- '...saw him suddenly disappear and came up again at a great distance.'
- '[Saw him] lying on the surface of the water perfectly still.'
- '...he plainly saw the serpent lying still on the surface.'
- 'He lay perfectly still in the water for some time, as he rose and lowered in the water.'
- '...at times perfectly still on the water.'
- '...'brot to' abreast of me perfectly still.'
- 'He then lay perfectly still...I continued looking at him about half an hour, and he remained still and in the same position, until I was called away.'
- 'No one thought they saw the whole length of the body at a time, the tail seeming always to be considerably under water.'
- '[The body] would rise in the water with an undulating motion, and then all his body would sink except his head. This would be repeated.'
- '...at other times, he lay nearly still...he sometimes darted under water, with the greatest velocity, as if seizing prey.
- 'He was lying on the surface of the water.'

With respect to the mysterious entangled animal of Gloucester Harbour, about all that can be deduced in terms of candidate screening is that: (1) one can be relatively certain that should a whale have been sleeping on the surface and breathing through its blowhole, its elongated train of entangled gear/debris nearby, that it would have been seen and recognized as such (e.g. as was the case for Inland in 2000); and (2) that an individual of a temperate zone, large fish species, such as a surface-swimming shark, would not have remained motionless long enough to produce the effect of the entangled gear/debris lying 'perfectly still'...with one caveat, however.

A second possibility to explain the string of gear/debris lying motionless on the surface is that the entangled animal to which it was attached was submerged and actively swimming about in a confined area. Certainly, the observed movement at times of the gear/debris, in terms of being pulled straight down beneath the surface 'like a rock' (Table 5.12), suggests that the entangled animal at the other end had plunged to greater depths 'with the greatest velocity, as if seizing prey,' as one observer noted. The duration that the string of gear/debris was observed to float motionless on the surface was described as varying from 'five to ten minutes,' to 'for some time,' to 'about half an hour.' Dolphins and porpoises are surface mammals who usually only hold their breath for about thirty seconds, with a maximum of no more than 6 to 10 minutes. In contrast, whales can remain submerged for periods of over an hour. Harbour seals (*Phoca vitulina*) routinely hold their breath for up to half an hour and actually exhale before submerging, relying upon oxygen stored in their blood and muscles while underwater. And upon surfacing, they can exchange as much as 90% of the air in their lungs with a single inhalation. The remarkable ability of marine mammals to stay underwater is due to the presence of positively charged oxygen-binding proteins, myoglobin, in their muscles upon which they can draw (Mirceta *et al.* 2013).

From these observations of lying motionless for extended periods, rapidly plunging down, and underwater swimming, the latter specifically noted by one eyewitness as occurring at shallow depths that could be traced by 'the motion of the water, on the surface,' we can make the following deductions about the identity of the entangled animal: (1) the disturbance created by a large whale submerging and swimming underwater would have been more noticeable then what was observed; (2) if the animal was a marine mammal, it was more likely to have been a seal than a porpoise; and (3) a large fish, submerged for the entire time, can be considered as a likely candidate.

The tightness of the curves of the string of fishing gear or maritime debris being pulled across the surface of the water, with both ends passing within a distance of less than a metre, and, in at least one instance, seeming 'almost to touch each other' (Table 5.13), is indicative that the entangled animal must have been moderate in size. Large whales simply cannot 'turn on a dime' as it were. Baleen whales and basking sharks feed by filtering large quantities of water through their mouths, entrapping plankton and small fishes. This is obviously nothing like the behaviour displayed by the entangled animal observed herein. Whatever that entangled animal was, it was probably actively feeding underwater on the abundant herring that were present. Rapidly darting one way and then suddenly and completely switching direction 180° in the other, each swimming track separated by only a metre apart, while engaged in the underwater pursuit of a school of prey, as was reflected by the surface marker of the entangled gear/debris, indicates a much more agile and smaller animal than a whale. Examination of eyewitness accounts of the movement patterns of what were believed to have been sea serpents but which were almost certainly large baleen whales that had become entangled with fishing gear or maritime debris (France 2016b), are very different from the descriptions recorded for the Gloucester UMO. Some smaller marine mammals such as pinnipeds, although nowhere near as rapid swimmers as dolphins and porpoises, are more agile, and can indeed twist themselves to tightly turn when in pursuit of prey.

From this observed behaviour, although it may be possible to exclude large baleen whales as being likely candidates for the entangled Gloucester animal, other potential candidates, such as pinnipeds, small toothed whales, and large fish, remain as possible suspects.

5.4.3 What the observed estimated swimming speed can tell us about the identity of the entangled animal behind the Gloucester UMO

When the Gloucester UMO was observed pulling its train of entangled fishing gear or maritime debris, it did so at remarkable speed. Eighty percent of all estimates from eyewitnesses were of a velocity between 10 and 20 mph (16 and 32 kmph), with fully half being at the upper end of that range. Indeed, the speed at which the gear/debris was dragged along the surface of the water was often observed to create both a bow wave and a noticeable trailing wake. Clearly, whatever type of animal it was at the front end of

Table 5.13. Descriptions of the UMO, imagined as a sea serpent, observed between 1815 and 1824 in Gloucester Harbour and elsewhere, and indicative of an object displaying great lateral flexibility, such as might occur through a result of a string of fishing gear or maritime debris being pulled by an entangled animal of a relatively moderate size actively engaged in pursing underwater prey.

- '[I] saw him turn in the water and the parts pop in the opposite direction.'
- '...saw him draw himself up into a coil and extend himself again.'
- '...at times playing about in circles...he turned short and quick...when his head came parallel with his tail, his head and tail appeared near together.'
- '...his motion when he turned was quick...the first part of the curve that he made in turning was of the form of a staple, and as he approached towards his tail, he came near his body with his head, and then ran parallel with his tail, and his head and tail then appeared together.'
- 'He afterwards went in different directions, leaving on the surface of the water, marks like tho made by skating on ice. Then he would move in a straight line west, and would almost in an instant, change his course to the east, bringing his head, as near as I could judge, to where his tail was; or in fact, to the extreme hinder part visible.'
- 'He turns quick and short...they [his head and tail] appear almost to touch each other.'
- 'When he changed his course, he diminished his velocity but little – the two extremes that were visible appeared rapidly moving in opposite directions, and when they came parallel, they appeared not more than a yard apart.'
- '...saw him move in a semi-circular form, repeatedly.'
- '...saw him turn...in quick motion.'
- 'He was in rapid motion and turned suddenly his head going one way and his tail the other.'
- '...he put his head near his tail and could plainly see one part of him move one way and the other part in a contrary direction.'
- 'He turned very short, and appeared as limber and active as an eel...the form of the curve when he turned in the water, resembled a staple; his head seemed to approach towards his body for some feet; then his head and tail appeared moving rapidly, in different directions, and when his head and tail were on parallel lines, they appeared not more than two or three yards apart.'
- 'He turns himself around almost instantaneously.'
- 'Once, laying stretched out on the surface of the water like a spar: at another time, coiled in serpentine folds, in the manner of a common snake.'
- '...it coiled itself up, instantly uncoiling again...he moves in all directions with the greatest facility.'
- 'They saw him turn and bring his body into a letter S, the head being at right angles with the tail.'

the entangled material, it was a powerful enough swimmer to achieve such a rate of locomotion while fighting the notable drag that would have been produced by the sequence of material trailing behind.

In general, horizontal movement on the surface of the water for buoyant animals like marine mammals requires more energy than for completely submerged animals such as fishes. Dolphins and porpoises get around this by leaping out of the water, called 'porpoising', to save energy due to the lower friction in the air, even considering the extra effort it takes to jump (Au and Weihs 1980; Weihs 2002). Pinnipeds can also exhibit porpoising behaviour when in proximity to the shore, but do so for purposes of orientation, and if distant from the shore, as an anti-predator avoidance strategy. It is only while leaping, that dolphins and porpoises can achieve speeds in the range of 27 to 56 kmph. But had such behavior occurred in Gloucester Harbour and elsewhere in New England and New York during the time of the UMO sightings considered here, one can assume that it would have been noted. There are plenty of other sea monster sightings from around the world and over the centuries in which dolphins and porpoises have been misinterpreted as sea serpents, as well as other cases where they were identified in close proximity to the presumed serpent, thereby raising the possibility of entanglement. Nothing resembling this was seen by the Gloucester eyewitnesses. Seals are much slower, generally moving at a rate of about 6 to 10 kmph, but can reach higher speeds while diving and in pursuit of prey (Thompson and Fedak 1993). Some species are capable of short bursts of rapid horizontal swimming but these rates are closer to 16 than they are to 32 kmph, and thus seem unlikely to account for the high speeds observed for the Gloucester UMO. Sharks normally cruise at about 5 kmph but are capable of short bursts while in pursuit of prey at about 19 kmph. Several species, however, such as blue sharks (*Prionace glavca*), great whites (*Carcharodon carcharias*), and especially, the fastest, macko sharks (*Isurus oxyrinchus*), can reach top speeds above 32 mph, possibly through a consequence of having warmer blood to allow energetically sustainable higher metabolisms. It is worth noting that one account reported the presence of a shark seen in proximity to the Gloucester UMO. But it is the scombroid fishes, however, the tunas, marlins, swordfish and sailfish, that are the record-holding Usain Bolts of the marine world, capable of reaching speeds exceeding 64-80 kmph for very short dashes.

Bluefin tuna (*Thunnus thynnus*) – from the Greek word *thynnus*, meaning to race forwards – normally cruise at 14 kmph in search of prey but easily triple that rate when engaged in hunting, and are the most efficient and remarkable of all fishes in terms of locomotion (Ellis 2008; Safina 1997; Whynott 1995). Pectoral and dorsal fins fold into recessed niches on their torpedo-shaped bodies to reduce drag. Even the eyes are flush with the body surface for increased streamlining. Often likened to being polished torpedoes, their scales are miniscule. They are truly 'superfish', and have been referred to as the pinnacle of piscine evolution through consequence of being the epitome of hydrodynamic excellence (Ellis 2008; Safina 1997). Their deeply forked, lunate tail is attached to a very narrow caudal peduncle filled with tendons having the strength of cables, and can beat at a rate of thirty times per second. Only the blur of hummingbirds' wings are faster. Finlets line the back third of their bodies, producing little vortices to break up the laminar flow of the water, and thereby further reduce drag. As Safina (1997) states, the 'superlative Bluefin' does not so much swim through the water, as split it like a wedge. Bluefin have distinctive musculature and blood heat-exchange systems to maximize efficiency for sustained and short-burst swimming. They have a gill surface area that is thirty times higher than that for any other species of fish, and is the most efficient enzymatic muscle fuel system known in nature (Whynott 1995). A Bluefin's heart is the largest relative to its body size for any fish, and the haemoglobin concentration is on par with humans. And if all that were not enough, they are endothermic or warm-blooded, and therefore capable of higher metabolic activity than other fishes. Always on the move, Bluefin tuna must travel a distance equal to their body length every second to keep their bodies oxygenated. They can dive to depths of more than a three hundred metres, and travel distances of close to two thousand kilometres in several months, such that over the course of their 40-year lifespan they could have swum over three million kilometres. Of all the aforementioned possibilities for an entangled animal to reach the estimated speed of 32 kmph observed for the Gloucester UMO, Bluefin tuna or some similar fish are the most likely candidates.

Given all that, it is no wonder that Bluefin tuna command a near reverence amongst their admirers. For many, it is nothing short of criminal to kill such magnificent creatures for either food or sport as they are undeniably one of the most highly evolved of any species on the planet, Given that the population in the

Northwest Atlantic declined by ninety percent in the last two decades of the twentieth-century, Safina (1997) believes that Bluefin are sentinels of oceanic health.

5.4.4 What the observed location of the sightings can tell us about the identity of the entangled animal behind the Gloucester UMO

Gloucester Harbour is quite shallow, with many locations being less than seven metres deep, and some with depths of just over four metres. That the Gloucester UMO was frequently observed actively swimming about these areas limits the likelihood of it having been a large baleen whale. Furthermore, several accounts mention the proximity of the UMO to the shore: once in Gloucester Harbour near enough to provoke alarmed children into fleeing for the safety of their homes; another, off Nahant in Boston Harbour, so close that it could have been touched with a long pole; another off Plum Island beach near Newburyport, just a few metres from shore where people had been bathing; and of course, the charming account about Mr. Stover in Gloucester Harbour, wherein he doffed his hat so as not to frighten the creature that was so near the shore. Then there are the two accounts in which the UMO was seen right adjacent to, or actually upon the shore of, Gloucester Harbour: 'The animal was stretched out, partly over the white sandy beach, which had four or five feet of water upon it, and lay partly over the channel'; and 'something, resembling a large log of wood,...resting partly on the rocks, and partly in the water.' Clearly, whatever the animal was at the front end of the entangled train of gear/debris, mistaken by onlookers to be the body or tail of the Sea Serpent, it must have been of only moderate size given that the harbour does not have many deep drop-offs so near to the shore.

5.4.5 What the observed phenology of the sightings can tell us about the identity of the entangled animal behind the Gloucester UMO

What other information do we have from which to make an inference about the identity of the animal(s) behind the Gloucester UMO? That many eyewitnesses posit the creature to have been actively feeding upon small fishes may be limited help, since many of the likely candidates – seals, sharks, large fishes, dolphins, porpoises, and large whales, are all piscivorous. What may be useful is the *phenology* or seasonality of the sightings, in that they were restricted to the summer months, and for 1817 at least, give the impression that the creature in question was observed at several locations during its journey southward at the onset of autumn. So which possible candidate animals are year-round residents of New England waters, and which, alternatively, are seasonal migrants? Unfortunately, here we have a some of the same problem in that many of the likely candidates, including mysticete cetaceans such as humpbacks and northern right whales, large cartilaginous fishes such as great white (*Carcharodon carcharias*) and basking sharks, and large bony fishes such as white marlin (*Tetrapturus albidus*) and Bluefin tuna, are all summer visitors who depart in the autumn for warmer climes along the southern Atlantic coast on their way to the Caribbean. However, it *is* useful to know that the prevalent pinniped species of harbour and grey (*Halichoerus grypus*) seals, and the prominent small odentocete species of dolphins and the harbour porpoise (*Phocoena phocoena*) are year-round residents, and thus do not match the seasonal pattern of the Gloucester UMO sightings.

Armed with a much better recognition of the enormous breadth of marine biodiversity, modern zoologists have recently begun to examine old sea serpent sightings and propose known animals as explanations for the hitherto mysterious creatures. To counter the confirmation bias frequently displayed in the past by cryptozoologists who promote their own particular pet theories without consideration of possible alternatives, contemporary scientists have used a simple procedure to objectively screen potential candidates for UMOs. The approach used can be referred to as 'likely candidate analysis' (LCA) and is not dissimilar to the weighting of evidence, sometimes of a conflicting nature, used in legal proceedings to assess culpability of suspects. Four examples follow.

In the middle of the sixteenth century, an 'abominable creature' was found in the strait between Denmark and Sweden and became a widely discussed subject throughout the learned society of Europe, generating commentary by the leading naturalists of the day. The merman resembled a bare-headed, cowled monk, and had a human face, an upper body that looked like a chasuble composed of scales, through the sleeves of which, short fins emerged in the place of the upper arms, a ruffled waist garment, and a multi-lobate tail that splayed out like a robe. Drawings of the so-called 'sea monk' are a staple of cryptozoology books. Other specimens of the same type of mysterious creature were recorded. In the nineteenth century, it was proposed that the creature was a giant squid, something that has been generally accepted ever since, although some have suggested particular fishes or pinnipeds. Paxton and Holland (2005) undertook a LCA of 9 'alternative suspects,' comparing their sizes, 7 physical traits, and two elements of their known distribution ranges to such information known for the observed aquatic cryptid. Their conclusion is that 'certainly with one or two caveats, [the angelshark] *Squatina* [*squatina*] makes a better contender for the Renaissance sea monk than the giant squid, for [the nineteenth-century] explanation, ingenious though it is, has some fundamental flaws.'

In 1741, off the coast of Greenland, Poul Egede encountered 'an extraordinarily horrible creature,' the subsequent description and accompanying illustrations of such, as recounted by his well-known father, can be said to have kick-started the modern age of sea serpentism. Potential candidates as to the identity of the creature have included the very real giant squid (Lee 1883) or imagined cryptids such as a giant long-necked seal or marine otter (Heuvelmans 1968; Oudemans 1892) which seem off-the-mark since, as Paxton *et al.* (2005) state 'things that blow like whales are, all other things being equal, most likely to be cetaceans.' The authors therefore undertook a LCA of 8 'possible cetacean candidates,' comparing 6 of their physical traits with those described in several accounts of the encounter with observed aquatic cryptid. Their conclusion is that 'the species seen was most likely to have been a humpback whale, a North Atlantic right whale, or one of the last remaining Atlantic grey whales [*Eschrichtius robustus*] either without flukes or possibly a male in a state of arousal.'

For more than four decades, reports from the northwestern coast of North America have suggested the presence there of a large creature of unknown phylogenic affiliation (Leblond *et al.* 2014) which, based on eyewitness accounts and some believed remains, led several biologists to propose to be a species of ancient reptile (LeBlond and Bousfield 1995). Identification of this plesiosaur by its proponents as the new species *Cadborosaurs willsi* has been lambasted as an example of fringe science. In 1968, what was thought to be a juvenile of the species, was captured and examined by a fisherman before being returned to the sea. This specimen was just over a third of a metre long and slightly over three centimetres in diameter and was purported to resemble the adult form of the *Cadborosaurus*, and thus taken as evidence for a precocial strategy of life history that is characteristic of reptiles. Woodley *et al.* (2011) undertook a LCA of 14 'plausible candidate identities' (various small fishes, a generic pinniped, a decapod, as well as the imagined adult cryptid), comparing 29 morphological traits with those described for the small specimen in question. The authors developed a numerical scoring system based on whether the diagnostic traits were similar, somewhat similar, or dissimilar, and concluded that 'the bay pipefish [*Syngnathus leptorhynchus*] shares more characteristics in common with...[the] specimen than any other candidate species,' thereby allowing them to 'contend that the 'reptilian hypothesis' does...need to be seriously re-examined in light of the removal of a strong piece of evidence.'

Finally, I investigated four particular encounters with sea monsters, which at the time of the sightings (1877, 1890, 1905, 1959), had garnered worldwide attention and considerable fame (France 2017). I selected 9 morphological, one distributional, and four behavioural characteristics, and applied Woodley *et al.*'s (2011) numerical weighting system for a composite UMO assembled from all four accounts. These were compared to the characteristics of the generic groupings of one extinct and 12 existing candidates of marine animals that have been posited in the literature as explanations for many sea serpent sightings. Chelonians were found to be the most likely candidate animals to explain the bodies at the front end of the four UMOs.

One of the more astute observers of the Gloucester Sea Serpent was James Prince. In his letter to Judge Davis about his August 1819 encounter with the UMO off Nahant, following a recounting of the particulars about the sighting itself, he concluded by offering a logical exclusion of what might be considered to be the most likely candidate animals (Oudemans 1892):

> Certainly it is a very strange animal. I have been accustomed to see whales, grampuses, porpoises, and other large fishes, but he partook of the appearance of none of these. The whale and the grampus would have spouted – the shark never raises his head out of water, and the porpoise skips and plays; neither of these has such appearances on the back or such a head as this animal. The shark it is true, has a fin on his back, and often the fluke of his tail is out of water; but these appendages would not display the form, and certainly not the number of protuberances, which this animal exhibited; nor is it the habit of the shark to avoid a boat. The water was extremely smooth, and the weather clear: we had been so habituated to see him, that we were cool and composed – the time occupied was from a quarter past eight to half past eleven – a cloud of witnesses exceeding two hundred, brought together for a single purpose, were all alike satisfied and united as to appearances, and as to the length and size of the animal; but you must deduct the influence which his passage through the water and the manner he propelled himself might have as to the apparent protuberances on his back, and the ripple occasioned by his motion on his real length, all of which you can judge equally well and better than myself. I must conclude there is a strange animal on our coast – and I have thought an unvarnished statement might be gratifying to a mind attached to the pursuit of natural science, and aid in the inquiries on a controversial question, which I knew to have interested you.

That Prince is knowledgeable enough to recognize the difference between a grampus or dolphin and a porpoise – a distinction I would suggest is probably unknown to almost all non-coastal and even many coastal residents today – gives weight to his interpretations. Indeed, I would go as far to say that this quotation represents the best attempt by an actual eyewitness, based on personal experience in the light of knowledge of the day, to discern the typology of the UMO. It is significant that in addition to several behavioural traits, much of Prince's argument for the uniqueness of the 'strange animal' comes from the noted presence of the long tail of protuberances, in addition to the head, which earlier in the account he referred to as closely resembling that of a swimming horse. And therein lies the difficulty in trying to identify the animal behind the Gloucester UMO. Remove the so-called tail, as it is most obviously a train of entangled debris, probably floats from a fishing net, and then remove the head, as it is quite possibly to have been the first link in the chain of entangled material, and our ability to discriminate amongst likely candidates is substantially weakened. This is very different to the aforementioned sea serpent LCA examples in which discriminating morphological characteristics predominate. For other sea serpent sightings which I believe to be demonstrative of entanglement in pre-plastic maritime debris, recognition as to the probable identity of the entangled animal at the front end – be it a whale (France 2016b, 2018), a sea turtle (France 2016a, 2017), or a seal or sea lion (France, in prep.), for example – is not difficult. This is because, in those situations, eyewitnesses provide enough evidence from their observations for one to make educated inferences. Indeed, often the eyewitness themselves assume the UMO to be an easily recognizable animal, only be forced into having to second guess and reassess their initial conclusion upon discerning the incongruous presence of the long 'tail'. Such is not the case, however, with respect to the Gloucester UMO. Most eyewitnesses do not mention anything concerning the nature of the front end at all; and for those that do, this is limited to what they believe to have been the small head, a conclusion which, as mentioned, I agree with Fama (2012), as being problematic. And so, the subsequent LCA is therefore more uncertain than the previously noted examples, based as it is on behavioural and distributional characteristics in the absence of corroborating morphological attributes. Indeed, as will be seen, the thesis developed below is based more on the absence of seeing some things than it is on confirming the presence of certain other things.

Table 5.14 compares 11 characteristics composited from available information for the Gloucester UMO in relation to those for 9 generic animals often identified as being frequently mistaken for sea serpents or lake monsters. The LCA generated the following sequence of ranked (highest to lowest) likelihood: scombroid fishes ~ sharks >>> giant squid >> dolphins and porpoises ~ seabirds ~ baleen whales > seals > sea turtles > manatees.

An upper level distinction can be made based on whether or not the entangled animal needs to surface in order to breath. If one believes, as Fama (2012) and I do, based on the points noted above, that the so-called 'head' was simply an oddly shaped first link in a chain of entangled debris, then there is no evidence that the animal figuratively behind (actually physically in front) of the UMO needed to surface for air. Besides the paleontological record, another compelling line of evidence against the idea that lake monsters are plesiosaurs comes from the simple truism that should such prehistoric creatures still be in existence, then they should have been sighted much more frequently given their need to repeatedly stick their heads up out of the water in order to inhale. Given that eyewitnesses failed to describe anything else of the body of the UMO besides the presumed long tail and occasionally the small head, implies that the Gloucester Sea Serpent may have been almost as unlikely to have been a mammal or reptile, as the Loch Ness Monster a plesiosaur. The putative tail of the UMO was frequently observed for extended periods, simply floating motionless on the surface, oblivious of the surroundings, with no mention made of any head suddenly appearing to catch a breath. In fact, quite the opposite occurred, wherein following such a prolonged period of immobility, the 'tail' could suddenly be pulled underwater for another long period of time. Cryptozoologists, however, who have an answer to every line of skeptical reasoning, would counter that such shy creatures as plesiosaurs only need to project their nostrils just a few centimetres above the surface to accomplish this task. What is known is that seals and manatees can indeed sometimes breath this way, keeping most of their heads hidden under water. Harbour porpoises are also subtle surface breathers in that they do not spout. Furthermore, their dark gray backs blend in with the steely-coloured surface, so it can be difficult to catch a glimpse of them (Johnson 2005). But

Table 5.14. Diagnostic characteristics of the composite UMO assembled from the accounts of eyewitnesses, and likely candidate analysis for 9 generic groups of animals that might explain the identity of the marine creature behind the sightings. Y = yes (numerical score of '1') or N = no (numerical score of '0') of trait in selected western Atlantic species within each candidate group, whereas P = perhaps or sometimes (numerical score of '0.5') for demonstration of trait.

Composite UMO	Seabirds	Giant squid	Baleen whales	Turtles	Dolphins/ Porpoises	Mana-tees	Seals	Sharks	Bony fishes
No emerged head seen	N	Y	N	N	N	N	N	Y	Y
No emerged body seen (beside putative tail)	N	Y	N	N	N	N	N	Y	Y
No prominent dorsal fin or bilobate tail seen	N	Y	N	N	N	N	N	N	Y
No spouting or breathing seen	N	Y	N	N	N	N	N	Y	Y
Demonstrable speed	Y	Y	Y	N	Y	N	N	Y	Y
Tight turning radius	Y	N	N	Y	Y	N	Y	Y	Y
Presumed piscivorous	Y	Y	Y	N	Y	N	Y	Y	Y
Extended time underwater	P	Y	Y	P	P	P	P	Y	Y
Seasonal migrant	Y	N	Y	P	P	P	N	Y	Y
Strength to tow entangled debris	N	P	Y	N	Y	N	P	Y	Y
Presence in shallow water	Y	N	N	Y	Y	Y	Y	Y	P
Total score similar	5.5	7.5	5.0	3.0	6.0	2.0	4.0	10.0	10.5

again, there are other characteristics that suggest that these particular marine mammals are not the most likely candidates for the observed UMO. For example, anyone who has observed the playful behaviour of dolphins and porpoises firsthand would concur with Prince that it is extremely unlikely that such an animal would never break the surface to demonstratively show itself during the many cumulative hours of observation. Likewise, even putting aside all other characteristics that argue against the UMO being a large whale, the absence of any notable exhalation of vapor, or 'spouting' – long the tell-tale trait that led to their decimation by the whaling industry – despite the observations of hundreds of eyewitnesses, is also suggestive that the animal must have been something else. Once again, Prince's reasoning in this regard seems sound.

Giant squids (*Architeuthis*) are extremely rare in near-shore coastal waters, as are open-water billfish such as Atlantic blue marlins (*Makaira nigricans*), which also often display their dorsal fins above the surface (Ross 1991). In contrast, only occasionally does the dorsal fin of a giant Bluefin tuna poke up out of the water during swimming. Swordfish (*Xiphias gladius*) are known to bask on the surface but are also mostly offshore, open-water hunters on the continental shelf (Ross 1991). Nurse (*Ginglymostoma cirratum*) and sandbar (*Carcharhinus plumbeus*) sharks can come into shallow water, and like almost all sharks, are powerful and tremendously strong, thereby capable of pulling a train of maritime debris.

And 'supersharks' such as the shortfin macko (*Isurus oxyrinchus*) can achieve amazing speeds like those noted for the UMO in question. For sharks, however, the dorsal fins are easily spotted during surface swimming, as is also the case for many cetaceans.

Sea monsters are often invented when eyewitnesses simply do not recognize mundane animals that are rarely seen through either their low abundance, elusive nature, or in consequence of them seldom straying beyond their normal distributional range. The latter are termed 'vagrants'. Manatees, identified by some as being the animal behind the 'Chessie' sea serpent in Chesapeake Bay (Cheezum 2007), occasionally travel all the way to New England (Annear 2016). Walruses, though at one-time numbering in the tens of thousands in Atlantic Canada (including on Sable Island off the coast of Nova Scotia) before their extirpation, were occasionally known to wander to northern Gulf of Maine waters in New England during the eighteenth century (Allen 1930). And then there are beluga whales (*Delphinapterus leucas*), a population of which survived glaciation by establishing a southern residence in the St. Lawrence River, from where occasionally individuals, often adventurous gray-coloured adolescents, will wander down to New England and even further south during the summer (including several seen in Gloucester Harbour – Jacobs 2014). Could any of these vagrant species have been the unrecognized animal entangled at the front end of the Gloucester UMO? I've observed the latter two in the arctic on several occasions and find it difficult to imagine how walruses at least could ever be likened to the descriptions of the Gloucester UMO. And beluga, though they do not have dorsal fins due to evolving to swim under the ice, and can swim along just beneath the surface with their admittedly horse-like heads just above the water, they do still exhale vaporous clouds, something that was not characteristic of the Gloucester UMO. Vagrant individuals of both walruses and belugas also move northward at the end of their occasional New England summer visitations, not southward into Boston Harbour and Long Island Sound. Finally, other traits of the Gloucester UMO, such as speed and manoeuvrability, seem to preclude the possibility that a vagrant manatee which would be southward migrating was the animal behind the monster.

For all these reasons, in addition to those enunciated earlier, I consider giant Bluefin tuna to be the most likely candidate for the entangled animal behind the Gloucester UMO. However, I must admit to being much less certain about what the UMO actually was, than I am about, in at least in one respect, about what it was not: namely, an entangled humpback whale.

Let's examine Fama's (2012) argument for the UMO being an entangled humpback. She admits to being moved by the 'magic' of old tales of improbable species inhabiting the reaches of the ocean, but then becomes aware of the dangers of ghost nets after watching a video of a humpback whale being cut free from its entanglement. Going back to some of the eyewitness testimonies Fama finds the answer 'hidden in the plain language of their own descriptions.' Some parts of the described encounters are then selected which offer support for her contention. In response to the fact that the UMO arrived coincident to the start of the mackerel fishing season, we have her comment that 'whales eat schooling fish like herring and mackerel.' In response to one account where the UMO was seen 'battling' the Sea Serpent, Fama comments 'Proximity of a whale to the serpent.' In response to one eyewitness who mentions white foam under the chin and a large wake as being indicative of the rapid speed of the UMO, she comments 'the foam suggests something is pulling the object, and the strand of kegs elongates when towed.' And finally, in response to the frequently made observation that the UMO could stay underwater for durations of many minutes, is her simple comment 'like a whale.' As discussed above, with the exception of the sighting of a nearby cetacean, all of this evidence is by no means unique to baleen whales. Through her research Fama, a novelist with a book about dangerous mermaids, finds 'a possible scientific solution had yielded to my effort':

> Between 1817 and 1819 (more likely much longer) I believe the 'sea serpent' was in fact the same poor humpback whale, entangled in a new or rope lined with keg or cork buoys, migrating to the Gulf of Maine every summer, powerful enough to survive the massive drag of its entanglement, and even to submerge the length of its torment into the depths with it, giving the illusion of the snake sinking. It's likely that just by chance, the first keg or buoy in the line was different than the others, or was made of multiple objects lashed together, to create the illusion of a head raised above the body.
>
> Monica Pepe, the Project Supervisor at the Whale and Dolphin Conservation Society in Plymouth, MA, confirmed that there have been many long-term entanglements, citing a North Atlantic right whale named 'Necklace' who had a fishing net wrapped around her tail stock for a decade. In most contemporary instances, disentanglement teams try to free the animals, but according to Ms. Pepe, 'If it doesn't appear to be life threatening they usually will try to let the animal free itself.'
>
> Perhaps the 'sea serpent' whale eventually freed itself. But given very similar sightings well into the 1830s (after which the descriptions are more varied), I believe instead that it spent its life inadvertently bringing science and magic together along the shores of New England.

The above interpretation is spot on about the causality behind the legend, erring, I think, for the reasons laid out in Table 5.14., only in suggesting it to have been a humpback whale that was the 'poor' animal so entangled. The presence of the UMO in shallow, near-shore water, and observation of a very tight turning radius reflected by the trailing chain of entangled debris, denotes an animal of a smaller size than the humpback whale Fama proposes; and the absence, except in one case, of an observed large body that could be likened to a cetacean (it is interesting to note that this particular account is not from the harbour of Gloucester but rather occurred in deep water, fully 10 kilometres off the coast), as well as the absence of any indication of exhalation, thereby denotes an animal that need not have had to surface in order to obtain oxygen.

But perhaps that dismissal is too simplistic. What about a minke whale (*Balaenoptera acutorostrata*) as being a likely candidate, or at least one more so in this regard than a humpback? There is some merit to such an argument. Minkes are much smaller, rarely reaching 10 metres long. They are also unobtrusive loners whom are 'not much to look at from a whale-watch boat, offering only a fleeting glance of a curved fin, no flukes, no spy-hopping, and hardly ever a breach' (Johnson 2005). Sounds like a promising possibility until one learns that when entangled, minkes tend to roll over and over, wrapping themselves up further in the gear, to which point they are rarely able to escape due to also being much less powerful swimmers than larger baleen whales (Johnson 2005). That said, remember, however, that nineteenth-century net material was not as strong as modern-day synthetic lines.

However, the strongest argument against the entangled animal behind the Gloucester Sea Serpent, or at least that one observed within the harbour during those halcyon days of August 1817, in being a large whale, is that, as was the case with respect to Ibis and Inland this century, one would expect that this would have been something easily recognized. Again, there are plenty of examples of sea serpents that clearly read as misidentified baleen whales from the eyewitness accounts (France 2016b). An even more compelling case against the entangled whale-cum-sea serpent theory for the August 1817 sightings arises from discussions I had in 2017 with staff members from both the coast guard and the harbour master services for Gloucester, as well as a series of fishermen I chatted with along the wharves, all of whom had grown up in the city with familial roots going back generations. The bottom line is that due to the very shallow harbour, which in consequence prevents normal feeding behaviour of baleen whales – which usually occurs in the vertical plane – whales almost never enter the harbour. Ibis became entangled outside the harbour, and Inland's presence within the harbour is believed among old-timers to have been the first large whale seen there in more than half a century.

Large fishes are particularly susceptible to becoming entangled in fishing gear, something that is no great surprise given that is specifically what the equipment is designed for after all. Entangled species include salmon sharks (*Lamna ditropis*), smalltooth sawfish (*Pristis pectinata*), manta rays (*Manta* spp.), blacknose sharks (*Carcharhinus acronotus*), dogfish (*Squalus acanthius*), skates, shortfin mako sharks, tiger sharks (*Galeocerdo cuvier*), and others (Laist 1997; NOAA 2014; Schwartz 1984; Seitz and Paulakis 2006; Wegner and Cartamil 2012), including tuna.

Among the most migratory of animals, Bluefin tuna arrive in New England and Maritime Canada in June and depart for the subtropics in late autumn (Ellis 2008; Safina 1997; Ulanksi 2008). Adults can reach 'monstrous' sizes (Safina 1997) of over four metres and weight more than seven hundred kilograms (Figure 5.34). Often hunting in packs like wolves, giant Bluefin are opportunistic feeders demonstrating fidelity to feeding sites were their primary prey, herring, congregate (Ulanski 2008). Pelagic animals, giant Bluefin are also a nearshore, and at times when attracted to food patches, even an inshore, species, something that was capitalized by early fishermen. For example, whether in the Bronze Age Mediterranean by Phoenicians, or in ancient (5,000-3,000 BCE) northeastern North America by Paleoindians (Ulanski 2008, 2013; Whynott 1995), capture methods were similar in that fish would be sighted by shore spotters, and along a section of shoreline with a steep drop-off, they would be lured into set traps, or more frequently captured in thrown nets, and beaten with clubs or speared. In what might be called the last medieval occupation (Whynott 1995), until recently, giant Bluefin tuna were caught by harpooning (Figure 5.35; Decker 2011).

Remember Captain Rich's admonishment of eyewitnesses of the Gloucester UMO, wherein he believed them to have been deceived into believing the presence of a sea serpent's long tail rather than the long wake trailing behind a tuna? In point of fact, detection of tuna for harpooning, whether through eagle-eyed shoreline spotters in antiquity, or those positioned in the elongated bowsprit of high-speed boats or even from accompanied aircraft today (note, the cost of an accompanying aircraft is small given that Bluefin are the world's most commercially valuable finfish, with single individuals capable of garnering truly astronomical prices such as two recent (2017, 2018) sales in Japan for the equivalent of $1.76 million and $ 3.1 million USD; that is for individual fish!; i.e. they are the planet's most expensively sought after animals), is based on discernment of swimming behaviour and the wake made from a distance (Decker 2011). Recognition of the presence of a giant Bluefin in advance of closing in for a capture attempt by screening out other contenders, offers a parallel useful for screening out candidates for the Gloucester UMO. During normal swimming, the giant Bluefin's fluid manoeuvers are 'excruciatingly beautiful' (Safina 1997). There is no column of exhaled vapor like whales, nor any show-off cavorting into the air like dolphins. Bluefin are very subtle, except when charging into a school of prey, when they move by quick bursts along the surface spraying water about their heads (Whynott

Figure 5.34. Examples of giant Bluefin tuna caught off southern Nova Scotia and in the Mediterranean. Photos taken at the Wedgeport Tuna Museum and Interpretive Centre and from a postcard handed out at the Tonno Restaurant in Gloucester (see Appendix One).

1995), and sometimes by *milling* (swimming about in a circle as a group in order to herd their prey). Individual tuna may also swim around in circles to maintain a minimum speed needed for respiration as a way to stay in one area. Might the latter explain the rapid twisting and turning about of the string of debris pulled behind the Gloucester UMO? For commercial spotters, giant Bluefin can be easily discerned from cetaceans (Safina 1997; Whynott 1995). Pilot whales (*Globicephala* spp.) move with their bodies above the surface. Humpback whales frequently break the surface for loud pursed exhalations, pitch-pooling (bobbing up and down), fin slapping of the water, or breaching, the latter two perhaps as ways to stun prey or communicate with other whales. And finback whales (*Balaenoptera physalus*) also frequently surface, sometimes to bubble-feed, wherein they blow air-rings to herd mouthfuls of prey into a circle. None of these behavioural traits, typical of cetaceans, were mentioned by observers of the Gloucester UMO. In contrast, such characteristics are often recorded with respect to other sightings of sea serpents which were likely to have been misidentified whales (France 2016b).

Figure 5.35. Harpooning giant Bluefin tuna. Photos taken at the Wedgeport Tuna Museum and Interpretive Centre.

There is no evidence that the Gloucester UMO, whatever it might have been, was anything other than a single animal seen at any one time. Bluefin tuna are considered to be the most socially advanced of all fishes. But whereas young and juvenile Bluefin occur in schools, and moderate-sized adults often hunt in small, cooperative packs, it is thought that giant adults assume a more solitary lifestyle.

Also of significance to my thesis is the fact that tuna tend to aggregate near or under floating objects such as buoys, logs, drifting algae, and fishing nets. A 1998 report, quoted in Ellis (2008), notes this attraction to 'floating rafts or buoys anchored in deep water which, for reasons not yet fully understood, cause tuna and other types of oceanic fish to gather around them.' Indeed, the best location for trophy fishing of tuna 'is to station near one of the herring nets set in every inlet by the inshore fishermen. Very often the angler's boat is moored to one of the nets; in fact many guides set herring nets especially for use of their parties, well-anchored and conspicuously buoyed' (Heilner *in* Ellis 2008). 'If it were not for the presence of these nets,' one sport fisherman noted (Ellis 2008), 'the tuna, in all probability, would not be there, or if they were, it would not be possible to hook them [with a fishing line].' Immediately following his discussion of a sea serpent seen in St. Margaret's Bay, Nova Scotia in 1849, Ambrose (1867) comments upon *T. vulgaris*, or tunny or Albicore, as Bluefin tuna were then called:

> The harpooner holds [the harpoon iron secured to 'a good strong rope'] in his hands in the bow of the boat, whilst an assistant paddles around the herring nets, occasionally throwing overboard a fresh herring, which is soon noticed and seized by the Albicore, prowling

> below. These fish take their prey with a rush, and so swiftly, that although not more than six or eight feet below the surface, their presence is only known by a white streak, like a flash of light through the water, and the sudden disappearance of the bait. The harpooner darts his weapon at the head of this streak, and thereby strikes the fish as near as possible to the pectoral fin; for if the weapon strikes any further back, it generally tears its way out.

Indeed, this latter is exactly what must have happened with respect to Captain Rich's first unsuccessful capture of what he believed to be the Gloucester Sea Serpent.

In nineteenth-century New England, this attraction of horse mackerel, as tuna were then also called, to fish-aggregating devices led to the fish being regarded as a nuisance by coastal fishermen. Tuna would follow small fishes into the nets, where, after eating their fill, would then tear their way out (Ellis 2008) or, I might add, if not successful in completely freeing themselves of entanglement, would swim off with attached netting material.

Pulling about a string of non-lethally entangled net and floats is certainly well within the realm of possibility for a fish as a powerful a swimmer as a giant Bluefin. Even the account of Captain Rich, wherein he and his crew were dragged about Gloucester Harbour, mirroring a 'Nantucket sleigh ride' with respect to harpooned cetaceans, is credible. Once harpooned, a tuna will run out a rope attached to a buoy and flag, sometimes until it is lost to sight. Not only extremely powerful, giant Bluefin are also very tough. Safina (1997) recounts one story by an old-timer: 'Some fish, Christ, they can just keep going and going and going. One time we struck a fish and his buoy flag just disappeared. One week later, it come sailing back into the bay again. The fish was still alive on it.' One cannot but help to wonder if anyone in the interlaying week who may have spotted this fishing gear, racing along the surface of the water as if on its own accord, may have thought they had seen a sea serpent. Heilner (*in* Ellis 2008) tells of an 1871 encounter with a tuna: 'He hooked to a monster tuna which towed his boat across the harbor and hurtled it merrily into a fleet of herring netters, swamping one and creating havoc with the rest.' Also, in 1909, he mentions that a tuna 'as big as hogsheads [barrels]' towed a boat all over a Cape Breton harbour for nineteen hours before 'the exhausted sportsman admitted defeat and cut his line.' According to Rivkin (*in* Ellis 2008), giant Bluefin of the Northwest Atlantic, 'often tow the small boats for miles.' It is the incredible strength of these tuna that makes them such a prize-worthy prey sought after by sportsmen. Heilner (*in* Ellis 2008) recounts six men switching off on a rod for sixty-two hours to land a nearly eight hundred pound (363 kg) giant Bluefin. And Rivkin (*in* Ellis 2008) describes a hooked tuna bolting at 'breathtaking speed' before settling down to a prolonged tug-of-war: 'If the tackle didn't fail, if the angler was resolute, and if the fish managed to avoid all the entanglements on its way out to sea, a weary but happy crew would eventually catch it.'

While once hunting a tuna near an abandoned wreck in the fog, Ulanski (2008) mentions mistaking the fish for a hideous sea serpent. There are two UMO sightings in eastern North America where passing reference is made to tuna. Off Nahant Beach Boston Harbour in 1849, 'a singular-looking fish in the form of a serpent' was seen moving in an undulating motion. 'I have seen all sorts of fishes and hundreds of horse mackerel, but I never say anything like this,' the eyewitness recounts (*in* O'Neill 1999), thereby adding yet another dig at Captain Rich's conclusions from across a span of three decades. Another sighting occurred the same year in Nova Scotia (Ambrose 1867), the body of the 20 metre UMO being 'proportioned like an eel, i.e. tapering towards the extremities, with no caudal fin perceptible, but one very high fin, or row of spines, each of about an inch in diameter at the base, erected along its back, serving indeed for a dorsal fin, like the folding fin of the *Tynnus vulgaris* or albacore.' The description continues to mention that this odd 'spinal erection' extended about a third of the length of the UMO, and looked like 'the sail of a skiff.' Various tuna species do share an ability to retract their fins into grooves so as to reduce drag, but this is the only time I am aware that this so-called folding ability is likened to something that might have been an actual piece of sail to which the animal was attached. Finally, from my notes, I have the following quotation which appeared as a letter to the *American Journal of Science and Arts* concerning a sighting that occurred in the Southern Ocean: 'I saw what was first supposed to be a fish called an Albicore [i.e. a tuna]; but, on further examination it was discovered to be a serpent of which I cannot give a clearer description.' Intriguing? – yes; but also, like so much else on the topic of mysterious UMOs, frustrating in its lack of clarity and conclusiveness.

Such nineteenth-century association of giant Bluefin tuna with sea serpents might not be that surprising given the lack of contemporaneous information about the species. As late as the end of the sixteenth century, the animal could still find its way into a book about fantastical fishes (Figure 5.36; Kingdon 2017). The species was only formally described at the end of the eighteenth century, with only a rare single specimen on exhibit in a museum in addition to no formal mention being made in the natural history literature until the middle of the nineteenth century (Hancock and Reilly 2017). Bluefin tuna

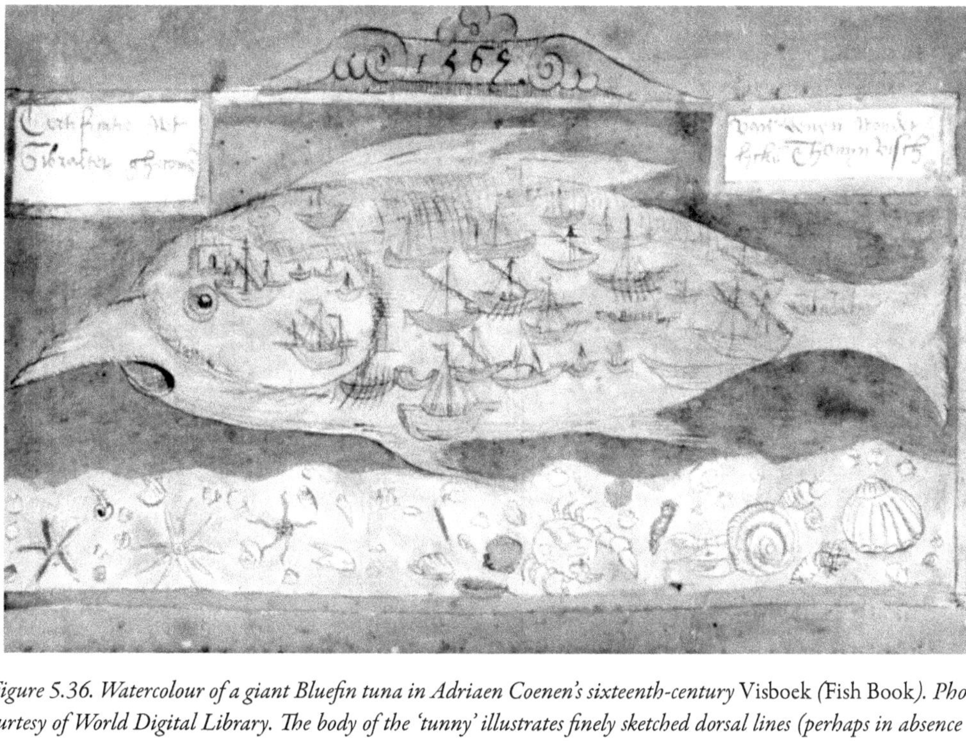

Figure 5.36. Watercolour of a giant Bluefin tuna in Adriaen Coenen's sixteenth-century Visboek (Fish Book). *Photo curtesy of World Digital Library. The body of the 'tunny' illustrates finely sketched dorsal lines (perhaps in absence of normal fish scales) as well as marking depicting 'ships Coenen imagined the fish had seen on its travels' before being caught in the Mediterranean Sea (Kingdon 2017).*

therefore would have been little known to landlubbers at the time when Captain Rich's example was displayed in Boston. In this light, perhaps some gullibility might be forgivable.

There have been a handful of accounts of UMOs that read as if they were misidentified fish, some entangled in maritime debris. Take for example, an encounter in 1875 off the Massachusetts coast south of Boston where a swordfish was, depending on the account, either being pursued or attacked by a sea serpent (O'Neill 1999). According to one eyewitness, the head of the monster was in constant motion, appearing above and beneath the surface, while the 20 metre-long body, black and white striped and shaped like a fish barrel (possibly a series of separated floats?), undulated 'like that of a caterpillar.' The description by another eyewitness is more dramatic:

> When the swordfish first attacked him [i.e. the sea serpent] he [again the serpent] reared his head at least ten feet above the water, and then dove down once more. These actions he kept repeating so that we had a fine opportunity to scrutinize him. His head was rather flat and closely resembled that of a turtle. The fin we first observed on the back, several feet from the head, while small fins protruded on each side. The body was at least eighty inches in diameter, and presented a shiny surface, covered with large coarse scales. When he moved his head the water seemed to boil as he rapidly clove his way through the wave, so that by far the largest portion of his body must have been under the water. We estimated his length to be at least sixty feet.

This encounter reads remarkably similar to those in which whales were thought to have been locked in mortal combat with sea serpents, but instead were most likely unfortunate animals attempting to thrash themselves free of entangled debris (France 2016b).

In 1884, off the coast of Maine, a sea serpent with a head 'like a shark' grasped a length of rope and 'dragged a boatload of fishermen through the water at tremendous speed' (O'Neil 1999). The parallel to Captain Rich's encounter with the Gloucester UMO six decades before is striking. Then along the northern border of Maine, in 1907, several fishermen reported seeing what they first thought to be a sea serpent but later concluded to be a big fish, probably a shark (O'Neill 1999 suggests it might have been a basking shark) repeatedly 'playing around a buoy.' 'Playing', or rather entangled, we may ask?

For several encounters from the region, initial contentions of UMOs as sea serpents have subsequently been altered to them being identified as rare fishes that went unrecognized by the fishermen. One example of fish-inspired sea serpent comes from 1910:

> Has his majesty, the great and mighty sea serpent permitted himself to be captured, after countless years of tumbling through the sea, and frightening the wits out of enthusiastic amateur fishermen, to say nothing of proving a great asset for summer hotel keepers along the Atlantic coast?
>
> The sea serpent was brought to Boston yesterday from Yarmouth [Nova Scotia] on steamer *Prince George*.
>
> The monster was seven feet long, with a flattened head over a foot long, and its great mouth filled with a toothlike structure. On its back was a dorsal fin composed of 28 spines, nearly a foot long, and connected by a thick, almost transparent skin. A huge, fan-shaped tail helped propel the marine monstrosity when it was alive, for be it known that the sea serpent was dead.

> A Boston commission merchant was the consignee of the fish, which had been carefully crated and shipped to Boston. The merchant did not appear to be overjoyed when he was presented with the freight bill and also learned that he would have to pay customs duties. there was a bargain sale of a sea serpent and it was bid in by an Atlantic avenue fish dealer for 25 cents, with the understanding that he would pay the duty and freight. The dealer, in turn, disposed of the curiosity to a representative of the Museum of Natural History, where casts will be taken and it will probably be mounted.
>
> The monster was captured after a terrific fight by four fishermen off Baccaro, Shelburne county, N. S. It is said that 30 years ago a similar specimen was brought to Boston, but no one could be found yesterday who had ever seen anything like it.
>
> One of the oldest dealers declared it was a species known as *Alepido-saurus ferox*. Others were of the opinion that it was a saw fish, which lives in 150 fathoms of water and is dreaded by every other fish that swims.
>
> The sea serpent flourished around this locality several years ago, but the unbiased witnesses who got a glimpse of him solemnly asserted that they saw more than 100 feet of length, with other portions of the great fish buried beneath the surface of the water.
>
> The New England specimen of the sea serpent has not been seen around here for years. He usually made his presence known to fishermen who went out for a day's outing in a small boat, with a bait and other things to keep up their courage and enthusiasm in case the fish failed to bite.
>
> He was the subject of many thrilling stories, both of prose and verse, and many of those who saw him formed what they called the Sea Serpent Club, which flourished for many years and for some time held annual reunions at Marblehead.

In an inspired bit of detective work, Hebda (2015) tracked down the fate of this specimen, to find out that, according to Boston Society of Natural History's *Museum and Library Bulletin* of 1911, a cast had been made of the animal, which indeed is labelled as a being a handsaw fish (*Alepisaurus*). Furthermore, the fact that there was a club of Massachusetts sea serpent enthusiasts whom met regularly shows that regional interest in the mysterious creature persisted long after the initial events of 1817.

In 1880, Maine fisherman S.W. Hanna reported catching an 8 metre-long, eel-like 'marine monster' in his net. Months later he was consulted, first by J.M. Allen, a scientist from Connecticut, and again by no less a personage than S. F. Baird, the US Commissioner of Fish and Fisheries. As Fritts (2017) notes: 'The idea of the most important fisheries scientist in the United States badgering a small-town fishermen for his recollections of a long-dead fish now seems absurd. However, scientists at the time had to rely on such chance encounters to find and identify new species. Without the technology that allows today's scientists to observe sea creatures in their own environment, any strange sighting by a fisherman warranted attention.' In response to Commissioner Baird's questioning as to whether the serpent could have been a misidentified shark, Hanna retorts that as an experienced fisherman 'acquainted with about every species of fish' – offering proof of his capabilities in the form of an itemization of every shark that the UMO was not (Fritts 2017) – he is adamant that what he saw was unusual, providing a sketch to the effect. The animal portrayed is clearly recognizable as a giant oarfish (*Regalecus glesne*), a rarely-seen, anguilliform fish for which there is a long history of being the basis for sea serpent sightings (Oudemans 1892). Of interest is that Fritts believes the legacy of the 1817 Gloucester Sea Serpent to be so pervasive, even six decades after the investigation of *Scoliophis atlanticus* by the Linnaean Society, that it explains the particular line of questioning put forward by Dr. Allen to fisherman Hanna concerning details about the head, mouth and gills of the 'snake'. So even if the Gloucester Sea Serpent itself did not live on, the methodology used to investigate its initial appearance seems to have done so. This then, was the lasting legacy of the Linnaean Society's 1817 *Report.*

If there is a single fish that comes closest to matching society's imagination as to how a classic sea serpent should look, it is the frilled shark (*Chlamydoselachus anguineus*). Growing to more than two metres and having an eel-like body with short pectoral and dorsal fins, frilled sharks, from the time of their formal identification, have been posited as the inspiration behind sea serpents (e.g. including a natural history article published in the *Bulletin of the Essex Institute,* a scholarly journal from the same region of Massachusetts visited by the Sea Serpent decades before). Ubiquitously distributed, they are occasionally observed on the surface where, one glimpse of its gaping, tooth-filled mouth and short rounded snout (Figure 5.37) would be all that it would take to raise the cry of 'sea serpent!' Furthermore, as Howard (2015) states, the fact that apparently an 8 metre-long related animal was caught in 1880 'suggests there may be some giant frilled sharks in the sea that could be taken for sea serpents.' The seas do indeed hold all manner of real creatures of such remarkable nature that there is no need to invent fictitious ones.

Figure 5.37. The mysterious and seldom-seen frilled shark. The most sea serpent-like of all fishes? Photos from Mother Nature Network Earth Matters and Awashima Marine Park (Noel Kirkpatrick).

As mentioned previously, the traits described by eyewitnesses for other UMOs from around the world, which today can be inferred to have likely been large cetaceans (France 2016b, 2018; Galbreath 2015; Paxton *et al.* 2005), marine chelonians (France 2016a, 2017), or pinnipeds (France, in prep.), are very different from the descriptions made by observers of the Gloucester UMO. James Prince was by no means the only eyewitness to have emphatically stated that to his experienced eyes the Gloucester UMO of circa 1817 was dissimilar to commonly observed fauna. Furthermore, some sightings of what were believed to have been other sea serpents in New England waters made throughout the remainder of the nineteenth century (O'Neill 1999) are notable for mentioning associations (either through resemblance or from contemporaneous and same-location sightings) with recognized animals, something missing in almost all the 1815 to 1824 sightings examined herein. For example, in 1875, off Nahant, a UMO was observed for two hours and shot at 'without apparent effect' at least twenty times. So far, this reads as if it could be another sighting of the Gloucester UMO, if it were not for the fact that this particular UMO was observed in a school of blackfish (pilot whales of the genus *Globicephala*), as well as clearly displaying a dorsal fin at the back of the neck 'resembling that of a blackfish.' Also in 1875, this time off

Cape Cod, an UMO was seen 'chasing porpoises.' In 1886, a 27 metre-long, vertically undulating UMO was spotted off Cape Ann 'showing some ten to fifteen [dorsal] ridges at once.' One eyewitness reported (*in* O'Neill 1999): 'I am well acquainted with the movements of porpoises, puffers, sharks, and other sea animals common on this coast and feel full sure that it was not one of them. After it disappeared a school of porpoises passed, rolling in their peculiar way and showing their fins.' And another eyewitness stated: 'After the serpent had disappeared, and while we were still looking, a school of porpoises passed, so that we had a chance to compare their appearance with that of the serpent. I speak of this, as it has often been said that the former was mistaken for the latter,' before concluding 'I shall never doubt that the sea serpent is fact.' Based on such incidences, I find it difficult to believe that a porpoise or dolphin could have been the entangled animal responsible for the 1817 Gloucester UMO sightings. In the same location, ten years later, yet another UMO was spotted. This time, however, one discriminating observer would have none of it, identifying the unusual creature as being nothing more than a wayward and regionally exotic beluga whale (*Delphinapterous leucas*). It is possible that all of these may have been sightings of entangled, known animals misconstrued as sea serpents.

Given the history of misidentifying whales for sea monsters (France 2016b), it comes as no surprise that the same pertains to New England and New York as well. In 1877, two miles off the mouth of Gloucester Harbour, a 'shapeless creature of enormous bulk' was observed which had first been thought to have been a rock ledge 'covered in kelp' (netting, perhaps?) before it raised its eyeless, head out of the water, looking 'like a gigantic alligator' that was 'covered with humps of varying sizes, some being as large as a two-bushel basket' (O'Neill 1999). Compare this with a modern description of a cetacean, and the mystery disappears: 'A humpback's head bears dozens of bumps, each with a single stiff hair and a mass of sensory nerves. The scientists call these humps tubercles. They give the animal a keen sense of touch and the look of a knobby alligator when they rest with their narrow rostrums at the surface of the water' (Johnson 2005). Add to this, the fact that when the 1877 Gloucester UMO disappeared 'the water closed over it with a tremendous roar and surge and spray, many feet into the air,' and one can see why the likelihood that the 1817 Gloucester UMO to have been a humpback whale must be regarded as remote. Also in 1877, an UMO was seen in New York whose head 'seemed to be enveloping in white sea foam as if in the act of spouting'; i.e. obviously another whale and again a behaviour quite unlike that described by eyewitnesses of the 1817 Gloucester UMO. An 1898 sighting of what can be inferred to be an entangled animal – i.e. described as having 'a form resembling a seventy-five-feet string of hogsheads,' and a body that was either 'creased in sections or the color markings camouflages as such,' with 'an absence of fins and the eel-like shape of the tail' – certainly reads like a sighting of the 1817 Gloucester UMO. Until that is, one notes that the first sighting of the UMO was of 'a black object having the appearance of a ledge just awash, or a whale's back loomed up' (O'Neill 1999); i.e. obviously an entangled whale. By the turn of the century, it seems some skeptics and the press may have had enough of UMO sightings, publishing an illustration (Figure 3.19) in 1902 that contrasted the sea serpent that some believed they had seen off Nahant with a more traditional whale-like creature suggested to have been what they probably actually *did* observe.

5.5 LINGERING QUESTIONS CONCERNING THE NATURE OF THE BEAST

Short of having a time machine to travel back to Gloucester thereabouts, circa 1817, we will never know, beyond reasonable doubt and credence given to logically argued hypotheses, one such as I posited in the previous section, as to what type of animal the unusual UMO might actually have been, and as to whether it was indeed, as I believe to have been the case, an example of an animal entangled in fishing gear or other maritime debris. There are of course other questions about the phenomenon that intrigue just as much as they frustrate attempts made at historic forensic interpretation. The most pressing of these concerns as to whether all the sightings pertain to the same individual animal, irrespective of what type of creature it might have been.

Cryptozoologists writing about UMOs that have been repeatedly observed in lakes such as Loch Ness, Lake Champlain, and Lake Okanogan, or along the British Columbia and the Pacific Northwest coast, all too often fall into the trap of assuming that the entire corpus of sightings deal with the same individual, if the timeframe is one of years, or type of individual, if the timeframe is decadal or longer. And if it is the latter, these 'scholars' face the sticky wicket of having to invent a self-sustaining population of reproducing animals, the bulk of whom are seldom if ever seen over time, in order to keep the species alive. In so doing, cryptozoologists open themselves up to easy parody by legitimate aquatic biologists (e.g. Scheider and Wallis 1973; Sheldon and Kerr 1972).

Up until this point, the impression conveyed has been that the UMO sightings detailed in Chapter 4 pertain to a single individual animal observed in a regionally constrained location over a limited period of nine years. Do I believe this to have truly been the case, however? The answer is, probably no. My lumping together of what I more realistically think to have been a set of grouped sightings concerning disparate individuals was made for purposes of convenience. Table 5.15 summarizes the sightings covered in the present book. As we progress through this spatial-temporal sequence, I will offer a rating of my own confidence about what is being discussed, ranging from high to moderate assurance, and progressing on to moderate and then major uncertainty.

With a high degree of assurance, I believe that the four clusters of tightly coupled (i.e. in terms of time and space) sightings – those 44 in Gloucester Harbour during August 1817, those 5 near and in Long Island Sound between September and November 1817, those 10 around Gloucester from July to September 1818, and those 10 in Boston Harbour during August and September 1819 – to represent a single individual for *each* one of the clusters. Whether those four clusters in turn pertain to the same individual or are instead independent of one another is more difficult to access. Nevertheless, I am moderately assured that the two clusters of the Gloucester August 1817 and Long Island October-November 1817 sightings can be linked via the September 1817 sighting that occurred in-between those two locations. The overwinter linking of the two clusters of Gloucester Harbour sightings between 1817 and 1818 can also be deduced with moderate assurance. Whether these sightings can in turn be linked over the next winter to those in Boston Harbour in the summer of 1819 is something about which I have moderate uncertainty.

The recurring appearance of the UMO over the next five years between 1820 and 1824 (probably sighted more frequently but only sporadically reported as the newsworthiness of the phenomenon had by then been waning due to rising skepticism), can only be inferred about with moderate uncertainty, tempting though that it is to suppose that the 7 midsummer sightings that repeatedly occurred in Boston Harbour over this time period suggest a single animal with a predilection for that specific location year after year. Finally, whether all of these 1820-1824 sightings, together with those from other locations over the same time period, as well as the single sighting in 1815, are related to the bulk of the total sightings that occurred between 1817 and 1819, is something about which there is major uncertainty.

Table 5.15. Spatial-temporal sequence of sightings of the reputed Gloucester Sea Serpent between 1815 and 1824. Undated sightings identified as only 'summer' were ascribed to August. Bracketed numbers indicate frequency of sightings.

Year	North	Cape Ann	In-between	Boston	In-between	Long Island
1815					July (2)	
1816						
1817		Aug (44)				
					Sept (1)	
						Oct (2)
						Nov (2)
1818	May (1)					
		July (7)				
		Aug (1)				
		Sept (2)				
1819			July (2)			
				Aug (9)		
		Aug (1)				
				Sept (1)		
1820				Aug (2)		
1821	Aug (1)			Aug (2)	Aug (1)	
1822				Aug (2)		
1823				July (1)		
		Aug (2)				
	Aug (1)					
1824	Aug (1)					

In conclusion, I believe that, with few exceptions, the sightings that occurred in Gloucester Harbour and Long Island Sound during the summers of 1817 and 1818 represent a single individual animal, possibly a large fish like a giant Bluefin tuna, that was entangled in fishing gear or maritime debris. And I may be willing to extend that illation, although with more reticence, to also include those sightings that occurred in Boston Harbour the following year, 1819. That reticence arises, in part, due to confusion surrounding Captain Rich's encounter with what he believed to be the UMO, but also due to the lower chance that entanglement could last for such a duration. Certainly, the longevity of a nineteenth-century hemp rope attached to an array of floats can make conceivable the idea that a single, migrating animal may have been seen one year and then again the next, thereby being responsible for the Sea Serpent craze. Can such reasoning also be extended to include a third year? Perhaps, but I would argue with some less likelihood. We can therefore be more certain of the following conclusion: given what is known about the rate of breakdown of natural fibre material, as well as the debilitating effects of non-lethal entanglement, there is a near impossibility that it could have been the same individual animal that was observed over the entire nine-year period. In other words, the UMO sightings that occurred in 1815 and in 1820-1824 may very well represent other entangled animals. Several of the eyewitness accounts, for example, mention the proximity of known animals, such as a whale or sharks, to the UMO.

One entangled animal of course in no way precludes another falling victim to the same. Indeed, we now know that it is in fact the opposite that is more common; i.e. in those situations wherein one animal becomes entangled, there is a greater likelihood that another will suffer the same fate; as witness to the sad events that transpired during the summer of 2017 in the Gulf of St. Lawrence with the multiple entanglement-related deaths of Atlantic right whales. Certainly there was no shortage of fishing gear scattered about New England and Maritime Canadian waters in 1817 following the Tambora eruption, for many animals of different species to become entangled. In the end, it is impossible to reach a definitive conclusion. And therein lies the reason why scientific ecologists, comfortable in their realm of falsifiable predictability, find it frustrating to be satisfied with the implicit uncertainties frequently germane to in the realm of natural history (France in prep.; Peters 1980, 1991).

Can the mundane be invoked to explain the mysterious? Well certainly, as that is the entire thesis behind this book. But can the same parsimonious reasoning be applied to sorting out the enigma associated with Captain Rich in the saga? For the second unanswered question we face concerns whether or not Rich's captured tuna bears any relationship at all to the so-called Gloucester Sea Serpent. Here too, the best we can do is only surmise.

Rich's best defence in building the case for ascribing the captured giant Bluefin to being the UMO is that eight of his eleven crewmembers whom were familiar with the latter (indeed these individuals were actually hand-picked for just that very purpose as Rich had himself not observed the Gloucester Sea Serpent) 'all agreed to a man' that what was seen and later captured was indeed 'the supposed Serpent.' But how much credence can be placed in this argument? Note, we do not have any direct testimonies from the crewmembers themselves. All we have to go on comes second-hand from their captain. Even if we had first-hand accounts from the crew that confirm the Captain's statement, could these really be believed? After all, what poorly paid sailor would turn down his share of the prize money (not to mention bragging rights) to contradict his captain in the name of the lofty goal of preserving historical fidelity? The irony is that it is preciously here where those members of the legal profession who made up the ranks of the Linnaean Society could have made a solid contribution to the elucidating the phenomenon by using their professional skills rather than going off, as they had previously done, into realms of natural history fantasy. Would that they had pursued and obtained independent crewmember testimonies about the tuna pursuit and capture, even if they were expressed only anonymously, with the same rigor with which they had sought out eyewitnesses of the UMO the previous year. Finally, what are we to make of the statement by Dr. Samuel Mitchell, raised a decade later, that 'the persons [presumably the crewmembers?] who, when they saw him [the fish] in the water, declared positively that he was the sea-serpent, now changed their minds, and swore he was not.' Oh for a good lawyer to have grilled the 'persons' for the truth, as to whether Captain Rich was legitimate or was rather just a (sea-serpent)snake-oil salesman.

For many, the fact that the Sea Serpent was seen the following year, 1819, was proof positive that Captain Rich's tuna had nothing to do with the monster. This reasoning assumes of course that the 1819 sightings, most of which were made in Boston Harbour, were in fact of the same UMO seen the preceding two years around Cape Ann and in Long Island Sound. As discussed above, however, in relation to the longevity of natural fibre ropes and the debilitating effects of long-term entanglement, I am hesitant to definitively equate the two groups of sightings. And so, the fact that sightings of *the* or another entangled animal continued after Rich's capture of the giant Bluefin, is neither here nor there in terms of providing insight as to what the relation of said captured tuna might have been to the UMO.

Rich's next piece of evidence that he had captured the Gloucester Sea Serpent, whose size he offers to swear upon oath to be 'not less than 100 feet in length,' is his insistence that those who had 'described his [the Serpent's] body as being like kegs fastened together' were simply deceived due to the tuna's 'peculiar movement and his velocity.' From this we surmise he meant that it was the wake produced by the rapidly swimming giant Bluefin that fooled everyone into believing that what they were viewing was an extended tail. Is this argument credible? Remember that Charles Sargent certainly did not think so, concluding his 1818 pamphlet by dismissing as preposterous the very idea that anyone could confuse a wake for a tail. But on the other hand, even the most cursory review of the literature on the Loch Ness Monster reveals that such confusions have been frequently made before, some even honestly admitted to by cryptozoological believers in Nessie. As well, recollect that similar arguments were also made about wakes and waves being behind the mystery of Chessie in Chesapeake Bay. The problem with this conclusion of Rich's is that, as often as not, the Gloucester UMO was observed at rest, its so-called tail lying strung out motionless upon the surface of the water. In such cases it is simply impossible to equate such a 'tail' resembling a string of herring floats or 'kegs fastened together' to be anything other than what it actually was, rather than being a deception created by a motion-induced wake. It is odd that critics of Captain Rich did not make more of this discrepancy. Instead, they dwell upon the physical differences irrespective of locomotion. 'M,' for instance, focuses his attention on Rich's theory being unable to explain the presence of the head that had been seen by some of the eyewitnesses. And the editor of the *New York Daily Advertiser* questioned how it was physically possible 'to place so many of them [i.e. the 'twenty-three bunches...of the size of a ten gallon keg'] upon the back of a fish not more than ten feet in length.' Well, as I hope I have convincingly argued, the answer to the latter query is entanglement, and in relation to the former concern is that, as Fama also believed, the first piece of debris in the string by coincidence probably looked like a horse's head. And of course, the deception underlying the imagined tail is due to a combination of things; i.e. pull along an entangled string of maritime debris bouncing up and down on the surface of the water within a churning wake produced by the rapid movement of an animal and, presto, you have what would look to many, even the most discriminating observer, to be a real tail wiggling about. In other words, one theory does not preclude the other; i.e. the perception of a tail was no doubt due to *both* the trailing debris and the wake.

What is certain, however, is that the specific giant Bluefin that was eventually captured and displayed did not in any way have a physical tail that resembled that of the observed UMO. One can imagine that had Rich caught the *actual* animal, whatever it might have been, tuna or otherwise, that was entangled in the debris that gave rise to the Gloucester Sea Serpent, he would have not only mentioned but would have also championed this in the face of the onslaught of criticism that was subsequently heaped upon him. The fact that he did neither indicates that the giant Bluefin that was captured and paraded about bore no relation to what I believe to have been the actual UMO. This conclusion does not, however, preclude my supposition that the most likely candidate for the entangled animal behind the legend of the Gloucester Sea Serpent was a giant Bluefin, or possibly another large fish. It is important to note that this supposition, reasoned through to a degree of some likelihood as outlined in Chapter 5, is made independent of Captain Rich's tuna-cum-serpent proposal. And so, in the end, we have what might be an historical coincidence wherein the intrepid captain, though he did not capture the very individual responsible for the sightings in 1817 and 1818, may in fact have captured a member of the same species behind the legend. An intriguing possibility.

ADDENDUM: My lingering unease concerning the uncertainty about my hypothesis that the entangled animal behind the legend of the Gloucester Sea Serpent was a giant Bluefin tuna was not alleviated by any number of books and scientific articles read. And so, it was at the end of a two year-long process of research and writing and rewriting, that I decided I needed to bounce my ideas off experts. Richard Ellis, he of comprehensive book *Tuna: Love, Death, and Mercury*, abruptly hung up the phone when thrice I contacted him. This bad experience led to me abandon soliciting opinions from published experts. In consequence, opting for local wisdom over scholarly knowledge, I decided to obtain feedback from tuna fishermen. Unfortunately, attempts to do so with those from the *National Geographic* television show 'Wicked Tuna' did not bear fruition during my visit to Cape Ann. Because of this, it was not until mid-autumn 2017 when the first draft of this manuscript was near completion, that I drove four hours from my home at the top of the Bay of Fundy down to the southern tip of Nova Scotia overlooking the Gulf of Maine. There, in the small French-speaking Acadian village of Wedgeport, the self-professed 'Historic Sport Tuna Fishing Capital of the World,' frequented in its heyday by the rich and famous, such as the likes of Franklin D. Roosevelt, Amelia Earhart, and Ernest Hemingway, I visited the Tuna Museum and Interpretive Centre. More importantly, I had arranged to meet with experienced locals whom had participated in Bluefin fishing in the International Tuna Cup Matches held in the surrounding waters during the 1970s.

After touring the museum, I sat down with the fishermen, all in their seventies and eighties, gathered as they always do on Saturday mornings to drink coffee and reminisce about the great hunts of decades past. I barraged them with questions before providing a synopsis of the 1817 events in and around Gloucester, particularly Captain Rich's exploits, and then asked for comments about my tentative conclusions regarding the putative identity of the entangled animal.

Some of the information was reassuringly supportive of my thesis:

- Bluefin tuna indeed have a very noticeable and distinctive wake due to their large, wide bodies and propensity to swim near the surface. Nothing else produces a comparable wake.
- Marine mammals, both frequent visitors like pinnipeds, dolphins, and young cetaceans, as well as occasional vagrants such as belugas, are easily discernable to even the most inexperienced observer.
- Giant Bluefin can swim extremely close to the shore in complicated terrain, in some cases surprisingly far inland up estuaries.
- Generally the larger the individual adult tuna, the more likely it will be a solitary wanderer.
- When swimming after prey, sometimes plunging to depths while so engaged, giant Bluefin zip about at lightning speed, frequently changing direction on a dime, and are observable only if one happens to be directly above them, looking down through clear water.
- Harpooning tuna is a difficult enterprise, with many struck individuals escaping.
- Giant Bluefin are extremely powerful and are known to be able to pull a dory for many hours. Indeed, after being struck with a harpoon, such small boats were often used as easily observable surrogate buoys for subsequent locating and capturing the exhausted animal. The persistence of harpooned tuna may be due to their diving deep and the consequent water pressure acting to constrict the open wound and so by limit bleeding.

But other information, however, was problematic, if not outright contrary, to my thesis:

- There is never a problem with tuna becoming entangled in fishing nets or lobster traps. There are two reasons for this: first, the animals are remarkably smart and can easily avoid the structures; and second, giant Bluefin can easily break free from the nets. Upon further questioning, however, it seemed that the reason for the latter is that once a tuna is caught on a fishing line and darts away in panic into nearby deployed fishing gear, the abrasion from the taught piece of heavy-duty fishing line easily severs the net material, enabling the fish to force its way through the net (whether such an easy escape could be effected in absence of the abrasive fishing line is unclear).
- Because of these two reasons, giant Bluefin are never seen swimming with any piece of entangled material in tow.
- Another reason for this is that the tuna have a series of sensitive nerves in their caudal peduncle which if pinched by a piece of rope seriously impedes the mobility of the animal. Notably, this debilitating effect is capitalized upon during the procedure of hauling aboard a captured animal. As a result, one never sees a giant Bluefin towing a train of fishing gear entangled around its tail.
- Sharks, however, are seen towing pieces of fishing gear. As well, sharks are often much more subtle in their swimming behaviour than is commonly believed, in many cases with only a few inches of their dorsal fins protruding above the surface.

The bottom line from talking with these old salty dogs was that as a group they did not give much credence to my hypothesis as to the likelihood of the Gloucester Sea Serpent being an entangled giant Bluefin tuna. During the long drive home I tried to reconcile my deflated mood brought about through the dismissive opinions of the fishermen by telling myself that the major purpose of my ethnozoological investigation, and the one about which I remained confident in the accuracy of my conclusions, was the assertion that the Gloucester UMO, the most sighted and studied sea serpent in history, was an entangled animal. What the specific identity of that entangled animal might have been, was of secondary importance, and not worth unduly fretting about.

But to the scientist in me, motivated by investigating the unknown, it still nagged as an unsatisfying bit of unfinished business. Some of this changed upon settling into bed that evening and reading Ken Fraser's *Possessed*, about his line-fishing world record for Bluefin tuna (Figure 5.34), which I had purchased earlier that day at the tuna museum. This self-published book was basically a year-by-year summary of which sport fisherman in the Canadian Maritimes had successfully caught which tuna of which size, culminating in description of his own capture of a 1496-pound (679 kg) behemoth in 1979, which still holds the world record. It was getting late and my attention was waning as I dutifully made my way through the pedestrian book. And then, in a remarkable twist of fate, I bolted upright in excitement at what was written in the very last paragraph of this, the very last book to be read by me for this ethnozoology research project. For here, in direct contrast to what I had been told by the Wedgeport fishermen, was at least some credence for my thesis. Here is that final paragraph:

> Naufrage Harbour is the next harbour west of North Lake [in the northwest corner of Prince Edward Island], and is home to about a hundred fishing boats. David Jamieson and Orville Jackson were commercial fishermen from St. Charles and fish out of Naufrage with their boat the 'Aquarius I'. Late in the season on October 8, 1985 they were hauling their nets east of Naufrage when they noticed a tuna in one of their nets. The tuna apparently had been feeding out of the net when it got tangled up and couldn't get free. It was still very much alive and the fishermen had their hands full trying to catch it. When they finally got it tied to the side of the boat they noticed it had a rope and two buoys attached to a leader with a hook still in its mouth. One of the buoys had the name

> 'Vic Ann' written on it. When they got ashore the tuna weighed 1130 lbs and was in good shape. They checked the name of the boat with the Department of Fisheries and it was registered in New Brunswick. In conversation with the owner of the boat it was learned they lost the tuna around the middle of September and it had been towing the two buoys around ever since.

So here we have an example of an apparently healthy giant Bluefin tuna pulling a length of fishing gear that had become caught in its mouth, for a duration of three weeks, and over a distance of at least one hundred and sixty kilometres, before becoming entangled in a fishing net. Perhaps my secondary supposition concerning the tuna-cum-sea serpent mystery may not be so far-fetched after all. On the other hand, it is certainly a very big, and possibly untenable, leap from these statistics in terms of entanglement time, distance, and number of towed floats, to the particulars of the historical situation examined in the present investigation. Therefore, conscious of the *fata morgana* allure of sea serpents and wary of the seductive pull of unrealistic cryptozoological certitude in this regard, I feel it is time for me to stop indulging in such wishful thinking and bid adieu to the Gloucester UMO... the mystery intact as to *whatever* the entangled animal may have truly been. Bowing out with one enigma being potentially explained and another left unresolved is in some respects the perfect time to leave the investigation. For to live in a world – particularly beside a sea – devoid of *all* wonder and mystery, would indeed be a pretty dull existence. For as Einstein said 'the most beautiful thing we can experience is the mysterious. It is the source of all true art and science.'

Literary hypograph and final note

Excerpted from the penultimate chapter in Sarah Perry's masterpiece of literary fiction, *The Essex Serpent*, (Serpent's Tail, Profile Books Ltd., London, 2016):

> 'What if it's still here,' said Naomi, only half-teasing: 'What if it's still here after all and has come back for us?' Out of a shameful desire for revenge she withdrew her arm and stepped backward into the mist, and cupping her hands to her mouth gave a kind of beckoning call. 'I'll summon it, shall I?' she said, frightening herself, but not wanting to stop. 'Watch out! Here it comes!'...
>
> 'Nomi – stop playing games, when you know full well it was just a great big fish...'...
>
> Then the pale pall lifted, and there it was – fifty yards distant, no more: black, blunt-tailed, with an ugly lumpish surface, not the sleek lapping scales of fish or serpent. Naomi half-screamed, half-laughed...
>
> The strengthening sun cast a clear light on the shingle, so that each girl saw, at the same bright moment, what has been cast up. It was a black boat, small and clinker-built, long sunk in the Blackwater and thick with barnacles which gave it the look of uneven flesh, coarse and battle-scarred. Its upturned hull had rotted and begun to sink, so that there was the impression of a blunt snout nosing at the shore; it moved in the last lap of the receding tide, causing its wood to grind against the shingle, and now and then its timber groaned in distress. It was possible to make out, beneath the draping of bladderwrack and sugar-kelp, the name GRACIE picked out in blue-white paint: Banks's boat, long since given up for lost, all the while casting up on the marsh on the whim of the tides, sending a village clean out of its wits.
>
> They clutched each other, not knowing whether to laugh or weep...

A final note on the elusive nature of sea serpents

In 2017, during the two hundredth anniversary of the Gloucester sightings, I went in search of *The Essex Serpent*, a pub located near Convent Garden in London, England. Sadly, even here, where one might have romanticized about sharing pints with locals while engaging in cryptozoological discussions, one is met with disappointment. For the onetime pub has now been replaced by a shop called 'Urban Decay'. So it seems that catching a glimpse of even nothing more physically tangible than an illustrated sign about a famous serpent, much less the creature itself, is thwarted by the progress of modernity.

Sea serpents, however, will always continue to exist in our imagination. Even though entanglement has been recognized by ecologists in specialized publications from the late 1980s onward, and by the general public due to televised newscasts probably since about the turn of the present century, it is likely that many today can still be fooled into believing that they have seen an UMO, if of a cautious disposition, or a definitive sea serpent, if prone toward an overt imagination. For example, in reference to the *Pauline* sighting of a sperm whale locked in mortal combat with a sea serpent (France 2016b), Philip Hoare (2010), in perhaps the most sensitive book ever written about whales, penned the following: 'I must confess I have seen whales look like sea monsters, rolling in the waves. My childhood desire to believe in a lost world (Arthur Conan Doyle, on honeymoon in Greece, claimed to have seen a young ichthyosaur in the sea) seeks to create something palpable out of the apparently incredible; to conjure an abyssal nightmare out of the pages of scientific certainty.' Sea serpents have always been, and will always be, conceptual constructs. Long may 'His Snakeship' so reign supreme in this regard, continuing its...OK, for this time only...*His* swimming about the margins of our romantic, childhood imaginations.

REFERENCES

Aiken, W.R.G. and J. Purser. 1936. The preservation of fibre ropes for use in sea-water. Plymouth Laboratory New Series 20:643-654.

Alexander, K.E., W.B. Leavenworth, T.V. Willis, C. Hall, S. Mattocks, S.M. Bittner, E. Klein, M. Staudinger, A. Bryan, J. Rosset, B.H. Carr and A. Jordaan. 2017. Tambora and the mackerel year: Phenology and fisheries during an extreme climate event. Science Advances 3:1-18.

Allen, G.M. 1930. The walrus in New England. Journal of Mammalogy 11:139-145.

Alverson, D.L., M.H. Freeberg, S.A. Murawski and J.G. Pope. 1994. A global assessment of fisheries bycatch and discards. Fisheries and Oceans Fisheries Technical Paper 339. FAO.

Alves, R.T. and U.P. Albuquerque (eds.) 2017. Ethnozoology: animals in our lives. Elsevier.

Ambrose, J. 1867. Some account of the petrel-the sea serpent-and the Albicore – as observed at St Margaret's Bay, – together with a few observations on a beach mound or kitchen midden, near French Village. Proceedings and Transactions of the Nova Scotia Institute of Science 1:34-44.

Anderson, E.N., D. Pearshall, E. Hunn and N. Turner (eds.) 2011. Ethnobiology. Wiley-Blackwell.

Anderson, M.T.G. and B. Ibatoulline. 2005. The serpent came to Gloucester. Candlewick Press.

Angenot, M. and N. Khouri. 1981. An international bibliography of prehistoric fiction. Science Fiction Studies 8:38-53.

Annear, S. 2016. Strange visitor from the South (a manatee) is spotted off Cape Cod. The Boston Globe, August 29.

Anonymous. 1817. A particular account of a monstrous sea-serpent, the largest ever seen in America. Which has made its appearance in Gloucester Harbor, Cape Ann, (Mass.). And has been seen by hundreds of respectable citizens. With an exact representations of that wonderful monster. Brattleboro Book Store.

Anonymous. 2014. How to unmake a sea serpent: the case of the Scoliophis Atlanticus. Available at: https://tinyurl.com/ybnob4ow

Arment, C. 2004. Cryptozoology: Science & speculation. Coachwhip Publications.

Au, D. and D. Weihs. 1980. At high speeds dolphins save energy by leaping. Nature 284:548-550.

Babson, J.J. 1860. History of the town of Gloucester, Cape Ann. Peter Smith. Available at: https://archive.org/details/historytownglou00changoog/.

Baker, P.F. 1960. Objects seen in Loch Ness. The Scotsman, 13/09/60.

Barber, L. 1980. The heyday of natural history 1820-1870. Doubleday & Company.

Barber, R. 1999. Bestiary. The Boydell Press.

Barnes, D.K.A., F. Galagani, R.C. Thompson and M. Barlaz. 2009. Accumulation and fragmentation of plastic debris in global environments. Philosophical Transactions of the Royal Society B 364:1985-1998.

Bartholomew, R.E. 2012. The untold story of Champ: A social history of America's Loch Ness monster. Excelsior Editions.

Bauer, H.H. 2013. Cryptozoology and the troubles with 'skeptics' and mainstream pundits. Journal of Scientific Exploration 27:690-704.

Beckman, N. 2016. Swimming in the medicinal leech. Model Systems in Neuroethology. University of Illinois.

Bekker-Nielsen, T. and D.B. Casola. (eds.) 2010. Ancient nets and fishing gear. Aarhus University Press.

Benson, K.R. 1988. From museum research to laboratory research: The transformation of natural history into academic biology. In: Rainger, R. (ed.) The American development of biology. University of Pennsylvania Press.

Benton, T. 1991. Oceans of garbage. Nature 352:113.

Berger, C. 1983. Science, God, and nature in Victorian Canada. University of Toronto Press.
Bigelow, H.B. 1927. Physical oceanography of the Gulf of Maine. United States Government. Available at: https://archive.org/details/physicaloceanogr00bige/page/516.
Bigelow, J. 1820. Documents and remarks respecting the sea-serpent. American Journal Science and Arts II:147-154.
Binns, R. 1984. The Loch Ness mystery solved. W.H. Allen.
Binns, R. 2017. The Loch Ness mystery reloaded. Zoilus Press.
Bolster, W.J. 2012. The mortal sea: Fishing the Atlantic in the age of sail. Harvard University Press.
Bowen, H. (ed.) 1817. A monstrous sea serpent. Broadside version of Anonymous (1817).
Brigham, C. 2007. Gloucester's bargain with the sea. Whale's Jaw Publishing.
Brink-Roby, H. 2008. Siren canora: The mermaid and the mythical in late nineteenth-century science. Archives of Natural History 35:1-14.
Brongersma, L.D. 1968 The Soay beast. Beaufortia 15:33-46.
Brown, C.M. 1990. A natural history of the Gloucester sea serpent: Knowledge, power, and the culture of science in Antebellum America. American Quarterly 42:402-436.
Brown, J. and G. Macfayden. 2007. Ghost fishing in European waters: Impacts and management responses. Marine Policy 31:233-241.
Buel, J.W. 1887. Sea and land: An illustrated history of the wonderful and curious things of nature existing before and since the deluge. Philadelphia Historical Publishing Company. Available at: https://archive.org/details/sealandillustrat00buel.
Bunge, M. 1984. What is pseudoscience. The Skeptical Inquirer 9:36-42.
Bunting, W.H. 1971. Portrait of a port. Boston 1852-1914. The Belknap Press.
Burnett, D.G. 2007. Trying Leviathan: The nineteenth-century court case that put the whale on trial and challenged the Order of Nature. Princeton University Press.
Burns, E.I. 2014. Monster on the margin: The sea serpent phenomenon in New England, 1817-1849. Ph.D. Thesis, Department of History, University of Buffalo.
Burton, M. 1960. The Soay beast. The Illustrated London News 04/06/60.
Campbell, S. 1997. The Loch Ness monster: The evidence. Prometheus Books.
Capone, L. 2011. Inland: The true story of a North Atlantic humpback whale. North Country Press.
Cardamone, J.M. 2001. The aging, degradation, and conservation of historic materials made from cellulosic fibres. In: Cardamone, J.M. and M.T. Baker. (eds.) Historical textiles, papers, and polymers in museums. Washington Museum Publication.
Carlson, J.K. and G.R. Parsons. 2001. The effect of hypoxia on three sympatric shark species: Physiology and behavioral responses. Experimental Biology of Fishes 61:427-433.
Carr, A. 1987. Impact of nondegradable marine debris on the ecology and survival outlook of sea turtles. Marine Pollution Bulletin 18:352-356.
Champagne, B.A. 2001. A preliminary evaluation of a study of the morphology, behavior, autoecology, and habitat of large, unidentified marine animals, based on recorded filed observations. Crypto (Draconology Special Issue) 1:93-112.
Champagne, B.A. 2007. A classification system for large unidentified marine animals based on the examination of reported observations. In Heinselman, C. (ed.) Elementum bestia. CRYPTO.
Chase, O. 1999. The wreck of the whaleship Essex. Harcourt Brace.
Cheezum, E.A. 2007. Discovering Chessie: Waterfront, regional identity, and the Chesapeake Bay sea monster, 1960-2000. Ph.D. Thesis, University of South Carolina.
Chen, J., W.O. Friessen and T. Iwasaki. 2011. Mechanisms underlying rhythmic locomotion: Body-fluid interaction in undulatory swimming. Journal Experimental Biology 214:561-574.
Chevalier, T. 2010. Remarkable creatures. Dutton.

Clifford, D., A. Warwick, E. Wadge and M. Willis. (eds.) 2006. Repositioning Victorian sciences: Shifting centres in nineteenth-century scientific thinking. Anthem Press.
Cogswell, C. 1847. A plea for the North Atlantic sea-serpent. Zoologist 1847:1841-1846.
Coleman, L. and P. Huyghe. 2003. The field guide to lake monsters, sea serpents, and other mystery denizens of the deep. Putman.
Conybeare, W. 1824. On the discovery of an almost perfect skeleton of Plesiosaurus. Transactions of the Geological Society of London, number 2.
Crafts, W. 1819. The sea serpent, or Gloucester hoax: A dramatic jeu d'esprit in three acts. A.E. Miller.
Cronon, W. 2011. Changes in the land: Indians, colonists, and the ecology of New England. Farrar, Straus and Giroux.
Cronon, W.J. and M.J. McDonnell. (eds.) 1993. Humans as components of ecosystems: The ecology of subtle human effects and populated areas. Springer.
Crumley, C.J. (ed.) 1994. Historical ecology: Cultural knowledge and changing landscapes. School for Advanced Research Press.
Csanady, G.T. 1973. Transverse internal seiches in large oblong lakes and marginal seas. Journal of Physical Oceanography 3:439-448.
Da Silva Vieira, K., W.L. Da Silva Viera and R.R. Alves. 2017 Imaginary zoology: Mysterious fauna in the reports of ancient travelers and chroniclers. In: Alves, R.T. and U.P. Albuquerque (eds.) 2017. Ethnozoology: Animals in our lives. Elsevier.
Das, N. 2009. Occam's razor and cryptozoology. The Scienta Review. MIT.
Davis, J., F.C. Gray and J. Bigelow. 1817. Report of a committee of the Linnaean Society of New England relative to a large marine animal, supposed to be a serpent, seen near Cape Ann, Massachusetts, in August, 1817. Linnaean Society Private Publication.
De Camp, L.S. and C. Crook de Camp. 1985. The day of the dinosaur. Bonanza Books.
Dean, B.P. 1993. Faith in a seed. Island Press.
Decker, C. 2011. A hard way to make an easy living. Trafford Publishing.
Deedy, A. 2017. Hark! A sea monster! (Oh, no, just a dying whale). All those strange sea monster sightings in days of yore? This may be the best explanation yet. Hakai Magazine 08/02/17.
DeMelo, R., R. France and D. McQueen. 1992. BIomanipulation: Hit or myth? Limnology and Oceanography 37:192-207.
Dendle, P. 2006. Cryptozoology in the medieval and modern worlds. Folklore 117:190-206.
Denny, M.W., B.L. Lockwood and G.N. Somero. 2009. Can the giant snake predict paleoclimate? Nature 457:E3-E4.
Derraik, J.G.B. 2002. The pollution of the marine environment by plastic debris: A review. Marine Pollution Bulletin 42:842-852.
Dewdney, S. and K.E. Kidd. 1973. Indian rock paintings of the Great Lakes. University of Toronto Press.
Dexter, R.W. 1986. Cape Ann visits of the great sea-serpent (1639-1886). American Neptune 46:213-220.
Dinnerstein, D. 1976. The mermaid and the minotaur: Sexual arrangements and human malaise. Harper & Row.
Dixon, D.P. and S.M. Ruddick. 2013. Monsters, monstrousness, and monstrous nature/s. Geoforum 48:237-2328.
Drake, S.A. 1884. A book of New England legends and folklore in prose and poetry. Roberts Brothers. Available at: https://archive.org/details/bookofnewengland00dra.
Drazin, P.G. and R.S. Johnson. 1989. Solitons: An introduction. Cambridge University Press.
Drinnon, D. 2010. Extractions from Heuvelmans, In the wake of the sea-serpents, 'mistaken observations' category. Available at: https://tinyurl.com/y8bk7ya8.
Eagle, L., M. Hamann and D.R. Low. 2016. The role of social marketing, marine turtles and sustainable tourism in reducing plastic pollution. Marine Pollution Bulletin 107:324-332.
Ehrman, B.D. 2017. Jesus before the Gospels: How the earliest Christians remembered, changed, and invented their stories of the Savior. Harper One.

Ellen, R. 1993a. The cultural relations of classification, and analysis of Nuaulu animal categories from central Seram. Cambridge University Press.
Ellen, R. 1993b. Nuaulu ethnozoology: A systematic inventory. Cellar Book Shop.
Ellis, R. 1994. Monsters of the sea. Alfred A. Knopf.
Ellis, R. 2008. Tuna: Love, death & mercury. Vintage Books.
Elman, R. 1977. First in the field: America's pioneering naturalists. Van Nostrand Reinhold.
Emling, S. 2009. The fossil hunter: Dinosaurs, evolution, and the woman whose discoveries changed the world. Palgrave Macmillian.
Erickson, E. 1956. When New England saw the serpent. American Heritage 7:26-27.
Evans, H.E. 1993. Pioneer naturalists. Henry Holt.
Evans, M. 2010. The roles played by museums, collections and collectors in the early history of reptile palaeontology. Geological Society London Special Publications 343:5-29.
Fama, E. 2012. Debunking a great New England sea serpent. Available at: https://tinyurl.com/y8qadzz4.
Feldkamp, S.D., D.P. Costa and G.K. DeKrey. 1989. Energetic and behavioral effects of net entanglement on juvenile northern fur seals (*Callorhinus ursinus*). Fishery Bulletin 87:85-94.
Feldman, A., N. Sabath, R.A. Pyron, I. Mayrose and S. Meiri. 2016. Body sizes and diversification rates of lizards, snakes, amphisbaenians and the tuatara. Global Ecology and Biogeography 25:187-197.
Fenimore Cooper, J. 1846. Lives of distinguished American naval officers. Cary & Hart.
Fish, P.P. 2007. Antique houses of Gloucester: The families who built them, the mayor who moved them and the changing face of the harbor village. History Press.
Folke, C. 2006. Resilience: The emergence of a perspective for social-ecological systems analysis. Global Environmental Change 16:254-267.
Forman, R.T.T., D. Sperling, J.A. Bissonette, A.P. Clevenger, C.D. Cutshall, V.H. Dale, L. Fahrig, R. France, C.R. Goldman, K. Heanue, J. Jones, F. Swanson, T. Turrentine and T.C. Winter. 2002. Road ecology: Science and solutions. Island Press.
Forth, G. 2011. Why the porcupine is not a bird: Exploration in the folk zoology of an eastern Indonesian people. University of Toronto Press.
France, R. 1983. Life history response of the crayfish Orconectes virilis to acidification in the Experimental Lakes Area, northwestern Ontario: A laboratory and field study. M.Sc. Thesis, University of Manitoba.
France, R. 1989. Recommended use of the terms 'production', 'production dynamics', and 'production ecology'. Bulletin of the North American Benthological Society 6:196-199.
France, R. 1992. Garbage in paradise. Nature 355:504.
France, R. 1993. Gaian integrity: A clarion precept for global preservation. The Trumpeter 9:159-163.
France, R. 1994. Nitrogen isotopic composition of marine and freshwater invertebrates. Marine Ecology Progress Series 115:205-207.
France, R. 1995a. Carbon-13 enrichment in benthic compared to planktonic algae: foodweb implications. Marine Ecology Progress Series 124:307-312.
France, R. 1995b. Source variability in ð15N of autotrophs as a potential aid in measuring allochthony to freshwaters. Ecography 18:318-320.
France, R. 1995c. Stable nitrogen isotopes in fish: Literature synthesis on the influence of ecotonal coupling. Estuarine and Coastal Shelf Science 41:737-742.
France, R. 1996a. Stable isotopic survey of the role of macrophytes in the carbon flow of aquatic foodwebs. Vegetatio 124:67-72.
France, R. 1996b. Scope for use of stable carbon isotopes in discerning the incorporation of forest detritus into aquatic foodwebs. Hydrobiologia 325:219-222.

France, R. 1997. Land-water linkages: Influence of riparian deforestation on lake thermocline depth and possible implications for cold stenotherms. Canadian Journal of Fisheries and Aquatic Sciences 54:1299-1305.

France, R. 2001a. Conservation biology: The newest tool in the biodiversity toolbox. Biodiversity 2:17-21.

France, R. 2001b. Reflecting heaven: Thoreau on water. Houghton Mifflin.

France, R. 2003a. Deep immersion: The experience of water. Green Frigate Books.

France, R. 2003b. Profitably soaked: Thoreau's engagement with water. Green Frigate Books.

France, R. (ed.) 2008. Handbook of regenerative landscape design. CRC Press.

France, R. 2010. High Arctic extreme science: Environmental research from the trans-Ellesmere Island ski expedition. Green Frigate Books.

France, R. 2011a. Evidence for tracing incorporation of sewage-derived organic matter into an urbanized food web with stable nitrogen isotopes. International Journal of Environmental Science 1:1047-1055.

France, R. 2011b. Veniceland atlantis: The bleak future of the world's favorite city. Libri.

France, R. 2013. Feeding on the edge: σ13C analysis of a macrofauna food web from disappearing eelgrass beds in a polluted urban harbour. Aquatic Conservation: Marine and Freshwater Ecosystems 25:81-90.

France, R. 2016a. Historicity of sea turtles misidentified as sea monsters: A Case for the early entanglement of marine chelonians in pre-plastic fishing nets and maritime debris. Coriolis: An International Journal of Maritime Studies 6:1-24.

France, R. 2016b. Reinterpreting nineteenth-century accounts of whales battling 'sea serpents' as an illation of early entanglement in pre-plastic fishing gear or maritime debris. International Journal of Maritime History 28:686-714.

France, R. 2016c. From land to sea: Governance-management insights from terrestrial restoration research useful for developing and expanding social-ecological marine restoration. Ocean and Coastal Management 133:64-71.

France, R. 2017. Imaginary sea monsters and real environmental threats: Reconsidering the famous Osborne, 'Moha-moha', Valhalla, and 'Soay Beast' sightings of unidentified marine objects. International Review of Environmental History 3:63-100.

France, R. 2018. Illustration of an 1857 'sea-serpent' sighting re-interpreted as an early depiction of cetacean entanglement in maritime debris. Archives of Natural History 45(1):111-117.

France, R. and M. Sharp. 1993. Comparison of avian species richness and abundance for localized oases on Ellesmere Island. Polarfuschung 61:147-152.

France, R. and M. Sharp. 1995. Polynyas as centers of organization for structuring the integrity of Arctic marine communities. In: Ehrenfeld, D. (ed.) To preserve biodiversity – an overview: Readings in conservation biology. Blackwell Science Inc. Publ., Cambridge, MA.

France, R. and R. Steedman. 1996. Energy provenance for juvenile lake trout in small Canadian Shield lakes shown by stable isotopes. Transactions of the American Fisheries Society 125:512-518.

France, R., J. Holmquist, M. Chandler and A. Cattaneo. 1998a. ∂15N evidence for nitrogen fixation associated with macroalgae from a seagrass-mangrove-coral reef system. Marine Ecology Progress Series 167:297-299.

France, R., M. Chandler and R. Peters. 1998b. Mapping trophic continua of benthic foodwebs: Body size – σ15N relationships. Marine Ecology Progress Series 174:301-306.

France, R., N. Yan, P. Olesiuk, R. Nero, P. Del Giorgio and W. Keller. 1995. Secondary analysis of zooplanktivorous macroinvertebrates and phytoplankton and water clarity associations. Freshwater Biology 34:255-261.

Franklin, A. 1990. Cherokee folk zoology. Garland Science.

Franklin, A. 1999. Animals and modern cultures: A sociology of human-animal relations in modernity. SAGE Publications.

Fraser, K. undated. Possess: World record holder for Bluefin tuna. Self-published. (copy obtained from the Wedgeport Tuna Museum and Interpretive Centre).

Fritts, R. 2017. The mystery of the 19th-century Maine marine monster. Hakai Magazine, Feb. 6.

Fry, B. 1988. Food web structure on Georges Bank from stable C, N, and S isotopic compositions. Limnology and Oceanography 33:182-190.

Fuller, J. 2001. Before the hills in order stood: The beginning of the geology of deep time in England. Geological Society London Special Publications 190:15-23.

Gage, S.H. 1896. Zoology as a factor in mental culture. Science 4:207-213.

Galbreath, G.J. 2015. The 1848 'enormous serpent' of the *Daedalus* identified. The Skeptical Inquirer 39(5):42-46.

Gardner, D. 2007. Whale survives harpoon attack 130 years ago to become 'world's oldest mammal'. The Daily Mail. Available at: https://tinyurl.com/y9b2vdbk.

Geller, L.D. 1970. Notes on sea serpents of coastal New England. New York Folklore Society 24:153-160.

Gibson, J. 1890. Monsters of the sea: Legendary and authentic. T. Nelson.

Goode, G.B., J.A. Allen, H.W. Elliot, F.W. True, E. Ingersoll, J.A. Ryder, R. Rathbun, R.E. Earll, W.A. Wilcox, A.H. Clark, F. Mather, J.W. Collins, M. McDonald, S. Stearns, D.S. Jordan, T.H. Bean, L. Kumlien, N.P. Scudder, C.G. Atkins, C.H. Gilbert, J.T. Brown and J.H. Swan. 1884. The fisheries and fishery industries of the United States. National Oceanographic and Atmospheric Administration. Available at: https://tinyurl.com/ycbuk6nr.

Gosse, P.H. 1861. Romance of natural history. Sheldon and Company.

Gould, C. 1886. Mythical monsters. W.H. Allen.

Gould, R.T. 1930. The case for the sea-serpent. Philip Allen.

Gould, R.T. 1934. The Loch Ness monster. Geoffrey Bles.

Graham, J.B., W.R. Lowell, I. Rubinoff and J. Motta. 1987. Surface and subsurface swimming of the sea snake *Pelamis platurus*. Journal of Experimental Biology 127:27-44.

Greener, M. 2010. The golden age of sea serpents. Fortean Times. March 2010.

Gregory, M.R. 2009. Environmental implications of plastic debris in marine settings – entanglement, ingestion, smothering, hangers-on, hitch-hiking and alien invasions. Philosophical Transactions of the Royal Society B 364:2013-2025.

Hall, M.A., D.L. Alverson and K.I. Metuzals. 2000. By-catch: Problems and solutions. Marine Pollution Bulletin 41:204-219.

Halsted, T. 2017. Mysteries of the deep. The Cape Ann Museum Perspectives Magazine, Summer 2017.

Hancock, E.G. and M. Reilly. 2017. Possibly the oldest stuffed tuna in the world. Archives of Natural History 44:170-173.

Harley, C.M., M. Rossi, J. Cienfuegos and D. Wagenaar. 2013. Discontinuous locomotion and prey sensing in the leech. Journal of Experimental Biology 216:1890-1897.

Harrison, P. 2001. Sea serpents and lake monsters of the British Isles. Robert Hale Ltd.

Haury, L.R., M.G. Briscoe and M.H. Orr. 1979. Tidally generated internal wave packets in Massachusetts Bay. Nature 278:312-317.

Hawkins, T. 1840. The book of the great sea-dragons, Ichthyosauri and Plesiosauri. William Pickering.

Head, J.J., J.I. Bloch, A.K. Hastings, J.R. Bourque, E.A. Cadena, F.A. Herrera, P.D. Polly and C.A. Jaramillo. 2009. Giant boid snake from the Paleocene neotropics reveals hotter past equatorial temperatures. Nature 457:715-718.

Hebda, A.J. 2015. The serpent chronicles: Sea serpents and other marine creatures from Nova Scotia's history – a book about stories. Nova Scotia Museum.

Heinselman, C. 2001. Dracontology: Being an examination of unknown aquatic animals. CRYPTO Special Publication.

Henderson, J. 1914. Ethnozoology of the Tewa Indians. Smithsonian Institute Bureau of American Ethnology.

Heuvelmans, B. 1968. In the wake of the sea-serpents. Hill and Wang.

Heuvelmans, B. 1988. The sources and method of cryptozoological research. Cryptozoology 7:1-21.

Heuvelmans, B. 1990. The metamorphosis of unknown animals into fabulous beasts and of fabulous beasts into known animals. Cryptozoology 7:1-21.

Hill, S. 2011. Cryptozoology and pseudoscience. Skeptical Briefs 21:1-3.
Himmelman, J. 1990. Ibis: A true whale story. Scholastic.
Hoare, P. 2010. The whale: In search of the giants of the sea. Ecco.
Hoch, J.M. 2014. Review of abominable science! The Quarterly Review of Biology 89:198-199.
Holm, N. and S. Taffel, 2016. Ecological entanglemetns in the Anthropocene. Lexington Books.
Holm, P., T. Smith and D.J. Starkey. (eds.) 2001. The exploited seas: New directions for marine environmental history. Liverpool University Press.
Holmes, R. 2008. The age of wonder: How the Romantic generation discovered the beauty and terror of science. Harper Press.
Howard, B.C. 2015. Rare shark that inspired sea monster myths is caught. Frilled sharks are rarely seen denizens of the deep that resemble their dinosaur-era ancestors. National Geographic News, January 23.
Hoyle, W.E. 1902. Sea-serpent. Encyclopedia Britannica.
Humphreys, D. 1817. The letters of David Humphreys. Kirk Mercein.
Hunn, E.S. 1978. Tzeltal folk zoology: The classification of discontinuities in nature. Academic Press Ltd.
Hutchins, J. 1968. Discovering mermaids and sea monsters. Shire Publications.
Hutchinson, G.E. 1957. A treatise on limnology. Volume I. Geography, physics and chemistry. John Wiley & Sons.
Hynes, B. 2012. Here be dragons: Strange creatures of Newfoundland and Labrador. Breakwater Books.
Ishi, Y. and M. Shimada. 2010. The effect of learning and search images on predator-prey relationships. Population Ecology 52:27-35.
Jackson, J.R. 1833. On the seiches of lakes. Journal of the Royal Geographical Society of London 3: 271-275.
Jacobs, N. 2014. Two beluga whales have been seen floating about in Massachusetts waters in recent days. New England Cable News, June 27.
Jaffe, A. 2013. Sea monsters in antiquity: A classical and zoological investigation. Berkeley Undergraduate Journal of Classics 1:1-12.
Jaffe, M. 2000. The gilded dinosaur: The fossil war between E.D. Cope and O.C. Marsh and the rise of American science. Crown.
Johnson, A., G. Salvador, J. Kenney, J. Robbins, S. Kraus, S. Landry and P. Clapham. 2005. Fishing gear involved in entanglements of right and humpback whales. Marine Mammal Science 21:635-645.
Johnson, T. 2005. Entanglements: The intertwined fates of whales and fishermen. University Press of Florida.
Josselyn, J. 1674 (1865). An account of two voyages to New-England. Massachusetts Historical Society.
Jylkka, J. 2017. Is there a monster in Gloucester Harbor? The Other Cape, March 22.
Kaschak, E. 2015. Believing is seeing. The blindness of the sighted. Psychology Today 16/08/15.
Kierman, I.C. 1950. An exciting and authentic narrative of the visits of the sea serpent to Cape Ann in the years 1817 and 1886 including sworn depositions together with eye-witness accounts from Rockport Citizens, never heretofore published. Private Publication. Cape Ann Museum.
Kingdon, A. 2017. A Renaissance field guide to fantastical fish. Dutch fisherman Adriaen Coenen's visual masterpiece chronicled the ocean's denizens, real and imagined. Hakai Magazine, February 24.
Klingaman, W.K. and N.P. Klingaman. 2013. The year without summer: 1816 and the volcano that darkened the world and changed history. Macmillian.
Konar, G. 2009. Not your everyday animals: Applying Occam's Razor to cryptozoology. The Scienta Review. MIT.
Kristjonsson, H. 1971. Modern fishing gear of the world 3: Fishing gear, purse seining, aimed trawling. Food and Agriculture Organization.
Kurlansky, M. 1997. Cod: A biography of the fish that changed the world. Vintage Canada.
Kurlansky, M. 2008. The last fish tale: The fate of the Atlantic and survival in Gloucester, America's oldest fishing port and most original town. Riverhead Books.

Laist, D.W. 1997. Impacts of marine debris: Entanglement of marine life in marine debris including a comprehensive list of species with entanglement and ingestion records. In: Coe, J.M. and D.B. Rogers (eds.) Marine debris. Springer.

Lapierre, J.L., P.O. Kosenko, T. Kodama, J.H. Peever, L.M. Mukhametov, O.I. Lyamin and J.M. Siegel. 2013. Symmetrical serotonin release during asymmetrical slow-wave sleep: Implications for the neurochemistry of sleep-waking states. Journal of Neuroscience 33:2555-2561.

LeBlond, P., J. Kirk and J. Watson. 2014. Discovering Cadborosaurus. Hancock House.

LeBlond, P.H. and E.L. Bousfield. 1995. Cadborosaurus, survivor from the deep. Hordsal & Schubart.

Lee, H. 1883. Sea monsters unmasked. William Clowes and Sons.

Lehn, W.H. 1979. Atmospheric refraction and lake monsters. Science 205:183-185.

Lehn, W.H. and I. Schroeder. 1981. The Norse merman as an optical phenomenon. Nature 293:362-366.

Leland, C.G. 1884. The Algonquin legends of New England. Houghton Mifflin.

Lenik, E.J. 2010. Mythic creatures: Serpents, dragons, and sea monsters in northeastern rock art. Archaeology of Eastern North America 38:17-37.

Lewis, A.J. 2011. A democracy of facts: Natural history in the early republic. University of Pennsylvania Press.

Lillywhite, H.B., C.M. Sheehy, H. Heatwole, F. Brischoux and D.W. Steadman. 2018. Why are there no sea snakes in the Atlantic? BioScience 68:15-24.

Loftus, E.F. and J.C. Palmer. 1996. Eyewitness testimony. In: Baynard, P. and A. Grayson (eds.) Introducing psychological research. Springer.

Lotze, H.K. H.S. Lenihan, B.J. Bourque, R.H. Bradbury, R.G. Cooke, M.C. Kay, S.M. Kidwell, M.X. Kirby, C.H. Peterson and J.B.C. Jackson. 2006. Depletion, degradation, and recovery potential of estuaries and coastal seas. Science 312:1806-1809.

Loxton, D. and D.R. Prothero. 2015. Abominable science!: Yeti, Nessie, and other famous cryptids. Columbia University Press.

Lyamin, O.I., P.R. Manger, S.H. Ridgway, L.M. Mukhametov and J.M. Siegel. 2008. Cetacean sleep: An unusual form of mammalian sleep. Neuroscience and Biobehavioral Reviews 32:1451-1484.

Lyell, Sir C. 1850. A second visit to the United States of North America. Harper and Brothers.

Lyons, S.L. 2009. Species, serpents, spirits, and skulls: Science at the margins in the Victorian Age. SUNY Press.

Mackal, R.P. 1976. The monsters of Loch Ness. The Swallow Press.

Maddigan, M.J. 2014. Nemasket river herring: A history. Natural History Press.

Magin, U. 1996. St George without a dragon. Bernard Heuvelmans and the sea serpent. Fortean Studies 4:223-234.

Mangiacopra, G.S. and D.G. Smith. 2007. Does Champ exist? Notes on the historic lake monster conference held in Shelburne, Vermont, 29 August 1981. Coachwhip Publications.

Maritz, B., M. Kgaditse and G.J. Alexander. 2016. Snake body size frequency distributions are robust to the description of novel species. Ecosphere 7:1-7.

Mason, A. 2017. An ocean in the parlor: Home aquariums were the newest fad in the natural history-crazed Victorian era. Hakai Magazine, March 24. 12 pp.

Masse, W.B., E.W. Barber, L. Piccardi and P.T. Barber. 2007. Exploring the nature of myth and its role in science. Geological Society London Special Publication 273:9-28.

Matsuoka, T., T. Nakashima and N. Nagasawa. 2005. A review of ghost fishing: Scientific approaches to evaluation and solutions. Fisheries Science 71:691-702.

Mayor, A. 2000. The first fossil hunters: Paleontology in Greek and Roman times. Princeton University Press.

McCaskill, J. 2009. Conserving waterlogged rope: A review of traditional methods and experimental research with polyethylene glycol. MSc. Thesis, Texas A&M University.

McGowan-Hartman, J. 2013. Shadow of the dragon: The convergence of myth and science in nineteenth century paleontological imagery. Journal of Social History 47:47-70.

McKinney, B. 2013a. The Journal of Cryptozoology. Skeptoid Blog 02/01/13.
McKinney, B. 2013b. In response to Darren Naish: Why cryptozoology is pseudoscience. Skeptoid Blog 08/01/13.
McLendon, J. 1984. Deviant science: The case of parapsychology. University of Pennsylvania Press.
McPhee, J. 2002. The founding fish. Farrar, Straus and Giroux.
Medin, D.L. and S. Atran. (eds.) 1999. Folkbiology. MIT Press.
Meine, C. 1999. It's about time: Conservation biology and history. Conservation Biology 13:1-3.
Merchant, C. 2003. Reinventing Eden: The fate of nature in western culture. Routledge.
Meurger, M. 1988. Lake monster traditions: A cross-cultural analysis. Fortean Tomes.
Meville, H. 1851. Moby-Dick; or, the whale. Penguin Editions.
Miller Christy, R. 1885. Introduction. The flying serpent, or strange news out of Essex. Sole copy in the British Museum Library.
Miller, T.L. 2018. Tale of the super serpent. The Herald Bulletin, Aug. 19.
Mirceta, S., A.V. Signore, J.M. Burns, A.R. Cossins, K.L. Campbell and M. Berenbrink. 2013. Evolution of mammalian diving capacity traced by myoglobin net surface charge. Science 340:6138.
Mitchell, S.L. 1829. The history of sea-serpentism. American Journal of Science and Arts 15:351-356.
Moore, G. 1977. Moncester: The sea serpent that attacked New England. The Boston Herald American/Beacon 09/01/77.
Morison, S.E. 1921. The maritime history of Massachusetts 1783-1860. Houghton Mifflin Company.
Mowat, F. 1997. Sea of slaughter. McClelland-Bantam.
Muir, W. (ed.) 1939. New England blockaded, 1814: From the journal of Lt. Henry Napier, H.M.S. Nymphe. Whitehill.
Naish, D. 2006. Cadborosarus and the Naden Harbour carcass: Extant Mesozoic marine reptiles, or just bad science? Tetrapod Zoology Blog 02/09/06.
Naish, D. 2011. Cryptozoology at the Zoological Society of London. Cryptozoology: Time to come in from the cold? Or, cryptozoology: Avoid at all costs? Tetrapod Zoology Blog 19/06/11.
Naish, D. 2012a. The Cadborosaurus wars. Tetrapod Zoology Blog 16/04/12.
Naish, D. 2012b. The inaugural issue of The Journal of Cryptozoology. Tetrapod Zoology Blog 22/12/12.
Nash, R.F. 1982. Wilderness and the American Mind. Yale University Press.
National Oceanic and Atmospheric Administration. 2014. Report on the entanglement of marine species in marine debris with an emphasis on species in the United States. Silver Spring, MD: National Oceanic and Atmospheric Administration Marine Debris Program.
Nature. 2019. Attenborough and the sea dragon. Public Broadcasting Service.
Neilson, J.L., J.M. Straley, C.M. Gabriele and S. Hills. 2009. Non-lethal entanglement of humpback whales (*Megaptera novaangliae*) in fishing gear in northern southeast Alaska. Journal of Biogeography 36:452-464.
Nelms, S.E., E.M. Duncan, A.C. Broderick, T.S. Galloway, M.H. Godfrey, M. Hamann, P.K. Lindeque and B.J. Godley. 2015. Plastic and marine turtles: A review and call for research. ICES Journal of Marine Science. DOI: https://doi.org/10.1093/icesjms/fsv165.
Newman, E. 1848. The great sea-serpent. Zoologist 54:1604-1608.
Nigg, J. 2013. Sea monsters: A voyage around the world's most beguiling map. University of Chicago Press.
O'Gorman, E.J. and D.W.E. Hone. 2013. Body size distribution of the dinosaurs. PLoS ONE 8:10.1371.
O'Neill, J.P. 1999. The great New England sea serpent: An account of unknown creatures sighted by many respectable persons between 1638 and the present day. Down East Books.
Oudemans, A.C. 1892. The great sea-serpent. Coachwhip Publications.
Papadopoulos, J.K. and D. Ruscillo. 2002. A Ketos in early Athens: An archaeology of whales and sea monsters in the Greek World. American Journal of Archaeology 106:187-227.
Paterson, S.C. and C.C. Seaburg. 1991. Nahant on the rocks. Nahant Historical Society.

Pattazy, S. 2011. An introduction to ethnozoology: A case analysis. LAP Lambert Academic Publishers.
Paxton, C.G.M. 2009. The plural of 'anecdote' can be 'data': Statistical analysis of viewing distances in reports of unidentified large marine animals 1758-2000. Journal of Zoology 279:381-387.
Paxton, C.G., M.E. Knatterud and S.L. Hedley. 2005. Cetaceans, sex and sea serpents: An analysis of the Egede accounts of a 'most dreadful monster' seen off the coast of Greenland in 1734. Archives of Natural History 32:1-9.
Paxton, C.G.E. and R. Holland. 2005. Was Steenstrup right? A new interpretation of the 16th century sea monk of the Oresund. Steenstrupia 29:39-47.
Paxton, C.G.M. 1998. A cumulative species description curve for large open water marine animals. Journal of Marine Biological Association United Kingdom 78:1389-1391.
Paxton, C.G.M. and A.J. Shine. 2016. Consistency in eyewitness reports of aquatic 'monsters'. Journal of Scientific Exploration 30:16-26.
Peck, W.D. 1819. Some observations on the sea serpent. Memoirs of the American Academy of Arts and Sciences 4:86-91.
Perry, S. 2016. The Essex serpent. Serpent's Tail.
Peters, R.H. 1980. From natural history to ecology. Perspectives in Biology and Medicine 23:191-203.
Peters, R.H. 1986. The ecological implications of body size. Cambridge University Press.
Peters, R.H. 1991. A critique for ecology. Cambridge University Press.
Philo, L.M., J.C. George and T.F. Albert. 1992. Rope entanglement of bowhead whales (*Balena mysticetus*). Marine Mammal Science 8:306-311.
Pontoppidan, E. 1765. The natural history of Norway. A. Linde (online access).
Pope, P.E. 2004. Fish into wine: the Newfoundland plantation in the seventeenth century. University of North Carolina Press.
Posey, D.A. 2000. Cultural and spiritual values of biodiversity. Practical Advances Publications.
Pringle, J.R. 1892. History of the town of Gloucester, Cape Ann, Massachusetts. G.H. & W.A. Nichols.
Radford, B. and J. Nickell. 2006. Lake monster mysteries: Investigating the world's most elusive creatures. The University Press of Kentucky.
Rafinesque, C.S. 1819. Dissertation on water snakes, sea snakes, and sea serpents. American Monthly Magazine and Critical Review 1:431-435.
Raynal, M. 2001. Cryptoecology and mathematics: How many cetaceans remain to be discovered? Crypto (Dracontology Special Issue) 1:93-112.
Rea, T. 2001. Bone wars: The excavation and celebrity of Andrew Carnegie's dinosaur. University of Pittsburgh Press.
Regal, B. 2011. Searching for sasquatch: Crackpots, eggheads, and cryptozoology. Palgrave Macmillian.
Regal, B. 2012. Richard Owen and the sea-serpent. Endeavor 36:65-68.
Regier, H.A. and R. France. 1990. Perspectives on the meaning of ecosystem integrity in 1975. In: Edwards, C.J. and H.A. Regier (eds.) An ecosystem approach to the integrity of the Great Lakes in turbulent times. Great Lakes Fishery Commission Special Publication 90-4.
Reyes, J.C., J.G. Mead and K. van Waerebeek. 1991. A new species of beaked whale *Mesoplodon peruvianus* sp. N. (Cetacea: Ziphiidae) from Peru. Marine Mammal Science 7:1-24.
Reynolds, D.S. 1988. Beneath the American renaissance: The subversive imagination in the age of Emerson and Melville. Harvard University Press.
Ritvo, H. 1997. The Platypus and the mermaid and other figments of the classifying imagination. Harvard University Press.
Ritvo, H. 2010. Noble cows and hybrid zebras: Essays on animals and history. University of Virginia Press.
Robertson, D.R. and J.M. Sheldon. 1979. Competitive interaction and availability of sleeping sites for a diurnal coral reef fish. Journal of Experimental Biology 40:285-298.
Rosby, H.T. and P. Miller. 2003. Ocean eddies in the 1539 Carta Marine by Olaus Magnus. Oceanography 16:77-88.

Ross, M.R. 1991. Recreational fisheries of coastal New England. University of Massachusetts Press.
Rossi, L. 2016. A review of cryptozoology: Towards a scientific approach to the study of 'hidden animals.' In: Angelici, F.M. (ed.) Problematic wildlife. Springer.
Ruddiman, W.F. 2013. The Anthropocene. Annual Review of Earth and Planetary Science 41:45-68.
Rudwick, M.J.S. 1992. Scenes from deep time: Early pictorial representations of the prehistoric world. University of Chicago Press.
Rupke, N.A. 1994. Richard Owen: Victorian naturalist. Yale University Press.
Safina, C. 1997. Song for the blue ocean. Owl Books.
Saila, S. 1983. Importance and assessment of discards in commercial fisheries. FAO Circular 765.
Sanderson, I.T. 1947. Don't scoff at sea monsters. The Saturday Evening Post 08/03/47.
Sargent, C.L. 1818. The sea serpent. Evidence about it. Unpublished manuscript. Sole copy in Cape Ann Museum Archives.
Scheider, W. and P. Wallis. 1973. An alternative method of calculating the population density of monsters in Loch Ness. Limnology and Oceanography 18:343.
Schembri, E. 2011. Cryptozoology as a pseudoscience: Beast in transition. Studies by Undergraduate Researchers at Guelph 5:5-10.
Schneider, C. 2015. The monstrosity in the English novel. The Victorian 3:1-11.
Schwartz, F.J. 1984. A blacknose shark from North Carolina deformed by encircling monofilament line. Florida Science 47:62-64.
Schwarz, H.T. 1976. Tales from the smokehouse. Hurtig Publishers.
Scotti, A. and J. Pineda. 2004. Observation of very large and steep internal waves of elevation near the Massachusetts coast. Geophysical Research Letters 31:L22307.
Seitz, J.C. and G.R. Poulakis. 2006. Anthropogenic effects on the smalltooth sawfish (*Pristis pectinate*) in the United States. Marine Pollution Bulletin 52:1533-1540.
Sewell, L. 1999. Sight and sensibility: The ecopsychology of perception. J.P. Tarcher/Putnam.
Shattuck, B. 2013. The monstrous serpent was real! Did the Gloucester fisherman see a massive tuna, or a serpent? And what are we conjuring, when we imagine the sea? Available at: https://tinyurl.com/yc2rbxol.
Sheavly, S.B. and K.M. Register. 2007. Marine debris & plastics: Environmental concerns, sources, impacts and solutions. Journal of Polymers and the Environment 15:301-305.
Sheldon, R.W. and S.R. Kerr. 1972. The population density of monsters in Loch Ness. Limnology and Oceanography 17:327-340.
Shermer, M. 1997. Why people believe weird things: Pseudoscience, superstition, and other confusions of our time. Freeman.
Shermer, M. 2003. Show me the body. Scientific American 288:37.
Shermer, M. 2010. Hermits and cranks: Lessons for Martin Gardner on recognizing pseudoscientists. Scientific American 35:1-3.
Siebert, V. 2014. Michael Phelps: The man who was built to be a swimmer. The most decorated Olympian of all time certainly has a boat load of dedication and steely focus, but he is also a biomechanical freak of nature. The Telegraph 25/04/14.
Soini, W. 2010. Gloucester's sea serpent. The History Press.
Standiford, L. 2011. The man who invented Christmas. Broadway Paperbacks.
Starkey, D.J., P. Holm and M. Barnard (eds.) 2007. Oceans past: Management insights form the history of marine animal populations. Routledge.
Starkey, L.J. 2017. Why sea monsters surround the northern lands: Olaus Magnus's conception of water. Preternature 6:31-62.
Stearns, R.P. 1970. Science in the British colonies of America. University of Illinois Press.

Stommel, H. and E. Stommel. 1983. Volcano weather: The story of 1816, the year without a summer. Seven Seas Press.

Stothers, R.B. 2004. Ancient scientific basis of the 'Great Serpent' from historical evidence. Isis 95:220-238.

Strong, R. 1998. Did Glooscap kill the dragon on the Kennebec? New England Antiquities Research Association Journal 32:38-43.

Sutherland, B.R. 2010. Internal gravity waves. Cambridge University Press.

Sweeney, J.B. 1972. A pictorial history of sea monsters and other dangerous marine life. Nelson-Crown.

Szabo, P. and R. Hedl. 2011. Advancing the integration of history and ecology for conservation. Conservation Biology 25: 680-687.

Taylor, L.R., L.J. Compagno and P.J. Struhsaker. 1983. Megamouth – a new species, genus and family of lamnoid shark (*Megachasma pelagios*, family Magachasmidae) from the Hawaiian Islands. Proceedings of the California Academy of Sciences 43:87-110.

Teresi, D. 1998. Monster of the tub: Through the icy depths of Lake Champlain glides a mysterious beast, affectionately known as Champ. Maybe it's a plesiosaur left over from the age of dinosaurs, or maybe it's just a very peculiar water wave. Discover Magazine 01/04/98.

Thomas, J. 2010. Curiosity. A love story. McClelland & Stewart.

Thomas, L. 2011. Weird water: The lake and sea monsters of Scandinavia and the Baltic States. CFZ Press.

Thompson, D. and M.A. Fedak. 1993. Cardiac responses of grey seals during diving at sea. Journal of Experimental Biology 174:139-154.

Thoreau, H.D. 1962. The journal of Henry David Thoreau, 1855-1861. Dover Publications.

Tidermann, S. and A. Gosler (eds.) 2010. Ethno-ornithology: Birds, indigenous peoples, culture and society. Routledge.

Torrens, H.S. 1991. The dinosaurs and dinomania over 150 years. In: Sargent, A.S. (ed.) Vertebrate fossils and the evolution of scientific concepts. Routledge, pp. 255-283.

Tuan, Y.-F. 1990. Topophilia: A study of environmental perceptions, attitudes, and values. University of Columbia Press.

Tytell, E.D. and G.V. Lauder. 2004. The hydrodynamics of eel swimming. I. Wake structure. Journal of Experimental Biology 207:1825-1841.

Ulanski, S. 2008. The Gulf Stream: Tiny plankton, giant Bluefin, and the amazing story of the powerful river in the Atlantic. University of North Carolina Press.

Ulanski, S. 2013. The billfish story: Swordfish, sailfish, marlin, and other gladiators of the sea. University of Georgia Press.

Van der Schrier, G. and P.D. Jones. 2008. Daily temperatures and pressure series for Salem, Massachusetts (1786-1829). Climate Change 87:499-515.

Van Duzer, C. 2014. Sea monsters on medieval and Renaissance maps. British Library Publishing.

Vegter, A.C., M. Barletta, C. Beck, J. Borrero, H. Burton, M.L. Campbell, M.F. Costa, M. Eriksen, C. Eriksson, A. Estrades, K.V.K. Gilardi, B.D. Hardesty, J.A. Ivar do Sul, J.L. Lavers, B. Lazar, L. Lebreton, W.J. Nichols, C.A. Ribic, P.G. Ryan, Q.A. Schuyler, S.D.A. Smith, H. Takada, K.A. Townsend, C.C.C. Wabnitz, C. Wilcox, L.C. Young and M. Hamann. 2014. Global research priorities to mitigate plastic pollution impacts on marine wildlife. Endangered Species Research 25:225-247.

Verne, J. 1869. Twenty thousand leagues under the sea: An underwater tour of the world. Create Space Publishing Platform.

Vitalano, D.B. 2007. Geomythology: Geological origins of myths and legends. Geological Society London Special Publications 273:1-7.

Wabnitz, C. and Nichols, W.J. 2010. Plastic pollution: An ocean emergency. Marine Turtle Newsletter 129:1-4.

Wada, S., M. Oishi and T.K. Yamada. 2003. A newly discovered species of living baleen whale. Nature 426:278-281.

Walker, B., C.S. Holling, S.R. Carpenter and A. Kinzig. 2004. Resilience, adaptability and transformability in social-ecological systems. Ecology and Society 9:5.

Walsh, E. 2013. 'Seeing is believing' and 'believing is seeing'. Available at: https://tinyurl.com/y8hk45cv.

Wegner, N.C. and D.P. Cartamil. 2012. Effects of prolonged entanglement in discarded fishing gear with substantive biofouling on the health and behavior of an adult shortfin mako shark, *Isurus oxyrinchus*. Marine Pollution Bulletin 64:391-394.

Weihs, D. 2002. Dynamics of dolphin porpoising revisited. Integrative and Comparative Biology 42:1071-1078.

Westrum, R. 1979. Knowledge about sea serpents. In: Wallis, R. (ed.) On the margins of science: The social construction of rejected knowledge. University of Keele.

White, M. 2012. Atrocities: The 100 deadliest episodes in human history. W.W. Norton.

White, T.H. 1984. The book of beasts. Dover Editions.

Whynott, D. 1995. Giant Bluefin. Farrar Straus Giroux.

Wilcocks, J.C. 1884. The sea-fisherman, comprising the chief methods of hook and line fishing in the seas, and remarks on nets, boats, and boating. Forgotten Books reprinting.

Wilcox, C., G. Heathcote, J. Goldberg, R. Gunn, D. Peel and B.D. Hardesty. 2014. Understanding the sources and effects of abandoned, lost, and discarded fishing gear on marine turtles in northern Australia. Conservation Biology 29:198-206.

Williams, G. 2015. A monstrous commotion: The mysteries of Loch Ness. Orion Books.

Wilson, F.A. 1928. Some annals of Nahant Massachusetts. Nahant Historical Society Reprint, 1977.

Wiseman, R. and P. Lamont. 1996. Unravelling the Indian rope trick. Nature 383:212-213.

Wood, G.D. 2014. Tambora: The eruption that changed the world. Princeton University Press.

Woodley, M.A. 2008. In the wake of Bernard Heuvelmans: An introduction to the history and future of sea serpent classification. CFZ Press.

Woodley, M.A., C.A. McCormick and D. Naish. 2012. Response to Bousfield and LeBlond: Shooting pipefish in a barrel; or, Sauropterygian 'mega-serpents' and Occam's Razor. Journal of Scientific Exploration 26:143-147.

Woodley, M.A., D. Naish and C.A. McCormick. 2011. A baby sea-serpent no more: Reinterpreting Hagelund's juvenile 'Cadbororsaur' report. Journal of Scientific Exploration 25:497-514.

Woodley, M.A., D. Naish and H.P. Shanahan. 2008. How many extant pinniped species remain to be described? Historical Biology 20:225-235.

Wright, D.B. and E.F. Loftus. 2008. Eyewitness memory. In: Cohen G. and M.A. Conway (eds.) Memory in the real world. Psychology Press.

Zalasiewicz, J., M. Williams, A. Haywood and M. Ellis. 2013. The Anthropocene: A new epoch of geological time? Philosophical Transactions of the Royal Society A. 369:1938-1949.

APPENDIX 1

Visiting the Gloucester Sea Serpent at Age Two Hundred

In August 2017, I returned to my one-time home of New England to specifically visit the sites, from the southern Gulf of Maine to the edge of Long Island Sound, where the famous UMO had been seen early in the nineteenth century. Photographs of what these locations look like today are shown in Chapter 2. My visit to Nahant was personally nostalgic in that it was here where I had first conducted research on New England's coastal ecosystem (France 2011a, 2013; France *et al.* 1998a,b). Notwithstanding Gloucester itself, at all other locations, even at Nahant where the so-called Sea Serpent had assumed a second home, nowhere was there any mention on an historic plaque of that which had so captivated thousands of citizens and naturalists two centuries before (this is in contrast to the situation at West Point, Prince Edward Island, for example, which celebrates its legacy of sea serpent sightings through interpretive signage). At this point in my visit I realized that my modern quest for traces of the Gloucester Sea Serpent phenomena that had at once been such a global *cause célèbre*, looked like it might prove frustrating. As a way to uplift my spirits, I spent time at the home of a former Harvard colleague who lived in Loblolly Cove, teasing him with a shared British 'taking the Mickey' sense of humour. In particular, although he is regarded as a leading environmental scientist of global renown, was it not an interesting coincidence that the precise location where he had recently decided to move, was also the very site of early American science's most humiliating fiasco; i.e. the Linnean Society's misidentified snake-as-a-baby-sea serpent.

With a present-day bustling population of nearly thirty thousand, to which can be added hundreds of milling tourists on a sunny August day, coupled with a predominance of largely unexciting contemporary architecture, Gloucester comes across as a thoroughly modern city, far different from the small fishing town of fifteen hundred persons huddled about its harbour back in 1817. The buildings that were standing when the Sea Serpent came to town two hundred years ago (Figure A1.1), are few and far between, and difficult to find, given that the most locally accessible book about them (Fish 2007) does not even contain a map identifying their location. In consequence, one has to be diligent in one's search for historical ambiance. For a far more atmospheric glimpse into how the early nineteenth-century town can be imagined to have looked, far better to leave Gloucester itself behind and to travel to the nearby coastal town of Ipswich (actually not too far away from where Captain Rich caught his imagined monster). There one can find an architectural treasure-trove of houses contemporaneous to the 1817 sightings (Figure A1.2). Furthermore, that Ipswich was one of the locational inspirations behind H.P. Lovecraft's brilliant horror story about interbred human-merebeings, 'The Shadow Over Innsmouth,' adds an aura of spookiness appropriate for those searching for sea monster ambiance.

Gloucester's waterfront, though nearly completely contemporary, is still an exciting place to visit for a sea serpent aficionado. This is because it is still very much a true working harbour. Here, one can chat with fishermen and examine their boats up close and find a spot from which to watch them manoeuver past pleasure craft in the harbour, all while sombrely imagining those many brave and hardy souls, who in bygone days, went 'down to the sea in ships' (Figure A1.3). The tourist office has a brochure that identifies the location of historical interpretive signs, one of which is actually about the Sea Serpent (Figure A1.4). And then there is the replica model of a portion of the nineteenth-century waterfront in

Figure A1.1. Examples of early nineteenth-century buildings in Gloucester dating from the time of the UMO sightings in the harbour.

the Museum of Cape Ann (Figure A1.5). But if one really wishes to see what the Gloucester waterfront would have been like back in 1817, it is once again necessary to leave the modern city behind and this time travel to the Mystic Seaport Museum in Connecticut (near Long Island Sound where the sea serpent appeared all those years ago). There one can wander along a waterfront of period buildings cobbled together from around the region over which tower the majestic masts and rigging of tall ships (Figure A1.6), similar to how things looked back in the early nineteenth century when the Sea Serpent swam past the mouth of the Mystic River a few kilometres away.

Figure A1.2. Examples of early nineteenth-century buildings in Ipswich dating from the time of Captain Rich's nearby capture of the putative Sea Serpent.

After several years of reading hundreds of books about sea serpents and lake monsters, and the modern tourism hype that often accompanies such cultural phenomena, I was, and remain, disappointed in Gloucester's lackluster celebration of the bicentenary of the appearance of its most famous maritime resident. In early spring 2017, I placed calls to various city officials as well as to the city's Maritime Museum and the Cape Ann Museum. Remarkably, some individuals with whom I talked had never even heard of the 'His Snakeship,' much less be busy making any arrangements to capitalize on the anniversary of its appearance as a way to celebrate local history and bring in tourist revenue. Back

Figure A1.3. Gloucester's Fisherman's Memorial to those lost at sea.

in the 1990s, when I was living in Massachusetts and an occasional visitor to Gloucester, I do not remember seeing anything that made reference to the Sea Serpent. But again, those were the heady days right in the middle of the *Perfect Storm* phenomena, which then dominated local hucksterism. When visiting in 2017, temporally well beyond the eye of the perfect storm so to speak, things were not much different vis a vis the Sea Serpent. One had to search hard to obtain a glimpse of any representations of the elusive cryptid (Figure A1.7). So different from the Italian town of Gubbio, for example, whose shops are filled with tourist kitsch, and wherein statues and interpretive signs abound, all about the legendary wolf being tamed there by St. Francis. However, whereas it is of course not possible to consume wolf meat in that Umbrian town, a visitor to Gloucester can manage to eat sea serpent, or at least *purported* sea serpent, in an upscale restaurant, followed by visit a nearby shop linked with a locally-filmed television show about capturing sea serpents-cum-tuna (Figure A1.8). In order to witness any serious effort made in regard to the anniversary and consequent tourism about Gloucester's Sea Serpent in 2017, one would have needed to have visited the Cape Ann Museum (Figure A1.9).

Inside the museum, and relegated to a single small display case tucked away in a hallway in the basement near the Archives, where normally only scholars venture, was what amounted to the exhibition about

Figure A1.4. Historical interpretive sign near the waterfront about the Gloucester Sea Serpent, to my knowledge the only such one in New England.

Figure A1.5. Representations of Gloucester's nineteenth-century harbour. Accurate replica model of the 1890 waterfront produced for the 1893 World's Fair in Chicago; the surviving portion of the diorama in the Cape Ann Museum shows rigged fishing vessels, wharf sheds, fish drying flakes, a fish packing house, smokehouse, and the railway used to haul ships out of the water for repair.

the Sea Serpent (Figure A1.10) that had looked so promising from the outside signage and the front cover of the membership publication with its announcement of a feature article inside about 'Gloucester's Mythical Sea Serpent' (Halsted 2017). But what was lacking in terms of quantity of material displayed was made up for by the quality and novelty of the items shown; enough to make any visiting academic sea-serpentologist or lay cryptozoologist happy. For the former, it is the Archives proper where the Gloucester Sea Serpent phenomena can best be appreciated. Facilitated by a helpful staff, a scholar, through pre-arrangement, could handle an original copy of the 1817 Linnean Society *Report* (Figure 3.3) as well as *the* copy of Sargent's 1818 private notebook (Figures 3.4 and 3.5), the latter being temporarily removed from the aforementioned display case for my perusal. There too, sitting at a large table occupied, at least during my visit, by several octogenarian historians wrapped up in their own research, one can view the Archives' collection of news-clippings relating to New England sea serpents. This includes the notebook assembled by Woodbury that had been mined so effectively by O'Neill (1999) for her seminal work on the subject. And in so doing, it is here where one comes as close as is physically possible to the mysterious marine visitor to nearby harbour back in 1817.

On August 10, 2017, a tour of the harbour aboard a sailing ship was offered as a way to visit the specific sites where the Gloucester Sea Serpent had first appeared those fateful days two hundred years prior. The added bonus to the outing was that it was to be guided by the author of the only hitherto published book specifically devoted to the UMO: Wayne Soini. When I found out about the event, congratulating myself in that it seemed my earlier chiding of authorities about the absence of any celebratory activities had been heeded, I arranged my travel plans accordingly, planning to return via the ferry from Portland, Maine to Yarmouth, Nova Scotia, just so I could play cryptozoologist and possibly

Figure A1.6. Period buildings beneath a canopy of tall-ship rigging at the Mystic Seaport Museum, the most atmospheric present-day location in New England providing a sense of how the Gloucester waterfront would have looked in the days when the Sea Serpent came to town.

spot a descendent of His Snakeship along the way (alas, only several humpbacks would be seen as the Gulf of Maine was crossed). It had been my intention to wait until the harbour tour was over and once ashore to offer clientele an explanation of the entanglement theory. But it soon became apparent that the majority of the people who were on the tour (two-thirds of whom seemed to be visiting tourists) had no real interest in the history of the Sea Serpent at all. Many rudely ignored Wayne Soini's reading of the eyewitness testimonies at the precise locations where the creature had been seen, and it was only one or two who were curious enough to ask him any questions. But for a sea-serpentologist, it was a wonderful experience (Figure A1.11).

Figure A1.7. Artistic representations of the Gloucester Sea Serpent seen about the city in 2017: a map near the tourist visitor centre, a t-shirt in a local shop, and a painting on the outcrop of Stage Fort Park.

Figure A1.8. Images from a Gloucester restaurant where one can eat a meal of Captain Rich's putative Sea Serpent, and of a nearby shop where one can buy memorabilia from a locally-filmed television show documenting such 'sea serpents' being caught regularly today.

After the tour, I was fortunate to be able to sit down on the dock and have a long chat with the affable Mr Soini, to learn more of his thoughts and opinions about the UMO. Basically these can be summarized as follows: (1) the UMO was indeed a sea serpent of an unknown species that has since been largely extirpated from New England shores; (2) his book focused only on the events in Gloucester Harbour during August 1817, as he believed the other sightings there and elsewhere in subsequent years were unrelated and unreliable, and therefore unworthy of detailed attention; (3) Captain Rich's captured tuna had nothing to do with the Sea Serpent observed the previous year; (4) fired bullets did not strike the creature because in all cases they missed their target; and (5) this is what startled the cryptid, spooking him into finally departing Gloucester.

Soini was also deeply disappointed in regard to the Gloucester's missed opportunity to effectively promote the bicentenary, and expressed his gratitude when informed about my own spring-time conversations with city and museum personnel which may have been responsible for him being contacted in July about giving the harbour tour. Proud in his status as a local historian and not an academic, Soini was unconcerned about the harsh (and in my opinion, unfair) comments posted by a knowledgeable scholar in an Amazon review of his book, and was uninterested in reading more about the creature in Burns' (2014) thesis when I informed him of is existence and quality of scholarship. Likewise, although I mentioned that I had published several papers over the last year making specific reference to the Gloucester Sea Serpent, and was nearing completion of what would be the second book, after his own, on the subject, Soini neither asked for details about the publications nor inquired as to my identity. It was impossible not to reach the conclusion that this was an example of an incurious, well-meaning cryptozoologist, comfortable within his own blinkered mindset, freed from viewing information that might pose a challenge to his accepted dogma. And it is preciously this ambivalence or opposition to countenancing further, possibly conflicting, data that is at the heart of what distinguishes real, from pseudo, science. I thanked him for both his book and the tour, departing without mentioning entanglement, not wanting to break his comfortable bubble of belief.

Figure A1.9. Advertising and tourist mementos at the Cape Ann Museum for their exhibition to mark the two hundredth anniversary of the 1817 sightings.

For the week I spent in Gloucester in Augusts 2017, my favourite activity became an early morning ritual of climbing atop the cliff at Stage Head and staring out past Half Moon Beach and Ten Pound Island and Piper's Rocks to enjoy the view of the full expanse of the harbour. There, I would spend an hour or so watching the boat traffic while rereading, for the umpteenth time, the original eyewitness accounts, punctuated by looking through my binoculars to where each sighting of the UMO had taken place (all the while munching upon a square of imported Double Gloucester cheddar cheese, jokingly hoping that in so doing it might somehow increase my own potential palatability as a way to lure any hungry sea monster out of its sequestration for a breakfast nibble). Also, I spent much time simply relaxing and admiring the way the rising sun progressively illuminated the bay. And then, one morning, about

Figure A1.10. Cape Ann Museum Archive's exhibition display case and items therein to mark the two hundredth anniversary of the sightings of the Gloucester Sea Serpent.

mid-way through my trip, I saw it...make that *IT*...my very own illusory sea serpent created *de novo* through a happenstance coincidence of natural and anthropogenic phenomena intersecting, for a brief moment through an uncritically receptive mindset before the deception became duly recognized. A wake produced by a boat that had long since moved off toward Hammond Castle rebounded off the rocks below me and while moving back outward into the harbour had collided with a front of waves produced by a sudden and localized onshore sea breeze (which blew past me some moments later). And there it was, silhouetted against the rising sun: the multi-humped Gloucester Sea Serpent itself... *Himself*...before my very eyes that were believing what they were seeing...at least for a moment anyway. Alas, unlike the physically real UMO of 1817, my conceptual UMO of 2017 was short-lived, gone just as suddenly as it had appeared, only an imprint of its fleeting presence left upon my romantic heart rather than upon my rational mind. Still, I offer my own, hastily drawn at the time, contribution to be added to the pictorial canon of the world's most sighted and studied sea serpent (Figure A1.12).

Figure A1.11. Harbour tour undertaken on August 10, 2017 of the sites where the Gloucester Sea Serpent had been seen two hundred years before, guided by Wayne Soini, author of Gloucester's Sea Serpent *(2010).*

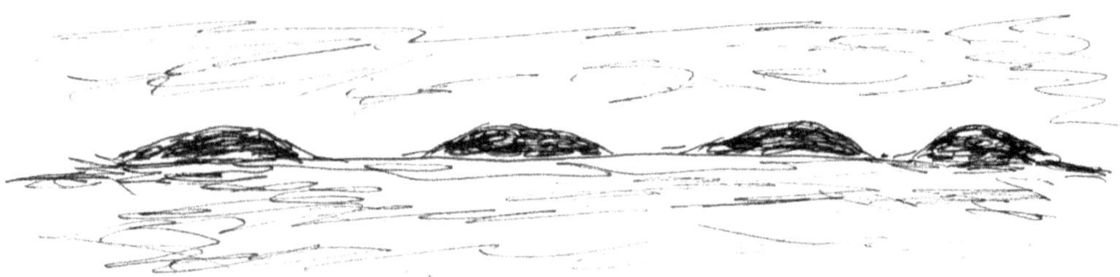

Figure A1.12. Drawing made by R. France based on his August 16, 2017 morning sighting of an UMO in Gloucester Harbour as observed from the vantage point of Stage Head cliffs.

APPENDIX 2

Sea Serpents: 'The Best Explanation Yet'

After reading a series of my papers and following telephone interviews with myself and several United States government experts, journalist Alexander Deedy wrote an article in *Hakai Magazine*, a publication that focuses on issues concerning coastal science and societies. The article had the following banner title and subtitle: 'Hark! A Sea Monster! (Oh, No, Just a Dying Whale). All those strange sea monster sightings in days of yore? This may be the best explanation yet.' This Appendix summarizes some of the case studies abstracted from those papers of mine dealing with what I believe to have been entangled whales (France 2016b) and sea turtles (France 2016a, 2017) misidentified at the time as sea serpents or monsters.

In 1898, while in a sloop off the coast of Maine, Albert Wass reported, in a letter written many years later (*in* O'Neill 1999), seeing what he and his companion imagined to be a sea serpent, but which I believe to be one of the best descriptions we have of a nineteenth century entangled cetacean:

> [We] were on board ...when a black object having the appearance of a ledge just awash, or a whale's back loomed up at less than a quarter of a mile ahead. We could see and hear what seemed to be a spar buoy with its base held stationary, spanking its length on the surface of the water with a circular motion, at varying intervals of from two to five seconds.
>
> At first we thought it must be a whale being attacked by a thrasher or swordfish, but, on nearer approach, we were convinced that it was something different from what we had ever encountered. When we had reached a proximity of about 100 yards (and, believe me, that was close enough!), the monster straightened itself out to a form resembling a seventy-five-feet string of hogsheads...The motion of its swimming was like an eel, and its speed, at least, twenty knots. The absence of fins and the eel-like shape of the tail settled all doubt that his could be any known type of whale.

In this UMO sighting, we have what seems to be a whale but one with an elongated, crenulated body that is specifically likened to a string of buoys or hogshead barrels, and which can alternatively curl up upon itself or move forward at great rapidity.

When, in 1903, the *Glengrant* returned to a Scottish port from fishing in the North Sea, they had a story to tell (in the *Daily Express*) about a fearful encounter with an angry UMO that captured the interest of readers following its publication in several newspapers. Following a cry of 'Whales boys! Look out for your nets,' the crew watched 'an enormous monster with a head like a Chinese dragon' (Harrison 2001) make its way toward their steamer. Parts of the creature were seen to be above the water, 'swaying' to and fro, before the 'monster' dragged its seventy metre 'long sinuous body' underneath the fishing vessel. The UMO moved at great speed, its undulations 'wriggling like a serpent,' and having a horse-like head that sported a long mane that flowed down the back of the creature. It is not hard to consider this as being a large whale or perhaps a basking shark (*Cetorhinus maximus*) swimming underwater, oblivious to what its dragged chain of 'long sinuous' material was doing upon the surface. The long, flowing 'mane' down the back of the UMO suggests an entangled fishing net.

Heuvelmans (1968) reports an 1898 sighting from south Asia in which the observers, to quote Polonius, believe the spotted UMO to be 'very like a whale' without, however, actually being so in their limited, Horatio-esque dreams of all that is possible: 'We saw it only once from close to, and there can be no doubt about it. The appearance of the beast is as follows: a round black fat body like that of a big whale, then a *sinuous* part, not emerging completely, but seeming to join the body to the head. The latter fairly large, continuing the neck, of rather oval shape, and with two gaping holes. Finally a set of dorsal spine-like saw-teeth' (Heuvelmans 1968). The description of 'a set of dorsal spine-like saw-teeth' attached to a whale-like animal suggests an entangled string of gear or debris.

In 1882, off the coast of Scotland, the crew of the *Bertie*, 'said to be men of more than average intelligence,' encountered an UMO with 'a huge head covered with barnacles as large as herring barrels; a prodigious square mouth, with whiskers seven or eight feet long' that was making straight towards them: 'On rising to the surface he had blown, but not like any ordinary whale, which these men were used to seeing. Three hillocks, each about the size of a six-oared boat, with water between them...was next seen passing under their craft. The next moment the monster rose and was approaching the craft with a cavernous mouth, so very wide open that it seemed capable of swallowing the boat and all on board in one gulp' (Harrison 2001). The encounter continues with the crew pelting the UMO with ballast stones which simply bounced off its surface 'like marbles' followed by firing a double shotgun into its mouth which did not dissuade it from following the craft for several hours. The crew reported that the UMO 'had two fins, apparently as large as their mainsail, 'stretched up from his back or sides' and the only resemblance that the creature bore to a whale was its tail, which they were sure would have measured forty feet across.' Clearly, this is a description of a large whale pulling some sizeable pieces of debris which were, of course, impervious to damage by stones or shot. Eight foot-long 'whiskers' suggest an entangled fishing net.

In 1912, off the coast of New England, the crew of the fishing boat *Philomena* had their nets ruined by a sea serpent. As initially run in the *Gloucester Daily Times* (*in* O'Neill 1999), the account is significant in being one of the few to use of the word 'entangled':

> A small school of mackerel in the seine boat were pulling in the seine when a commotion was noticed among the fish, and the sea serpent, which had evidently been under the seine, made its appearance alongside the boat to the alarm and disgust of the crew, who had never seen anything like it before...It became entangled in the seine, tearing it to pieces, and then started off at a 2.40 gait with the boat, seine and everything in tow, all the mackerel estimated at about 40 barrels, getting away...Captain McKinnon describes the sea monster as being from 50 to 60 feet in length, its body which resembled in size and shape an immense tree trunk being black with rough skin covered with barnacles. It had what the fishermen call a hammer head and an immense fin on the back resembling a leg of mutton sail and nearly as large.

After receiving assistance from other nearby boats, the UMO was stabbed and cut adrift. A leg-of-mutton sail is used on small dories. The size, colour, encrustation with barnacles, pursuit of mackerel, strength to pull a laden dory, and location in the cold waters of New England imply the UMO to have been a whale or large shark. O'Neill (1999), a believer in sea serpents, including one in this particular account, nevertheless hits the mark with her interpretation of the event: 'Perhaps the creature [whatever it might have been] is unable to see the net and becomes entangled as it swims up on a closely packed meal.'

The sea has always been a foreign, mysterious, and occasionally scary place, the abode of *Hic sunt dracones*. A common leitmotif of early maps and lithographs was the image of brave sailors and their vulnerable ships being locked in ferocious battle with all manner of denizens of the deep (Nigg 2013; Van Duzer 2014). Likewise, it is not difficult to find images of imagined combat between sperm whales and 'kraken' or giant squids (Sweeney 1972). And of course one of the most dramatic scenes in Jules Verne's widely popular 1864 novel *Journey to the Center of the Earth* is of battling sea monsters, particularly a prehistoric ichthyosaurus and a plesiosaurus. No surprise then that encounters between whales locked in mortal combat with sea serpents formed a popular staple of nineteenth-century maritime lore.

The *Ville de Rochfort* was crossing the Gulf of Mexico in 1840 when the captain reported an intriguing UMO (Oudemans 1892):

> [We saw something like a] long chain of enormous rings, resembling a number of barrels linked together, and in the form very like the back of a silkworm...As the ship approached...we presently saw the extremity of an enormous tail, longitudinally divided into two sections, white and black. This tail appeared to wind itself up, and repose on a part of the object itself. Then, at the other extremity, we saw a membrane rising to the height of about two metres from the water, and inclining itself at a considerable angle upon the mass (without leaving it, however) and this led me to conjecture that the monster before us was provided with an apparatus for the purpose of respiration...At last we perceived something like an antenna rising from the water, to the great height of nearly eight metres, terminated by a crescent at least five metres from one extremity to the other.

Heuvelmans (1968) backs away from supposing that it was a sea serpent that had been seen. He suggests, instead, that the 'membrane' was merely a fin, and not 'an apparatus for the purpose of respiration,' and goes on to state that 'the 'antenna' and 'crescent' could have been a big whale's tail, and the other tail which they saw 'wind itself up' could have been the tentacle of a giant squid attacking the whale.' It is worth noting, however, that incidences of giant squids observed 'attacking' whales are incredibly rare relative to the thousands of cases of cetaceans which have been observed entangled in fishing gear, or in this case, possibly maritime debris. The fact that the imagined serpent's tail, which resembled 'a number of barrels linked together' and which wound itself around and lay in 'repose' upon the body of the UMO all indicate entanglement (i.e. the noted length of the 'tail' precludes it being the alternative hypothesis of an aroused penis a là Paxton *et al.* 2005). The 'antenna rising from the water' is similar to the descriptions of similar 'horns' on the Gloucester Sea Serpent which were likened to a harpoon by one eyewitness. It is possible, as Heuvelmans contends, that the 'membrane' and 'crescent' were a fin or the tail of the whale. But given the case for entanglement noted above, one wonders whether this might be a case of a cetacean dragging along debris from derelict sails or large net traps. Today, when one consults various on-line galleries containing photos of entangled whales, it is possible to see unfortunate cetaceans dragging all manner of billowy hoops, odd-shaped buoys, or other large pieces of anthropogenic bric-a-brac.

In 1879, while aboard the steamship *Kirushiu-maru,* off the southern coast of Japan, the captain and chief officer saw what Heuvelmans (1968) describes as 'a new single combat between a big whale and what...[was] thought was a sea-serpent':

> The chief officer and myself observed a whale jump clear out of the sea, about a quarter a mile away. Shortly after it leaped out again, when I saw there was something attached to it. Got glasses, and on the next leap distinctly saw something holding on the belly of the whale. The latter gave one more spring clear of the water, and myself and chief then observed what appeared to be a large creature of the snake species rear itself about thirty feet out of the water. It appeared to be about the thickness of a junk's mast and after standing about ten seconds in an erect position, it descended into the water, the upper end going first. With my glasses I made out the colour of the beast to resemble that of a pilot fish.

Most interestingly, accompanying the original article published in the periodical *Graphic* were two sketches (Figure A2.1). Oudemans (1892) states elsewhere in his book that it was a porpoise and not a whale whose lateral fin was 'seized' by a sea serpent. Despite his unsupported belief that it was a whale 'of the smaller kind' that was so engaged with a sea serpent, it must be said that the drawing looks quite like a typical breaching baleen whale. Heuvelmans (1968) describes this incident as a whale being 'juggled' by a sea serpent. He cites an 1880 article in *Nature* wherein the author 'even thought that the whale's fluked tail was part of the same animal as the 'swan-neck' in front of it, and thence deduced that the sea-serpent was a cetacean.' He then goes on to advance a cryptozoological explanation for the observation: 'They [the two drawings] show a baleen-whale fighting a sort of monstrous worm that could be a swan-necked sea-serpent, an enormous conger, or the arm of a giant squid. Since the long-necked sea-serpents are usually described as unaggressive, I incline towards the last alternative. This would explain why Captain Davison 'saw something holding on to the belly of the whale'. An animal that fights with teeth and jaws like a conger would have found a better grip on a fin.' Rather than fighting a sea serpent, the drawings show the normal breaching behaviour of a whale with a piece of fishing gear, of a 'thickness of a junk's mast' or several feet in diameter, attached to its right fin. The accompanying

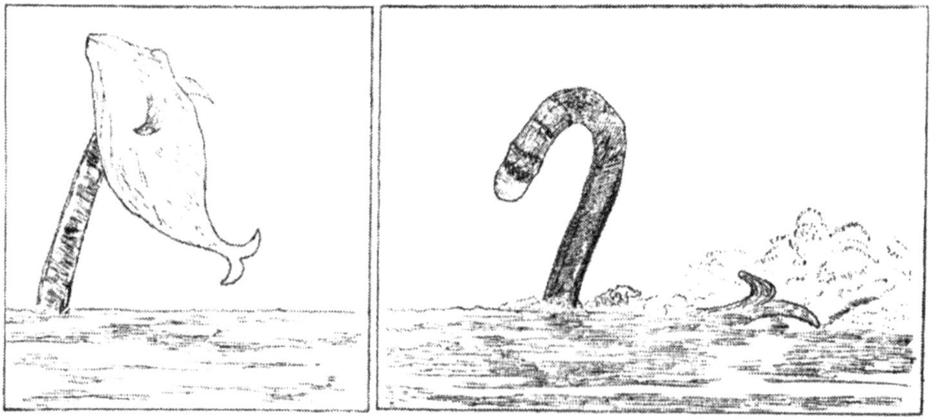

*Figure A2.1. Depictions of a purported sea serpent attacking a whale that resembles (left drawing) a normal breaching whale with a thick line of debris entangled over its right fin, and the same whale moments later heaving the entangled debris upward into the air while raising its tail to plunge back into the water (right drawing). Sighing from 1879 off the coast of Japan (*Graphic *illustration reprinted in Oudemans 1892, Heuvelmans 1968 and France 2016b).*

drawing is of a single creature, probably a cetacean as the *Nature* contributor correctly concluded, depicting, I believe, the tail of the whale and the end of the string of entangled gear being thrown up in the air following the cetacean's plunge back into the water after its breaching shown in the first drawing.

Figure A2.2. The most famous and notorious account of a cetacean observed to have been locked in mortal combat with a purported sea serpent, or simply a long string of entangling gear or debris wrapped around an unfortunate sperm whale? Sighting from 1875 off the coast of Brazil. Illustration from the Illustrated London News *reprinted in Oudemans (1892) and France (2016b). See France (2016b) and Deedy (2017) for other illustrated versions of this encounter.*

Captain Drevar and the crew of the barque *Pauline* were bound for Zanzibar in 1875 when, off the north-east coast of Brazil, they observed an epic combat – a 'duel to the death' as one commentator would later phrase it – between a sea serpent and a sperm whale. This would become the most famous imagined battle between a sea serpent and a whale in history. Upon reaching the British Naval Stores in Zanzibar, the observed battle became the talk of the wharf, so much so that it attracted the attention of Rev. E. Penny, who interviewed the witnesses and recorded details of the sighting and drew a sketch of the encounter (Figure A2.2), both of which he mailed off to the *Illustrated London News* for publication. With the subsequent translation and republication in Europe, the *Pauline* incident became the most famous sea serpent sighting of its day. One article, in the *Daily Telegraph,* was filled with purple prose worthy of Jules Verne, including statements that 'The ribs of the ill-fated fish [i.e. the embattled whale] were distinctly heard cracking one after the other, with a report like that of a small cannon; its bellowing ceased', causing the crew to be 'struck...aghast with terror.' Another article in the *Graphic* contained an additional rendering of the encounter (France 2016a). (In his retelling of my thesis for a popular audience, Deedy (2017) presents yet another rendition of this famous encounter, indicating it to be one of the most popular stories in nineteenth-century sea serpent lore.) By now, the barque was on its way to Burma, and Captain Drevar was forced to pen a retort in a Calcutta newspaper in which he blasts the *Telegraph* for its exaggerations which, as his relatives had written him, had ridiculed his reputation. Two years later the *Pauline* returned to England, from where the interest began anew with publication of details in more than half a dozen periodicals, including the journal *Nature* and the *Illustrated London News*, the latter including the dramatic sentence that 'The serpent whirled its victim round and round for about fifteen minutes, and then suddenly dragged the whale to the bottom head first.'

Immediately upon debarking in Liverpool, Captain Drevar and members of his crew sought out a Justice of the Peace and filed sworn testimonials about the truth of their observation of the battle between the whale and the sea serpent. The most detailed account of the famous encounter comes from the Captain's log-book which was published in the *Graphic* (Heuvelmans 1968; Oudemans 1892):

> At first glance I took all to be breakers as the sea was splashing up fountain-like about them, and the pillar a pinnacle rock, bleached with the sun; but the pillar fell with a splash, and a similar one rose. They rose and fell alternatively in quick succession, and good glasses showed me it was a monstrous sea-serpent coiled twice round a large sperm whale. The head and tail parts, each about thirty feet long were acting as levers, twisting itself and victim around with great velocity. They sank out of sight about every two minutes, coming to the surface still revolving; and the struggles of the whale and two

> other whales, that were near, frantic with excitement, made the sea in their vicinity like a boiling cauldron; and a loud and confused noise was distinctly heard. This strange occurrence lasted some fifteen minutes, and finished with the tail portion of the whale being elevated straight in the air, then waving backwards and forwards, and lashing the water furiously in the last death struggle, when the body disappeared from our view, going head downward to the bottom, where no doubt it was gorged at the serpent's leisure; and that monster of monsters may have been many months in a state of coma, digesting the huge mouthful. Then two of the largest spermwhales that I have ever seen moved slowly thence towards the vessel, their bodies more than usually elevated out of the water, and not spouting or making the least noise, but seeming quite paralysed with fear; indeed, a cold shiver went through my own frame on beholding the last agonizing struggle of the poor whale that has seemed as helpless in the coils of the vicious monster as a small bird in the talons of a hawk. Allowing for two coils round the whale, I think the serpent was about 160 or 170 feet long, and 7 or 8 feet in girth. It was in colour much like a conger-eel; and has the head, from the mouth being always open, appeared the largest part of its body.

Evidence, however, suggests that the death struggle was not fatal, since five days later and a several hundred kilometres along the coast, the *Pauline's* crew were 'astonished to see the same or a similar monster... throwing its head and about 40 feet of its body in a horizontal position out of water as it passed onwards by the stern of our vessel.' When the 'Leviathan' started 'looking angrily at our vessel' as if preparing to attack 'we kept ready all our axes, to give it a warm reception. But the animal dived and disappeared.' Given the widespread knowledge about the fate of the whale-ship Essex (Chase 1999) which became the foundation for Melville's *Moby-Dick*, one can perhaps understand the captain and crew's concern.

Many interpretations exist for the famous *Pauline* sighting of the sperm whale and sea serpent confrontation. One contributor to a periodical at the time insisted that the whale was in mortal combat with a hitherto unknown species of giant boa or python adapted to marine life. Lee (1883), however, makes a good point in dismissing such an idea that any serpent could elevate its body to one-third the height of the Monument to the Great Fire of London. He suggests that the 'supposed coils of the snake may have been the arms of a great calamari [i.e. giant squid], cast over and around the huge cetacean.' Oudemans (1892) notes that the circumference observed of the elongated object greatly exceeds that of any squid's arm (i.e. about half a metre), notwithstanding also that the terminus of the 'monster' was believed to have a mouth that was 'being constantly open,' as reported in several of the periodicals. Oudemans then enters into a discussion about the violent nature of sperm whales before turning the encounter on its head by siding with the serpent in a remarkable, and quite ridiculous, bit of prose:

> Knowing the wild and angry nature of the warlike and pitiless sperm-whale, and the rather harmless character of the sea-serpent, we cannot believe that a sea-serpent has ever or will ever attack such a formidable antagonist. Every one will rather believe that a sperm-whale, when meeting a sea-serpent, would suddenly attack it. Moreover, if the sea-serpent was the attacker, it would not have had 'its mouth always open', – an unfailable sign of great pain – but would have bitten repeatedly the whale! And so I firmly believe that one of the three spermwhales, has seized with its colossal mouth a sea-serpent by the trunk. The poor defenseless sea-serpent with its enormous flexible body wound round the upper jaw and forepart of the quadrangular head of the spermwhale...The sea-serpent in its violent pain raised its long neck high in the air and extended its jaws; it is even probable that it

> uttered a roaring sound or shriek, this is not mentioned; it may have been drowned out by the dreadful noise caused by the fight of these two monsters, for we may suppose that the sea-serpent was not destitute of muscular strength, and must have been a formidable antagonist.

Oudemans continues in a similar vein, before questioning whether the coils of the serpent were wrapped twice around the whale, dismissing Rev. Penny's drawings as 'not worth our attention' as they had been created months after the encounter. He then discounts the suggestion by the captain that the serpent used 'its extremities as levers' during the fight, and believes it is 'more probable that the sea-serpent, feeling itself free for a moment, suddenly escaped, dived and was followed by the sperm-whale.' Writing more than a half century later, Heuvelmans (1968) reviewed the evidence and concluded that the idea that the two encircling coils were the flexible neck of a prehistoric saurian or a mystery seal, as Oudemans had contended, was unfathomable. 'Personally I think,' he wrote, 'it was more likely some kind of giant eel,' proceeding into a discussion about the powerful muscles of a wounded moray.

Time for a bit of reality therapy. The observations and facts do not support the contention that the whale was in battle with any creature. No giant species of marine snake or eel are known to exist approaching the estimated length of the UMO. Nor could any such creature, if one even did exist, structurally manage to lift itself up to such a height above the water, much less be capable of thrashing about, up and down, so quickly. In the absence of evidence to the contrary, prehistoric pleisiosaurs or long-necked seals are known to exist only in the minds and writings of cryptozoologists. Nor does any extant vertebrate have a neck of such flexibility as to be capable of wrapping itself around the girth of a whale. And no known cephalopod has arms, nor a whale has a penis, both of which do display such flexibility, anywhere near the size – circumference with respect to the former, and length with respect to the latter – that were noted for the elongated UMO.

A much more likely explanation, and a parsimoniously satisfying one, is that what we have here, in 1875, is one of the first observations of a whale spotted in the process of trying to disentangle itself from a long string of fishing gear or other maritime debris which has become wrapped around its body. In this regard, Captain Drevar's empathetic language is much closer to the mark when he refers to the 'poor' whale being caught 'helpless in the coils of the vicious monster,' than is Oudemans' reference to the 'pitiless' whale attacking the 'poor defenseless sea-serpent.' Of all the commentaries and suggested hypotheses, one, from the German translator of several of the earlier articles reprinted in Oudemans (1892), comes closest to what I believe to have been the real circumstances underlying the observed event, when he posits that 'the whale was playing with a large tree or with a broken mast,' notwithstanding, of course, the sad fact that entangled cetaceans are not at 'play' with their trailing detritus. It is heartening to note that, contrary to Captain Drevar's supposition of the whale being caught in a death embrace before it was dragged down to be feasted upon by the sea serpent, the unfortunate creature seems to have survived its entanglement long enough to have been spotted the following week and many kilometres away, though, of course, its long-term survival in such a, possibly debilitating, state might have been problematic.

While fishing from their boat in 1934 off the Queensland coast of Australia, several individuals observed four dark objects in the water which, upon close inspection, submerged 'like a submarine' before reappearing five minutes later (*Victorian Naturalist* article reprinted *in* Heuvelmans 1968). 'After waiting half an hour and seeing no movements, excepting the head swaying from side to side, as if watching us, we decided to go back to town.' Heading out onto the water a short while later they continued to watch the 'monster': 'The head rose about 8 feet out of the water, and resembled a huge turtle's head...The colour was greyish-green. The eye...was small in comparison to the rest of the monster. The other part in view was three curved humps about 20 feet apart, and each one rose from 6 feet in the front to a little less in the rear. They were covered with huge scales about the size of saucers, and also covered in barnacles. We could not get a glimpse of the tail, as it was under the water.' Additional information included mention of the presence of a 'dark line along its spine,' barnacles being the size of soup-plates, and the scales shining in the sun and being described as 'butted and perpendicular' A week later, another, or more likely the same, UMO was observed nearby by another group of fishermen, who reported it to be about ten metres long, with a turtle-like head, and a body shaped like 'a huge armoured hose,' before continuing: 'Its head resembled a huge turtle more than anything else, and was slightly arched. Farther along three smaller dark objects were seen, giving the appearance of a Monster of the sea with a series of humps.'

The presence of 'a series of humps' is suggestive of a marine animal pulling a string of fishing gear or maritime debris. Likewise, a 'tail' composed of large, 'perpendicular' projecting 'scales' that resembles 'a huge armoured hose' is exactly what a chain of fishing floats looks like. That they were mentioned as 'shining' might indicate their composition to be blown glass rather than cork. Heuvelmans (1968), noting a parallel between this sighting and that of the famous 'Moha-moha' creature, which was almost certainly a turtle (Gould 1930), one I believe to have been entangled in fishing material or debris (see below), correctly mentions the physical impossibility of horizontal rows of scales upon the humps of any animal that must bend, stating 'If the monster was really humped and scaled as he [i.e. the observer] had depicted it, it is quite unlike any other sea-serpent, let alone any animal known to science.' Encrusting barnacles commonly occur on slow-moving whales, sometimes on the shells of sea turtles, and frequently, of course, as I suggest to be the case here, on submerged pieces of anthropogenic material. But then the large size noted for the barnacles here might indicate that they might instead have been sherds of glass floats. Furthermore, as has been observed elsewhere, which I suggest to be indicative of entanglement, the 'swaying' motion of what was mistaken for the head (due to the distance of the first sighting), may be nothing more than a piece of upright debris that is gently oscillating 'side to side' in the waves. The behaviour of an UMO lying motionless upon the surface of the water for extended periods is what would be expected when entangled debris is floating while the marine animal to which it is attached is submerged and possibly engaged in feeding upon a dense patch of prey in a small area.

In 1926, two fishermen were trawling off the coast of southern England when they 'caught themselves a sea monster,' as described in a local newspaper (Harrison 2001): 'It was twenty feet long, had an eight-foot tail and a beak, which they thought was 2 feet long and six inches wide. It also had four scaled legs, and a wide flat back, which was covered in a matted brown hair. The creature managed to escape by tearing the nets.' This seems to have been a doubly unlucky UMO. The 'beak,' 'wide flat back' and 'four scaled legs' certainly suggest a normal sea turtle. Whereas, the presence of a very narrow 'eight-foot tail' and 'matted brown hair' suggests an entangled fishing net draped over and trailing behind the shell. And then, as reported, the unfortunate creature gets caught, once again, albeit briefly this time, in a second net.

In 1925, a letter was published in *La Nature*, a French magazine popularizing science (Heuvelmans 1968), which describes a sighting in the China Sea:

> I was surprised to see on the surface of the water, hardly immersed...an animal looking like a 10-foot snake, with a maximum diameter of 7 to 8 inches, decreasing towards the tail, the body annulated from one end to the other, alternatively light yellow and black, the length of each ring being about 4 ½ to 6 inches. The rather small head was immediately followed by a swelling about 12 to 14 inches in diameter, as thick as the body which followed it, and with four feet, two each side, recalling the shape of a turtle, but without the shell, striped in yellow and black and with an annulated tail about 10 feet long. I could not see whether this body was round like a snake's, or, as I supposed, slightly flattened.

He continues in mentioning that the animal moved slowly with no discernable appendages observed along the extended length of the body. The accompanying sketch, redrawn and commented upon by Heuvelmans (Figure A2.3), 'shows what seems to be a sea-snake swallowing a turtle.' He believes such UMOs represent a large, unknown species of ray rather than what he refers to, somewhat dismissively, as 'the turtle-bodied snake.' This phrase, however, perfectly sums up my present thesis, as does the sketch suggesting itself as an appropriate schematic should a T-shirt emblem ever be designed about entangled chelonians.

On June 6th, 1877, when the royal yacht *Osborne* docked in Gibraltar, its captain, Commander Hugh Pearson, reported to the Admiralty that members of his crew had seen a sea monster four days before off the northern coast of Sicily. Three other crew members filed brief descriptions of the incident, of which Lieutenant William Haynes' was the 'most detailed' (*in* Heuvelmans 1968):

> On the evening of that day, the sea being perfectly smooth, my attention was first called by seeing a ridge of fins above the surface of the water, extending about thirty feet, and varying from five to six feet in height. On inspecting it by means of a telescope, at about one and a-half cables' [i.e. 100 metres] distance, I distinctly saw a head, two flappers, and about thirty feet of an animal's shoulder. The head, as nearly as I could judge, was about six feet thick, the neck narrower, about four to five feet, the shoulder about fifteen feet across, and the flappers each about fifteen feet in length. The movements of the flappers were those of a turtle, and the animal resembled a huge seal, the resemblance being strongest about the back of the head. I could not see the length of the head, but from its crown or top to just below the shoulder (where it became immersed), I should reckon about fifty feet. The tail end I did not see, being under water, unless the ridge of fins to which my attention was first attracted, and which disappeared by the time I got a telescope, were really the continuation of the shoulder to the end of the object's body.
>
> The animal's head was not always above the water, but was thrown upwards, remaining above for a few seconds at a time, and then disappearing. There was an entirely absence of 'blowing' or 'spouting'. I herewith beg to

Figure A2.3. A UMO described as a 'turtle-bodied snake' observed in 1925 in the China Sea. Illustration by Heuvelmans (1968) based on original eyewitness account, and reprinted in France (2016a, 2017).

> enclose a sketch [Figure A2.4] showing the view of the 'ridge of fins,' and [Figure A2.4] of the animal in the act of propelling itself by its two fins.

By mid-June the ship had returned to Portsmouth, England where the Commander filed his own official report, a synopsis of which was published, to much public interest, in the June 14th issue of *The Times*:

> At about five o'clock in the afternoon of [June] the 2nd, the sea being exceptionally calm, while the yacht was proceeding round the north coast of Sicily towards Cape Vito, the officer on watch observed a long ridge of fins, each about six feet long, moving slowly along...The fins were progressing in an eastwardly direction, and as the vessel more nearly approached them, they were replaced by the foremost part of a gigantic sea-monster. Its skin was, so far as could be seen, altogether devoid of scales, appearing rather to resemble in sleekness that of a seal. The head was bullet-shaped, with an elongated termination, being somewhat similar in form to that of a seal... Its features were only seen by one officer, who described them as like those of an alligator. The neck was comparatively narrow, but so much of the body as could be seen, developed in form like that of a gigantic turtle, and from each side extended two fins, about fifteen feet in length, by which the monster paddled itself along after the fashion of a turtle.

Figure A2.4. Upper: long row of serrated 'fins' of the 1877 Osborne *sighting of an UMO in the Mediterranean. Original illustration from* Land and Water, *reprinted in Oudemans (1892) and France (2017). Lower: front-end of the* Osborne *UMO which clearly resembles a marine chelonian.* Land and Water *illustration reprinted in Oudemans (1892) and France (2017). See France (2017) for other illustrated versions of this UMO.*

Gould (1930) gives considerable attention to the 'good deal of critical examination' generated from the *Osborne* encounter, including a body of opinions solicited by maritime and zoological experts published in *Land and Water*. One of those experts was Sir Richard Owen, who had, twenty-nine years earlier, published a tract dismissing the famous *Daedalus* sighting. But whereas in 1848 he offered a valid criticism of that sea serpent encounter, here, in old age, he assumed the role of a cranky curmudgeon in his outpouring of insults about anyone – by which he means anyone who is not as eminently qualified a naturalist as himself – who believes that they have observed a sea monster. In his opinion, the UMO was simply a cetacean. Heuvelmans (1968) is particularly peeved by Owen's legitimate scientific insistence that 'If the observer has not the zoological knowledge, or the opportunity for exact inspection, requisite for determining the species from the phenomena, he should abstain from giving the object any special name.' Heuvelmans also scolds Gould, whom 'would rather, it seems, call it 'object', or even 'thingmejig', than prejudice scientific research by using more appropriate terms like 'sea monster' or 'strange beast''.

And therein, lies the difference between the pseudoscience of cryptozoology and real science; hence my deliberate use of the purposely open-ended term 'UMO' (i.e. the equivalent of using 'UFO' rather than 'flying saucer').

The naturalist H. Lee, in his commentary in *Land and Water*, later reiterated in his 1883 book, also questions the previous hypotheses, before giving up and concluding that 'the combination of...[the imagined ridge of dorsal fins] with long flippers, and the turtle-like mode of swimming [of the body], forms a zoological enigma which I am unable to solve.' He states his belief that the creature is a species of marine animal that is 'either totally unknown to science, or which has hitherto been believed to be extinct.' In like fashion, A. Barrtlett, Superintendent of the Zoological Gardens, after ruling out the possibility that the observed creature could have been a whale, porpoises, sharks, or even a monster turtle – the latter excluded due to the noted presence of the long row of serrated fins – agrees that it was a specimen of an as yet unknown species.

Oudemans (1892) demonstrates a prodigious degree of confirmation bias in his quest to force fit all observations to support his pet theory that all sea serpents are long-necked seals. He undertakes an exercise in selective data screening, in one case discounting that the movements of the flappers were like those of a turtle, rather than being, as he believed, like those of a sea-lion, 'for a sea-turtle cannot possibly elevate its flappers so high above the surface of the water, while sea-lions are able to do so.' We now know this to be incorrect based on behavioural accounts and photos of sea turtles engaging in just such activities. Remarkably, Oudemans even goes as far as to actually invent wording for traits he believed the witnesses saw but neglected to report! He also selectively accepts one description of a trait while dismissing the drawings as being inaccurate, followed by doing just the reverse a few sentences later with respect to another trait. He is partially correct, however, about one thing, namely his 'firm conviction that the ridge of fins has nothing at all to do with the animal.' But for Gould (1930), this supposition is just one mystery too many:

> This merely gives us two enigmas instead of one. The fins – seven or eight in a line, and standing six or seven feet out of the water – were, in themselves, as unusual a sight, and one as difficult to explain, as the creature with the huge seal-like head, slender neck, and giant flippers, which replaced them. One such sight was remarkable enough: that two should occur – in precisely the same spot – at an interval not exceeding a few seconds, and the two should yet be completely independent, defies belief. However convenient theorists may find the 'ridge of fins,' these cannot, I suggest, be dismissed in this airy fashion, for they formed an integral portion of the evidence – and, presumably, of the 'sea-monster.'

In his 1890 survey of marine monsters, John Gibson comments on the hybrid fusion of the *Osborne* UMO, sporting as it did the dorsal fins of a basking shark, the head of a seal, and the flippers of a turtle. As to what the creature actually was, he too is at a loss: 'This conundrum of an animal has been fairly given up by naturalists as unexplainable by a reference to any known group of animals.' However, decades later, Burton (1960), through use of a side-by-side engraving of the *Osborne* creature and photographs of various turtles, some of which have dorsal serrations, which if magnified in size do resemble the UMO, makes the case that the creature was most likely to have been a turtle.

Heuvelmans is also puzzled by the fact that it is 'hard to reconcile the two drawings and apply them to the same animal.' He sides with the theory that the drawing of the serrated fins could be a school of sharks or dolphins, and suggests that the other end-view drawing indicates that the creature was fleeing being pursued and attacked by the school such that 'it might try in its terror to rise above the waters and so beat its fins in the air.' His alternative hypothesis is that it was indeed a single animal and that the ridge of fins was not dorsally situated but located along the sides of the animal as characteristic of his imagined 'great sea-centipede' category of sea monsters. And although he is off into flights of cryptozoological fantasy here, he is correct in rebuking the idea that the *Osborne* UMO could have been some type of prehistoric beast, believing that such thinking merely reflects the faddism of the day. 'Science,' he states, 'is as susceptible to fashion as hats, literature or dance-steps.' More recently, cryptozoologist Drinnon (2010) suggests that the large size of the *Osborne* UMO may be the result of either several whales in a line, or a 'mistaken view of a humpback whale from the rear, and the row of fins either a school of sharks or else the whale's pod of comerades [sic] in 'fin-up' position,' or simply the wake of a single cetacean.

The parsimonious explanation for the *Osborne* UMO is that of a sea turtle entangled in fishing gear or maritime debris. With little difficulty, the hitherto incongruous and mysterious ten-metre ridge of fins of irregular height shown in the illustration can be likened to a dragged length of fishing net and floats just as convincingly as can the body of the creature to being a chelonian based on the eyewitness accounts as well as the illustration. Of interest is the fact that the descriptions and drawing of the extended 'ridge of fins' lying upon the surface of the water matches closely with eyewitness accounts of the famous Gloucester UMO. Due to their life cycle and swimming, feeding and migratory behaviour, sea turtles are particularly susceptible to entanglement. As such, it is possible to view many on-line photographs of chelonians entangled in fishing gear including floats. It is easy to surmise how such net floats composed of cork or glass used in nineteenth-century purse-seines, if pulled in line strung out behind the unfortunate animal, and viewed from a distance as was the case for the *Osborne* UMO, could lead to the supposition of an elongated sea monster being seen.

A final example culled from my previously published articles recounts how, in June 1890, while on a beach on Great Sandy Island, Queensland, a schoolteacher, Miss. S. Lowell, accompanied by the lighthouse keeper and his two children and assistant, two other individuals, and an Aboriginal, spotted an UMO. Reported immediately in the local press, it was not until the 3rd of January 1891 that a letter was published in the well-known and internationally distributed magazine *Land and Water* (*in* Gould 1930):

> We have had a visit from a monster turtle fish. I send a sketch of it [Figure A2.5]. It let me stand for half an hour within five feet of it. When tired of my looking at it, it put its large neck and head into the water and swept round seaward, raising its huge dome-shaped body about five feet out of water, and put its twelve feet of fish-like tail over the dry shore, elevating it at an angle. Then, giving its tail a half twist, it shot off like a flash of lightning, and I saw its tail in the air about a quarter of a mile off, where the steamers anchor.
>
> It has either teeth or serrated jaw-bones. Native blacks call it 'Moha-moha', and they say they like to eat it, and that it has legs and fingers. I did not see its legs, as they were in the water. What I saw of it was 27 ft or 28 ft, but I think it must be 30 ft in all. Whilst its head was out of water it kept its mouth open, and, as I could not see any nostrils, I fancy it breathes through its mouth. The jaws are about 18 in in length; the head and neck greenish white, with large white spots on the neck, and a band of white round a very black eye and round upper and lower jaws.

> The body was dome-shaped, about 8 ft across and 5 ft high, smooth, and slate-grey in colour. Tail about 12 ft, the fish part wedge-shaped, and fin of chocolate-brown. Then beautiful silver shading to white scales size of thumb nail.

Subsequently, the assistant curator of the London Natural History Museum, William Saville-Kent, asked for a more detailed description from Ms. Lovell so he could attempt a taxonomic designation of the creature for the book he was writing, *The Great Barrier Reef of Australia*, to which she responded:

> I was (while walking on the Sandy Island beach) admiring the stillness of the sea, it being dead calm, when my eye caught sight of the head and neck of a creature I had never seen before.
>
> I went to the edge of the water and saw a huge animal, lying at full length, which was not at all disturbed by my close proximity to it, enabling me to observe the glossy skin of the head and neck, smooth and shiny as satin. Its great mouth was wide open all the time it was out of the water.
>
> In about a quarter of an hour or so it put its head and neck slowly into the sea, closing its jaws as it did so. I then saw what a long neck it had, as it moved round in a half circle, and also perceived that the head and neck were moving under a carapace. When the head was pointing out to sea it rose up, putting a long fish-like tail out of the water over dry shore, parallel to myself, and not more than five feet from me, not touching the sand, but elevated. I could have stood under the 'flukes of its tail.'
>
> The only part of the body that had marks like joints (like in size and shape to a common brick) was also on the dry shore, but *resting* on the sand; the great dome-shaped carapace, dull slate-grey, was standing quite five feet high, and so hid its long neck and head from my view, which before it rose I could see as a long shadow on the water. The carapace was smooth and without marks of any sort. The fish-like part of the tail was glossy and shiny as the head and neck, but of a beautiful silver-grey, shading to white, with either markings or large scales, not like those of a fish in position, as the fishes' scales lie horizontally, while the Moha's, if scales, lie perpendicularly, each the size of a man's thumbnail. It had a thick fleshy fin near the end, about three feet from the flukes, and, like them, chocolate-brown. The flukes were semi-transparent; I could see the sun shining through them, showing all the bones very forked. One of the girls asked me if a shark had bitten a piece out of its fail, and the other wanted to know if I thought it was an alligator! The fish-like part was quite twelve feet long.
>
> All the time the animal was on shore it was perfectly motionless; at last it gave a curious half-twist to the fluke of its tail, the movement only just reaching beyond the fleshy fin, and, without disturbing the water in the slightest degree, vanished. I seemed only to have taken one breath when I saw its tail out of the water about the place where the steamer anchors, sending a quantity of fish into the air. I then saw it give a twist of its tail, and it disappeared altogether. The black boy saw it on shore the previous Monday, the 9th inst.
>
> As I was so close to it for at least half an hour, I was able to study its shape and colouring. In moving about, head and tail were seen alternatively above water, but not even the shadow of its great body, and, from the length of that, a spectator could not guess that the head and tail belonged to the same creature, particularly as the colouring was so different. The parts I did not see were the legs. I stooped down and tried, but in vain, to

> see them, though the Moha was standing in only a foot of water, but the Black described them as being like an alligator.

The Moha-moha sighting occurred in 1890, too late to be included in Oudemans' seminal treatise. But that did not stop the world-wide flourish of speculation as to the nature of what certainly remains one of the strangest of all UMO sightings. This particular case is also of interest for its unseemly portrayal of male chauvinism and expert arrogance.

Following the first report in *Land and Water,* the editor, naturalist F. Buckland, remarked that he believed the creature to be a particular type of turtle found in New Guinea, neglecting to account for the marked disparity between Ms. Lovell's 10 metre-long animal and the fact that the purported candidate turtle is never larger than a single metre in size. Buckland does make the valid point that 'the length and shape of the tail, as described, are, however, impossible, as no turtles have caudal appendages corresponding to the description,' before concluding in a condescending manner that 'the fair observer must have been mistaken on this very important point' (when Gould reprinted this comment, he took it upon himself to emphasize the word 'impossible' by placing it in italics). Ms. Lovell retorted that 'You speak of the impossible length of its tail. I beg to state that this is a most astonishing statement from people who have never seen this monster, half fish, half tortoise. The tail was over the dry shore for half an hour, so close to me, that five footsteps would have enabled me to put my hand upon it.' She continues, informing that the translation of the local Aboriginal name for the creature was 'dangerous turtle'. Buckland, however, would not back down, replying in turn with condescension: 'With no wish to cast any doubt on our fair correspondent's veracity and powers of observation, we must adhere to the editorial remarks made...[earlier], and point out to her that from a scientific point of view, the existence of a fish, together with those of a tortoise, are absolutely impossible.' Again, he suggests another candidate turtle for the UMO, which Gould (1930), writing four decades later, has no trouble in dismissing. Buckland then recommends that Ms. Lovell should obtain a copy of a book entitled *Mythical Monsters* in which she will find 'all sorts of supposed monsters figured, including the sea-serpent.'

The *Land and Water* editor then goes on to state that the only solution to the half fish – half tortoise creature was that 'the monster tortoise had clasped by its nether limbs, and has pressed against its abdomen some very large fish, the upper part of whose body was thus concealed, while the tail projected behind, and gave the tortoise the appearance of having a fish's tail.' Gould begins by offering the valid comment: 'Why the 'tortoise' should clasp the (20-foot or so) fish, except to eat it; why the latter should endure this process with absolute calm for half an hour and then suddenly come to life; and why it should still remain so gripped by the 'tortoise', when the latter was swimming, as to give the impression of a single creature, he does not attempt to explain,' before concluding in a way to ensure his notoriety for evermore: 'But he [Buckland] makes one sensible remark: '...We lay no stress on the evidence of the blacks, as all coloured races are greatly given to exaggeration and invention.''

If Ms. Lovell thought she would be any better treated by the naturalist Saville-Kent, with whom she had corresponded, she was mistaken. He sarcastically suggests that she could become rich through marketing a new type of turtle soup, which he proposes to name *Chelosauria lovelli* after her.

Gould (1930), without hesitation, places the Moha-moha, along with the *Osborne* incident, within his 'giant' or 'monster' turtle category. Heuvelmans (1968), however, suggests that the Moha-moha might be a gigantic placoderm, a type of armoured fish surviving from the Devonian period, that was observed